Green Technologies for Pollutants Remediation and Recovery from Water and Soil

Green Technologies for Pollutants Remediation and Recovery from Water and Soil presents a comprehensive examination of sustainable and environmentally friendly methods for remediation and recovery from various types of pollutants in water bodies and soil. It discusses diverse remediation and recovery technologies, ranging from classic adsorption and membrane technologies to the most promising green technologies such as biocatalysts and microorganisms for obtaining value-added products for practical use. Through many case studies and best practices, this book explores current achievements and challenges and indicates future directions for global and cross-border collaboration.

Key Features:

- Provides a comprehensive overview of the achievements, challenges, and the current use of eco-friendly approaches for pollutant removal and recovery.
- Covers the role of biology, biotechnology, and chemical technology in designing green technologies for environmental applications.
- Presents innovative applications of microorganisms, enzymes, adsorption processes, and membrane technologies as a base for improvement in green technologies of environmental importance.
- Explains the basic assumptions, requirements, and mechanisms of green technologies used for pollutant conversion.
- Aims to better understand the importance of sustainable technologies in environmental protection.

This reference is a great resource for upper-level undergraduate and graduate students in environmental science and engineering, soil and water sciences, biotechnology, and chemical engineering, as well as researchers and professionals who work as environmental and technical engineers, ecologists, applied environmental scientists, and managers.

Emergent Environmental Pollution

Series Editor:
Jörg Rinklebe
University of Wuppertal, Germany

The nature of pollutants in the environment depends largely on their interactions in the soil, water, air, and organisms. Understanding the fate and transport of pollutants in the environment is essential for humans and animals, plants, water bodies, air, soils, and sediments. This series provides timely information on the biological, chemical, and physical processes governing the interactions of pollutants in the environment. Such knowledge is needed for minimizing the negative impact of pollutants on human health and the health of ecosystems. Books from this series are of value to professionals and researchers, students, industrial and mining engineers, government regulators, authorities, and consultants in the environmental arena.

Dynamics and Bioavailability of Heavy Metals in the Rootzone
Edited by H. Magdi Selim

Trace Elements in Waterlogged Soils and Sediments
Edited by Jörg Rinklebe, Anna Sophia Knox, and Michael Paller

Competitive Sorption and Transport of Heavy Metals in Soils and Geological Media
Edited by H. Magdi Selim

Nickel in Soils and Plants
Edited by Christos Tsadilas, Jörg Rinklebe, and Magdi Selim

Phosphate in Soils: Interaction with Micronutrients, Radionuclides, and Heavy Metals
Edited by H. Magdi Selim

Permeable Reactive Barrier: Sustainable Groundwater Remediation
Edited by Ravi Naidu and Volker Birke

Nitrate Handbook: Environmental, Agricultural, and Health Effects
Edited by Christos Tsadilas

Vanadium in Soils and Plants
Edited by Jörg Rinklebe

Maritime Accidents and Environmental Pollution - The X-Press Pearl Disaster Causes, Consequences, and Lessons Learned
Edited by Meththika Vithanage, Ajith Priyal de Alwis, and Deshai Botheju

Methylmercury Accumulation in Rice: Process and Regulation
Edited by Xinbin Feng, Jianxu Wang, and Jörg Rinklebe

Green Technologies for Pollutants Remediation and Recovery from Water and Soil
Edited by Jakub Zdarta, Teofil Jesionowski, and Jörg Rinklebe

For more information on this series, please visit: https://www.routledge.com/Emergent-Environmental-Pollution/book-series/CRCEEP

Green Technologies for Pollutants Remediation and Recovery from Water and Soil

Edited by
Jakub Zdarta, Teofil Jesionowski,
and Jörg Rinklebe

CRC Press
Taylor & Francis Group
Boca Raton London New York

CRC Press is an imprint of the
Taylor & Francis Group, an informa business

Designed cover image: iStock

First edition published 2026
by CRC Press
2385 NW Executive Center Drive, Suite 320, Boca Raton FL 33431

and by CRC Press
4 Park Square, Milton Park, Abingdon, Oxon, OX14 4RN

CRC Press is an imprint of Taylor & Francis Group, LLC

ISBN: 978-1-032-88121-8 (hbk)
ISBN: 978-1-032-88124-9 (pbk)
ISBN: 978-1-003-53627-7 (ebk)

DOI: 10.1201/9781003536277

Typeset in Times
by codeMantra

Contents

Preface

Still, a growing number of environmental pollutants of various origins, including hazardous metal ions, organic dyes, endocrine disruptor chemicals, personal care products, pharmaceuticals, and other xenobiotics, as well as new pollutants, such as microplastics and nanoplastics, appearing in uncontrolled quantities, cause growing social concerns. This in turn means that more and more efforts of scientists and engineers are needed, as well as financial outlays to develop effective methods not only for removing but also for neutralizing this type of pollution. The techniques currently used are not only ineffective in removing micropollutants but also often cost- and energy-intensive and generate significant amounts of by-products or waste. In accordance with current trends and efforts to improve the quality of the environment and increase its protection, it is therefore necessary to develop green technologies that will, above all, effectively remove the aforementioned pollutants, but additionally the remediation process itself will be sustainable and environment-friendly. Green technologies refer to the development and use of technologies that minimize the negative impacts of human activities on the environment and society. It encompasses a wide range of products, services, and practices that support a more sustainable future and usually also a circular economy. The concept of green technologies is gaining attention as the world seeks ways to address climate change and environmental pollution. The creators of green technology use scientific knowledge and innovation to support the protection of natural resources, reduce greenhouse gas emissions, and promote the use of renewable energy, as well as promote processes that generate no by-products and operate under mild conditions. Among green technologies for the removal of environmental pollutants, the most frequently used are biological methods based on microbes or enzymes that are capable of converting toxic compounds into less harmful substances. Further, the use of biomaterials as sorbents capable of partial biodegradation is also more and more common. All of these techniques are becoming increasingly important, but a lot of research and development work is still needed to implement these techniques into practical applications.

Nevertheless, the huge advantage in the development of green technologies for environmental applications is the fact that this approach might lead to the expansion of universal platforms enabling not only the simple removal/conversion of micropollutants but also the detection of hazardous compounds as well as remediation and their further recovery as a source of valuable compounds. Resource recovery is based on the use of wastes or toxic compounds as a starting material to create valuable products as new outputs. The main idea is to reduce the amount of waste generated and to reduce its toxicity, thereby reducing the need for landfill space and optimizing the values created from waste. An advantage of resource recovery is the fact that it delays the need to use raw materials in the manufacturing process. The application of green technologies for resource recovery will establish a recycling-oriented society towards sustainability. However, the wide-scale application of different technologies for pollutants recovery and reuse will, in many cases, require novel technologies to alter their practices. Due to the still progress in the development of new recyclable materials, green design, and energy conversion technologies, the recycling of organic and/or inorganic substances from wastes has become the focus and direction of numerous renowned labs around the world. Although various techniques for pollutants recovery could be used, the most promising are composting, pyrolysis, gasification, membrane processes, and sintering techniques. Currently, the development of solutions consistent with sustainable development and in accordance with green chemistry is one of the greatest challenges in the world of science, whereas their effective implementation remains an economic and political challenge. Hence, it is crucial to summarize and discuss recent achievements and existing challenges in sustainable development and environmental protection. It is also important to highlight the importance of future research directions to raise the government's awareness of the green approach towards a clean and safe environment.

This book covers different aspects of environmental applications of green technologies for both pollutants remediation and resource recovery from polluted streams. Chapter 1 addresses the state of the art regarding the use of various green techniques for remediation processes of pollutants from water and highlights common techniques for recovery of these compounds, showing also current limitations in these directions. Chapters 2 and 3 address the issues related to the presence of endocrine chemicals in surface and groundwater challenges and show that green technologies should be considered simple, inexpensive, efficient, and eco-friendly in the removal of endocrine disruptors. However, structure, chemical formula, and the presence of various functional groups strongly affect removal efficiency. In the next part of this book, a biotechnological approach to remediation and toxicity removal of persistent pollutants is presented. In Chapter 4, the importance of biological processes for pollutants recovery is supported by the great potential of these techniques on a wider scale. It is known that biological methods for green pollutants removal consist of microbial and enzymatic methods. The importance of the biocatalytic approach is shown in Chapter 5 followed by the strategies to improve enzymatic activity in environmental applications. By contrast, Chapter 6 presents the importance of microbial selection for pollutants removal, showing that microbial consortia usually act better than single-strain microbes. This book also demonstrates the progress in green technologies for the production of valuable compounds. Chapter 7 shows the great potential of converting waste biomass into value-added chemicals. Chapters 8 and 9 clearly show the virtually infinite potential of the microalgae for pollutants removal as well as show that microalgae, even after the removal of pollutants, might act as a promising and universal platform for the production of valuable compounds for further use. In the next part of this book, it moves to the more conventional, but also green, methods for environmental protection. In Chapter 10, advances in the use of membrane technology and membrane reactors for the efficient removal of pollutants from water bodies and the production of pure water streams are summarized. Chapter 11 shows the production of hydrochar from livestock manure wastewater and presents that such kind of hydrochar might be efficiently applied to acquire nutrient recovery. In Chapter 12, the importance of computer techniques in environmental chemistry is highlighted. Examples given in this chapter clearly show that computer modeling has to be considered a promising strategy for endocrine chemicals remediation. Finally, in Chapter 13, the possible use of iron oxide-based technologies for the removal of arsenic and chromium ions from the soil body is presented.

About the Editors

Dr. Jakub Zdarta is an associate professor in the Faculty of Chemical Technology at Poznan University of Technology. He has been selected as one of the 2% Highly Cited Researchers in 2022 and 2023 according to a ranking by Stanford University and Elsevier. His academic background covers biocatalysis, environmental sciences, chemical engineering, and organic chemistry. His main research is on the development of biocatalytic systems for pollutant removal and recovery and for the synthesis of active pharmaceutical ingredients with a special focus on green chemistry and circular economy. He also has a certain expertise in the field of nanotechnology.

Dr. Zdarta published plenty of scientific papers in leading international and national journals. According to the Scopus database (09.12.2024), he published 126 articles, which have been cited 5148 times with an *h*-index of 35. Also, he published numerous book chapters, and he has served as a leading editor in one book. He is a guest editor of international journals including *Environmental Technology & Innovations*, *Chemosphere*, *Materials*, *Molecules*, *AIMS Biotechnology*, and *Catalysis Communications*. He is also a member of editorial boards for several renowned journals. He was the chair of the fourth international conference "Green Technologies for Sustainable Environment 2024 (GTSE24)" in September 2024 and was a leading person in the organizing committee for the BioOrg 2022 conference in December 2022. He was also an invited speaker (plenary and keynote) at many international and national conferences.

For his outstanding scientific achievements, he received the following awards: Scholarship of the Minister of Science and Higher Education for Outstanding Young Scientists in 2019, Scientific Award of the 4th Division of Technical Sciences of the Polish Academy of Sciences in 2022, Award of the Prime Minister in 2022 for scientific achievements constituting the basis for conferring the degree of habilitated doctor, and Award of the Ministry of Higher Education and Science in 2023 for outstanding scientific achievements.

Dr. Zdarta visited numerous centres and scientific institutions in different countries. He collaborates mainly with MIT, the Technical University of Denmark, the University of Technology Sydney, and the University of Almeria. He is an expert at the National Science Centre, Poland; National Centre for Research and Development, Poland; Foundation for Polish Science; Polish Agency for Enterprise Development; Swiss National Science Foundation (SNSF); and Innovation and Technology Commission (Hong Kong).

Dr. Teofil Jesionowski is currently the rector of Poznan University of Technology (PUT), the honorary chair of the Conference of Rectors of Polish Technical Universities, and a member of the Board of the Conference of Rectors of Academic Schools in Poland. He has been a member of the Polish Academy of Science since 2020. He also serves as the head of the Chemical Technology Division and is a member of the Academy of Engineering in Poland.

Dr. Teofil Jesionowski received the title of professor of chemical sciences in 2013. His main areas of scientific interest are chemical sciences (chemistry) and engineering and technical sciences (chemical engineering, materials science, and environmental protection). His research interests include the synthesis (design and fabrication), characterization and applications of advanced functional materials, functional fillers and polymer composites, (bio)additives and eco-friendly fillers, biomineralization-inspired syntheses and extreme biomimetics, biocomposites and biomaterials, removal of wastewater pollutants via adsorption, photocatalysis or precipitation methods, pigment composites, enzyme immobilization, colloid chemistry and surface modification, hybrid systems, biopolymers, and biosensors.

He has published over 575 publications indexed by SCOPUS (2024-12-09). His publications received over 18,100 citations (*h* index = 62). He is also a co-creator of over 50 patents and patent applications. He is the co-author of chapters in numerous monographs, published in renowned publishing houses. He is also a member of the editorial boards in numerous renowned journals, e.g., *Dyes and Pigments*, *Physicochemical Problems of Mineral Processing*, and *Scientific Reports*. He served numerous issues as Guest Editor.

Dr. Jesionowski has visited numerous centres and scientific institutions around the world. He collaborates mainly with the MIT, Technical University of Denmark, University of Technology Sydney, and Kent State University. He has given lectures and speeches in numerous countries representing six continents. He is an expert at the National Science Centre, Poland; National Centre for Research and Development, Poland; F.R.S.-FNRS, Belgium; NWO, Netherlands; Swiss National Science Foundation (SNSF); and DFG, Germany.

Dr. Jörg Rinklebe is a professor of soil and groundwater management at the University of Wuppertal, Germany. He has been selected as an ISI Web of Science Highly Cited Researcher in 2019, 2020, 2021, and 2022. He was the vice president, the president, and now the past president of the International Society of Trace Element Biogeochemistry (ISTEB). His academic background covers environmental science, bioavailability of emerging contaminants, and remediation of contaminated sites. His main research is on soils, sediments, waters, plants, and their pollutions and linked biogeochemical issues with a special focus on redox chemistry. He also has a certain expertise in soil microbiology.

Dr. Rinklebe has published plenty of scientific papers in leading international and national journals. According to the SCOPUS database (25.11.2024), he published 573 articles, which have been cited 40,575.0 times with an *h*-index of 104. Also, he published five books entitled *Trace Elements in Waterlogged Soils and Sediments* (2016), *Nickel in Soils and Plants* (2018), *Soil and Groundwater Remediation Technologies* (2020), *Vanadium in Soils and Plants* (2022), and *Methylmercury Accumulation in Rice* (2024), as well as numerous book chapters.

Dr. Rinklebe has served as the editor-in-chief for *Environmental Pollution.* He was the editor of the *Journal of Hazardous Materials*; co-editor in chief of the international journal *Critical Reviews in Environmental Science and Technology* (*CREST*); and guest editor of the international journals *Environment International, Chemical Engineering Journal, Science of the Total Environment, Chemosphere, Journal of Environmental Management, Applied Geochemistry*, and *Environmental Geochemistry and Health.* Also, he is a member of several editorial boards (*Ecotoxicology, Geoderma, Water, Air, & Soil Pollution, Archive of Agronomy and Soil Science*).

Dr. Rinklebe was the chair of the first joint International Conferences of Biogeochemistry of Trace Elements (ICOBTE) and the International Conference on Heavy Metals in the Environment (ICHMET) in September 2023. He was an invited speaker (plenary and keynote) at many international conferences.

Dr. Rinklebe was appointed as the Honorable Ambassador for Gangwon Province, South Korea. Also, he got an appointment as an adjunct professor at the University of Southern Queensland, Australia; a visiting professor at Sejong University, Seoul, South Korea; and a guest professor at China Jiliang University, Hangzhou, Zhejiang, China.

Dr. Rinklebe has received the following awards:

- Award for Internationalization of Research and Teaching: Weltlöwe
- JIOMICS & Proteomass Scientific Society Career Awards
- IESGA Fellowship Award 2023

Contributors

Fatima Azeem
University of the Punjab
Lahore, Pakistan

Workisa Bacha
Haramaya University
Haramaya, Ethiopia

Gurunathan Baskar
St. Joseph's College of Engineering
Chennai, India

R. Baskaran
Jeppiaar University
Chennai, India

Tesfaye Kassaw Bedru
Wollo University
Kombolcha, Ethiopia

Murielly Fernanda Ribeiro Bihain
Federal University of Tocantins
Tocantins, Brazil

Aline Frumi Camargo
Federal University of
Fronteira Sul
Erechim, Brazil

Muhammad Mahroz Hussain
Yangzhou University
Yangzhou, China

Nazim Hussain
University of the Punjab
Lahore, Pakistan

Hussain Ibrabeem
National Institute for Biotechnology and
Genetic Engineering
Faisalabad, Pakistan

Katarzyna Jankowska
Technical University of Denmark
Kongens Lyngby, Denmark

Mani Jayakumar
Karpagam Academy of Higher
Education
Coimbatore, India

Teofil Jesionowski
Poznan University of Technology
Poznan, Poland

Ewa Kaczorek
Poznan University of Technology
Poznan, Poland

Michał Kapczyński
Poznan University of Technology
Poznan, Poland

Riti Thapar Kapoor
Amity University
Noida, India

Natalia Klanovicz
University of São Paulo
São Paulo, Brazil

Simone Kubeneck
Federal University of Fronteira Sul
Erechim, Brazil

Zhuoya Lv
China Agricultural University
Beijing, China

Maham
University of the Punjab
Lahore, Pakistan

Tazeen Malik
University of the Punjab
Lahore, Pakistan

Julaia Pieper Nerling
Federal University of Fronteira Sul
Erechim, Brazil

Muhammad Osama
University of the Punjab
Lahore, Pakistan

Douglas Henrique Pereira
Technological Institute of Aviation
São José dos Campos, Brazil

Manuel Pinelo
Technical University of Denmark
Kongens Lyngby, Denmark

Mohd Rafatullah
Universiti Sains Malaysia
Penang, Malaysia

Swati Rani
Diatom Research Laboratory
Amity Institute of Biotechnology
Amity University
Noida, India

Jörg Rinklebe
University of Wuppertal
Wuppertal, Germany

Aleksandra Rybak
Poznan University of Technology
Poznan, Poland

Agnieszka Rybarczyk
Poznan University of Technology
Poznan, Poland

Loisleini Fontoura Saldanha
Federal University of Fronteira Sul
Erechim, Brazil

Anna Karla dos Santos Pereira
Federal University of Tocantins
Tocantins, Brazil

Thamarys Scapini
Federal University of Paraná
Curitiba, Brazil

Hanna Sieracka
Poznan University of Technology
Poznan, Poland

Wojciech Smułek
Poznan University of Technology
Poznan, Poland

Tallha Sultan
Fujita Health University
Tokyo, Japan

Aleena Sumrin
University of the Punjab
Lahore, Pakistan

Huma Tahreem
University of the Punjab
Lahore, Pakistan

Archana Tiwari
Diatom Research Laboratory
Amity Institute of Biotechnology
Amity University
Noida, India

Helen Treichel
Federal University of Fronteira Sul
Erechim, Brazil

Shengsen Wang
Yangzhou University
Yangzhou, China

Jakub Zdarta
Poznan University of Technology
Poznan, Poland

Tao Zhang
China Agricultural University
Beijing, China

Xiaowei Zhang
China Agricultural University
Beijing, China

1 Progressing Forward
Sustainable Remediation and Recovery of Pollutants

Jakub Zdarta, Jörg Rinklebe, and Teofil Jesionowski

1.1 INTRODUCTION

Remediation and recovery of pollution are two essential pillars of modern environmental protection efforts, especially in the context of industrial, agricultural, and urban activity. Both processes aim to reduce the negative impact of human activities on ecosystems, but their goals, methods, and results differ significantly. Both remediation and recovery are responses to the growing problem of environmental contamination with toxic substances, although they differ in their approaches to these pollutants and how they transform contaminated materials in the removal or recovery of value (Figure 1.1) [1,2].

Pollution remediation refers to extensive activities that aim to remove, neutralize, or reduce the level of harmful substances present in the environment. It is a comprehensive process, involving multiple technologies and methods tailored to the specific conditions and characteristics of the contaminants. The primary goal of remediation is to protect human health and the stability of ecosystems by restoring

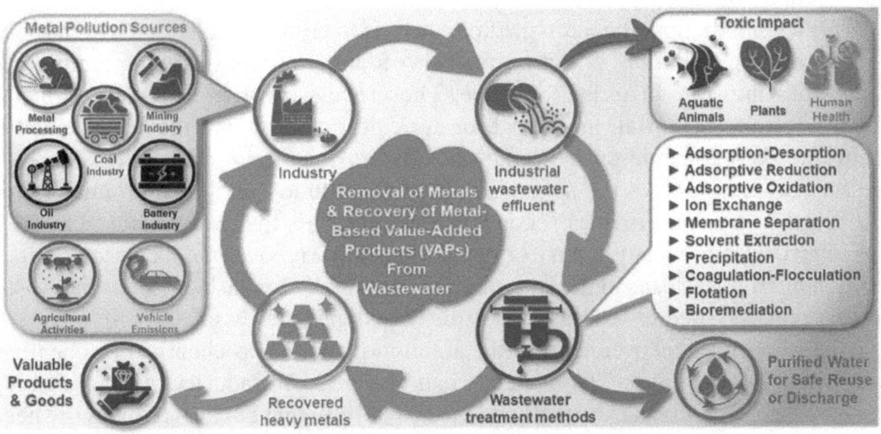

FIGURE 1.1 The general idea of the cycle for pollutants removal and recovery highlighting the green technologies in the example of metal ions. (Adapted from [3] with permission from Elsevier.)

DOI: 10.1201/9781003536277-1

degraded areas to a state that does not pose a threat [4]. Remediation processes can include physical, chemical, and biological methods, which are often used simultaneously in complex environmental cleanup projects. In the context of physical methods, remediation usually involves the removal of contaminated soil, groundwater, or river sediment [5]. An example is the mechanical pumping out of contaminated groundwater for further treatment using advanced filtration, adsorption, or ion exchange technologies. In cases of heavily contaminated soils, methods such as digging up and moving contaminated soil layers to special landfills, where they are subjected to further treatment, are used. Modern physical remediation methods also include thermal techniques, such as the use of hot air or microwaves to break down organic chemicals, reducing their toxicity. Equally important are chemical methods, which use chemical reactions to neutralize or transform contaminants [6]. For example, chemical oxidation, using powerful oxidants such as hydrogen peroxide, potassium permanganate, or ozone, can break down difficult-to-remove organic compounds including pesticides, industrial solvents, and petroleum-based solvents, transforming them into less toxic products. Chemical reduction, on the other hand, involves the introduction of chemicals that convert harmful pollutants into less mobile or toxic forms. In the case of hazardous metal contaminants such as lead or mercury, chemical methods may involve precipitating them out of solution as insoluble compounds that are easier to remove or store [7]. Bioremediation, on the other hand, which is one of the most advanced green approaches in remediation, involves using microorganisms or plants to break down or detoxify organic and inorganic contaminants. These organisms can metabolize or break down toxic substances into simpler, less harmful compounds. For example, bacteria and fungi are often used to break down petroleum hydrocarbons in soil, while specific plant species, such as the common reed, are used in phytoremediation processes to absorb hazardous metals. Bioremediation plays a particularly important role in natural environments, where complex interactions between microorganisms and soil or water can significantly accelerate the degradation of contaminants [8].

Unlike remediation, the main goal of which is to eliminate a hazard by removing or neutralizing pollution, pollution recovery focuses on converting contaminated substances into raw materials of value. The process involves extracting valuable materials from industrial, municipal, or agricultural waste and then reintroducing them into the economic cycle. Pollution recovery thus has a dual dimension – on the one hand, it reduces the negative impact of pollution on the environment, and on the other hand, it provides secondary raw materials that can be reused in various industrial processes [9]. An example of this is the recovery of hazardous metals from industrial waste, as in the case of leaching metals from ash or electronic waste. In this process, metals such as lead, zinc, or copper are dissolved with chemical reactants and then recovered from the solution using electrochemical or chemical precipitation techniques. These metals can be reused in industry, which not only reduces their potential harmful effects on the environment but also reduces the need for virgin raw materials [10]. Similarly, the recovery of phosphorus from wastewater, which is particularly important in the context of sustainable agriculture, allows it to be reused in the production of fertilizers while preventing the eutrophication of water bodies [11]. Pollution recovery is also finding applications in the renewable energy sector. Organic waste, such as plant residues, animal excrement, or food waste, can be subjected to anaerobic digestion to produce biogas. This process not only transforms waste into a valuable energy resource but also helps reduce methane emissions, one of the most powerful greenhouse gases. Biogas, which is a mixture of methane and carbon dioxide, can be used to produce electricity and heat, which fits in with the goals of a closed-loop economy [12].

While both remediation and pollution recovery seek to protect the environment, the differences between these processes are significant. Remediation focuses on the direct removal or neutralization of contaminants, with the goal of restoring balance to ecosystems and minimizing health risks to humans. It is a primarily environmentally oriented process and often requires significant financial and technological resources [13]. In contrast, the recovery of pollutants not only eliminates the hazard but also generates secondary raw materials, significantly influencing resource management. Recovery products, such as metals, biogas, or fertilizer ingredients, can be reused, making the process more economically viable. Despite these differences, both approaches play complementary roles in pollution management and, when used in the right contexts, contribute to sustainable development [14].

1.2 POLLUTANTS REMEDIATION

Pollution remediation processes include a wide range of technologies and methods to remove, neutralize, or reduce the concentration of harmful substances in the environment. Depending on the type of contamination, environmental conditions, and the scale of the problem, the selection of the appropriate remediation method can be crucial to the effectiveness of the action. These processes can be divided into three main categories: physical, chemical, and biological, each of which has its own specific applications and limitations [15,16]. Further, each of the above-mentioned approaches also represents green technologies (Figure 1.2).

Physical remediation processes primarily involve the mechanical removal of contaminants and containment techniques to limit the further spread of harmful

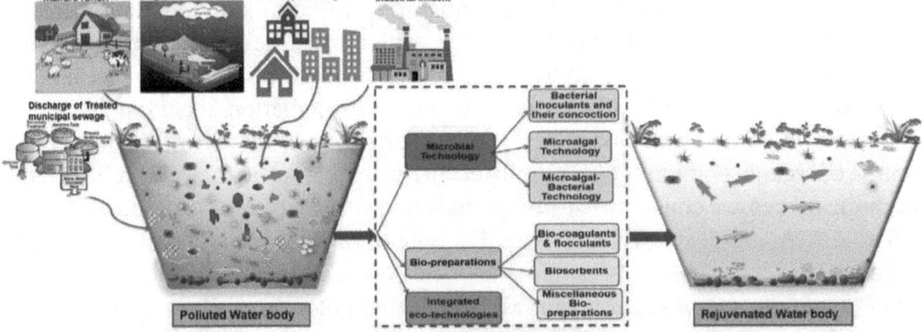

FIGURE 1.2 The general overview and idea of the green technology for the removal of pollutants from wastewater. (Adapted from [17] with permission from Elsevier.)

substances. One of the most commonly used physical processes is the "pump and treat" method, which involves pumping out contaminated groundwater and treating it using various technologies, such as adsorption filters (e.g., with activated carbon) or ion exchange processes, which might be considered green technologies [18]. Also popular in the context of soil remediation are excavation techniques, in which contaminated layers of soil are excavated and moved to special landfills or treated off-site [19]. Examples of such activities include work carried out on former chemical factory sites where the soil was heavily contaminated with hazardous metals and organic compounds, as is the case with the Love Canal factory site in the United States, which was a chemical waste disposal area for decades [20]. Another physical remediation method is the use of reactive barriers, which create physical or chemical obstacles that prevent the movement of contaminants in groundwater. Such barriers can be constructed of materials with high sorption capacity, such as zero-valent iron, which reacts with organic compounds, leading to their decomposition [21].

Chemical remediation methods are widely used when contaminants need to be neutralized or transformed quickly and efficiently. One of the most developed chemical processes is the *in situ* chemical oxidation (ISCO) of contaminants. It involves introducing powerful oxidants such as hydrogen peroxide, potassium permanganate, or ozone into the environment, which react with toxic compounds to convert them into less harmful substances. ISCO is particularly effective for volatile organic compounds (VOCs) such as benzene, toluene, and tetrachloroethene, which are common industrial pollutants [22]. Although this approach is not considered eco-friendly, it is highly efficient in the removal of pollutants. An example of the application of this method can be found in a study where the use of potassium permanganate made it possible to effectively clean chlorinated hydrocarbons from groundwater at an industrial site [23]. Another chemical method, known as stabilization and solidification (S/S), involves adding special substances (e.g., cement and lime) to contaminated soil to convert contaminants into less mobile forms. This method is mainly used for hazardous metals, which are immobilized in the cement matrix, reducing the risk of their further movement in the soil or groundwater. S/S has been successfully applied in a number of cases, including the cleanup of brownfield sites in Central and Eastern Europe, where historical industrial contaminants posed a serious threat to public health [24,25].

However, one of the most rapidly growing areas of pollution remediation is biological methods, particularly bioremediation, which uses the metabolic capabilities of microorganisms to degrade pollutants and is considered a fully green technology. Several technologies can be distinguished within bioremediation, including biodegradation, biostimulation, and bioaugmentation [26]. Biodegradation involves the decomposition of organic pollutants by naturally occurring bacteria and fungi, which, through metabolic processes, convert toxic compounds into simple, harmless products such as carbon dioxide and water. Examples of this type of process include numerous applications for the cleanup of oil-contaminated sites, where hydrophilic bacteria such as *Pseudomonas* sp. play a key role in the decomposition of aromatic hydrocarbons [27]. Biostimulation, on the other hand, involves adding nutrients, such as nitrogen and phosphorus, to the environment to stimulate the growth of natural populations of microorganisms capable of biodegrading pollutants [28]. This technique was widely used during the Exxon Valdez oil tanker disaster, where the addition of biostimulants accelerated the natural degradation process of the spilled oil [29]. Bioaugmentation, on the other hand, involves the introduction into the environment of selected strains of microorganisms that are more capable of degrading specific contaminants. This method is used especially for pollutants that are difficult to degrade, such as polychlorinated biphenyls and dioxins [30]. Within biological remediation methods, phytoremediation, which uses plants to remove contaminants from soil, groundwater, or air, is also particularly promising. Plants can accumulate hazardous metals such as cadmium, lead, or mercury in their tissues, allowing them to be safely removed later. Phytoremediation is used in areas where traditional remediation methods would be too costly or technologically unfeasible. This method not only cleans up soils but also contributes to the preservation of biodiversity and sustainable land use [31].

In summary, contaminant remediation processes include a wide range of technologies, from physical methods such as "pump and treat" to chemical oxidation to advanced biological methods such as bioremediation and phytoremediation. Each of these methods has its own unique applications and limitations, and their effectiveness depends on many factors, such as the type of contamination, environmental conditions, and the scale of the problem. Modern pollution remediation research is focused on developing green technologies that are more effective, less invasive, and less expensive, which is crucial in the context of global environmental and sustainability challenges [32].

1.3 POLLUTANTS RECOVERY

Pollution recovery processes include technologies that aim not only to remove pollutants from the environment but also to convert them into valuable secondary raw materials (Figure 1.3). In the context of today's environmental challenges and

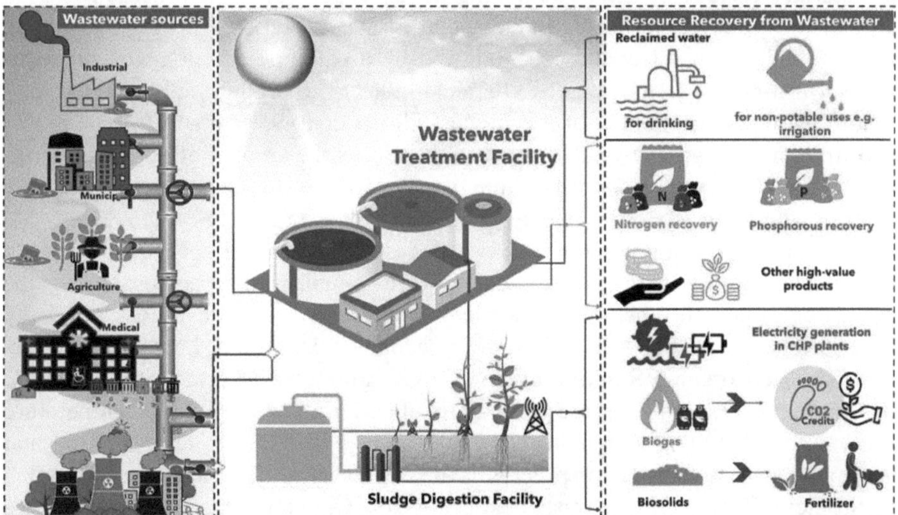

FIGURE 1.3 The representation of the resource recovery idea based on the biotechnological approach in wastewater treatment plant. (Adapted from [35] with permission from Elsevier.)

growing raw material needs, pollution recovery is becoming a key component of the circular economy [33]. Pollution recovery can be understood as processes that not only minimize the negative impact of pollution on the environment but also contribute to saving natural resources by reusing materials that might otherwise be considered waste. These processes can include a wide range of technologies, from physical and chemical to biotechnological methods, which vary depending on the type of pollution and the environment in which it occurs [34].

One of the most widely studied pollutant recovery processes is the recovery of hazardous metals from industrial water and solid waste. Metals such as lead, zinc, copper, and mercury are often released by industrial activities, posing a serious threat to ecosystems and public health. Instead of using traditional neutralization methods, which involve the removal of pollutants without recovering valuable materials, modern and green technologies seek to extract these metals and reuse them [36]. An example is biosorption technology, in which microorganisms or biomaterials selectively adsorb metals from aqueous solutions. Biomaterials, such as brown seaweed, are highly effective at adsorbing hazardous metals from wastewater, allowing them to be recovered as metal salts that can be reused in industry [37]. An alternative method for recovering metals is using membrane processes, such as nanofiltration or electrodialysis, which allow the separation of metals from solutions by penetrating selective membranes. These methods are increasingly used in industrial wastewater treatment plants, where they can effectively remove and recover metals in high purity [38].

Pollutant recovery is also crucial in the chemical and energy industries. One example is the recovery of sulfur from process gases. In many oil refining and natural gas production processes, sulfur occurs as an impurity in the form of hydrogen sulfide, which must be removed from process gases due to its toxicity and corrosion of materials. Instead of simply removing the hydrogen sulfide, a process has been developed to recover the sulfur in elemental form. The process involves oxidizing hydrogen sulfide to sulfur dioxide, followed by a catalytic reaction that produces pure sulfur that can be used in the chemical industry to produce sulfuric acid [39]. According to studies, the use of this process allows up to 95% of sulfur to be recovered from process gases, which significantly reduces the amount of pollutants emitted into the atmosphere and at the same time contributes to the production of a raw material with a high market value [40].

Another significant pollution recovery process is the recovery of industrial waste heat. Waste heat, which is generated during various industrial processes, such as the production of cement, metals, or chemicals, represents a significant loss of energy that could be used efficiently [41]. Modern technologies, such as heat recovery systems based on organic Rankine cycles (ORCs), allow waste heat to be converted into electricity. ORC systems are particularly effective at recovering low-temperature heat, making them useful in industrial applications where large amounts of thermal energy are lost as flue gas or process heat. ORC systems can achieve conversion efficiencies of 20%–25% for waste heat below 200°C, making them one of the more promising tools for improving energy efficiency in industry [42,43].

In the context of environmental protection and pollution recovery, the process of recovering phosphorus from wastewater is of particular interest. Phosphorus is an essential element for life, but its excess in aquatic ecosystems can lead to eutrophication, i.e., excessive growth of aquatic plants and algae, which leads to the degradation of water bodies [44]. At the same time, phosphorus resources are limited, and obtaining phosphorus from natural deposits is becoming increasingly expensive. Therefore, technologies have been developed to recover phosphorus from wastewater, such as precipitation of phosphate in the form of struvite (magnesium ammonium phosphate). Struvite can then be used as a fertilizer in agriculture, thus closing the phosphorus cycle and reducing the need to extract this element from primary resources. Examples of the implementation of such technologies can be found in numerous wastewater treatment plants across Europe, where phosphorus recovery is an integral part of wastewater management. These technologies are not only ecologically sustainable but also economically viable, as phosphorus recovered in the form of struvite has a high market value [45].

Pollution recovery is also crucial for e-waste, which contains valuable precious noble metals and rare earth metals such as gold, silver, palladium, platinum, and lithium. Due to the increasing consumption of electronic devices, the amount of e-waste is growing at an alarming rate, raising serious environmental and raw material issues [46]. Processes for recovering metals from e-waste are based on advanced separation methods, such as hydrometallurgical and pyrometallurgical processes. The use of cyanide leaching and electrochemical methods for the efficient recovery of gold and silver from e-waste makes it possible to transform waste materials into raw materials with a high market value. In addition, these technologies help reduce the consumption of primary raw materials and minimize the amount of hazardous waste that could end up in landfills [47].

Finally, the recovery of pollutants also includes the recovery of organic compounds such as VOCs from industrial processes where they are released as gases or vapors. VOCs can be recovered through adsorption methods, which use adsorbents such as activated carbon, or through condensation processes, which allow these compounds to be condensed and reused in industrial cycles. Recovery of VOCs is particularly important in the chemical and petrochemical industries, where their presence in the atmosphere poses a serious threat to human health and the environment [48].

1.4 CHALLENGES

Pollutant remediation and pollutant recovery processes, while essential for environmental protection and sustainable resource management, face a number of complex challenges that affect their effectiveness, cost-effectiveness, and potential environmental risks. These challenges are accompanied by technical, economic, regulatory, and social difficulties, making the successful implementation of these processes on a global scale require coordinated action, advanced research, and support from government and international institutions [49]. A deeper analysis of these challenges provides a better understanding of why, despite the growing interest in remediation and recovery technologies, their implementation often faces barriers that must be overcome to contribute to real environmental improvement [50].

One of the most important challenges in pollutant remediation and pollutant recovery processes is adapting the technology to the specific environmental conditions and diversity of pollutants. In practice, environmental pollutants are very heterogeneous chemically and physically, which means that one method is not a one-size-fits-all solution [51]. For example, hazardous metal contaminants such as cadmium, lead, and mercury are often found in soil, groundwater, and air in dispersed form, making them difficult to remove or recover using traditional technologies. The limited selectivity of technologies such as bioremediation or phytoremediation, which can only work effectively under certain conditions of pH, temperature, and humidity, creates additional difficulties. Phytoremediation, which involves the use of plants to remove hazardous metals from soil, is only effective when contaminant concentrations are low to moderate; otherwise, plants are unable to survive under toxic conditions. In contrast, in physicochemical technologies, such as ion exchange, membrane, or sorption processes, the selectivity of adsorption materials poses a challenge when contaminants are in low-solubility forms or in highly complex matrices, such as bottom sediments, which can block access to target molecules [52].

Pollutant recovery technologies are particularly challenging when it comes to processes involving the recovery of complex chemicals and organic compounds. An example is VOCs, which are difficult to recover due to their high degree of dispersion in the air and their tendency to escape quickly into the atmosphere. VOC adsorption processes that use activated carbon are often limited in sorption capacity and can lead to secondary contamination when the sorption beds reach a saturated state and require regeneration [53]. Activated carbon regeneration can be costly and the technology itself energy energy-intensive, increasing operating costs and making it less cost-effective in the long run. Moreover, some VOCs have low adsorption activity, which hinders their efficient recovery and leads to the need for more advanced and costly technologies, such as condensation or freezing [54].

Economic issues are another major challenge associated with pollutant remediation and recovery processes. Although these technologies have great potential for sustainable resource management, high installation and maintenance costs often discourage companies and local governments from implementing them. The problem is particularly pronounced in developing countries, where a lack of funding and technology often results in the problem of pollution being marginalized and access to modern treatment and resource recovery methods remains limited [55]. For example, in a study of industrial wastewater treatment and hazardous metal recovery in India, the high cost of membrane technologies such as nanofiltration is a major barrier to their widespread implementation, even though the technology is highly effective in removing and recovering metals from wastewater. In addition, waste-to-energy technologies, such as waste incineration plants and biomass processing technologies, are often seen as uneconomical from an economy-of-scale perspective, especially in low-population density regions where waste generation is insufficient to justify the cost of building and maintaining advanced facilities [56].

Another challenge is legal and regulatory constraints, which often have not kept pace with the development of pollutant recovery technologies. While many countries are implementing laws to promote recycling and the closed-loop economy, there is a lack of consistent global regulations that facilitate the cross-border movement of technology and waste and promote resource recovery internationally [57]. In the context of EU waste management legislation, regulations such as Waste Directive

2008/98/EC, while promoting the waste hierarchy, do not always provide sufficient financial and legal incentives for the development of advanced technologies for the recovery of metals from electronic waste. The lack of consistent technical standards and the diversity of regulations at national and regional levels often lead to chaos in waste recovery systems, limiting their effectiveness [58].

The risks of secondary pollution during remediation and recovery processes are also a major challenge. In pollution recovery processes such as pyrolysis and incineration, there is a risk of toxic emissions that can be generated as by-products. Pyrolysis of organic waste, while effective in converting it into fuels, can generate substances such as dioxins and furans, which are highly toxic and can cause serious health risks. For example, even in advanced thermal installations, the risk of dioxin emissions remains if appropriate flue gas cleaning technologies, such as activated carbon filters or catalytic reduction systems, are not employed [59].

Last but not least is the need for global coordination and technology transfer. Developed countries with advanced pollution remediation and recovery technologies should promote the transfer of these solutions to developing countries, where pollution problems are often greatest but technological and financial resources are most limited [60]. A globalized economic and trade system, with waste moving between countries, requires international cooperation to prevent the export of pollution and promote the recovery of raw materials in a sustainable manner. An example of positive efforts in this direction is the cooperation between European Union countries and China in recovering critical raw materials, such as rare earth metals, from electronic waste. Integrating global efforts in this area is essential to meet the challenges of pollution and ensure a cleaner and more sustainable environment for future generations [61].

1.5 SUMMARY

Summarizing the detailed analysis of pollutant remediation and pollutant recovery processes, it can be concluded that both these approaches play a key role in environmental protection in the broadest sense and are essential elements of sustainable development strategies aimed at reducing the negative impact of human activities on ecosystems. These processes, while complementary, differ in their nature, goals, and technologies used. Nevertheless, their integration into modern industrial and economic systems creates new opportunities in terms of both minimizing pollution and making rational use of resources. However, this requires overcoming numerous technological, economic, legal, and social challenges that currently limit their widespread implementation [62].

In the case of pollutant remediation, the key objective is to remove, neutralize, or degrade harmful substances from soil, water, or air, with direct implications for improving environmental quality and public health. These processes cover a wide range of methods, including biological, chemical, and physical technologies, each with its own specific applications and limitations. For example, bioremediation, which involves the use of microorganisms to degrade organic pollutants such as hydrocarbons, pesticides, or complex aromatic compounds, is one of the most promising approaches due to its ecological nature and relatively low cost of implementation.

However, as studies have shown, bioremediation is effective mainly under specific environmental conditions, such as appropriate pH, humidity, and temperature, which limit its universal applicability. Similarly, phytoremediation, which relies on plants to extract hazardous metals such as lead, mercury, and cadmium from soil, is only effective when contaminant concentrations are relatively low. High environmental toxicity can lead to weakened plants and their inability to accumulate metals, which is a serious limitation of this method. In addition, many remediation processes, especially for highly toxic or complex contaminants, require the use of more advanced physicochemical technologies, such as ion exchange processes, adsorption, or the use of membranes, which are much more expensive and technically complex [63].

Unlike remediation processes, pollutant recovery technologies are not limited to neutralizing or removing pollutants but aim to recover valuable raw materials from waste and pollution. This concept fits directly into the tenets of the circular economy, which promotes recycling and reuse of resources to minimize waste. Recovery of hazardous metals, such as gold, copper, cobalt, and rare earth metals, from electronic waste (e-waste) is one of the most growing areas in this field. As studies have shown, hydrometallurgical technologies such as leaching allow for efficient recovery of metals, which not only reduces the need for primary extraction of raw materials but also reduces the amount of waste. In this context, the recovery of raw materials is becoming a key component of sustainable natural resource management, especially with the growing demand for strategic raw materials such as lithium and neodymium, which are essential for batteries, smartphones, and other high-tech products. However, the technical challenges of recovering valuable raw materials are significant. Hydrometallurgical processes, while efficient, generate toxic liquid waste that must be properly neutralized, which in turn increases operating costs. In addition, recovery of raw materials from waste with low metal concentrations requires advanced separation methods, which are not only costly but also technically complex [33].

A significant challenge for both pollutant remediation and pollutant recovery processes is economic issues. The high cost of the technologies, including installation, maintenance, and required energy, is a significant barrier to their widespread implementation. For example, in the case of industrial wastewater remediation, membrane technologies such as nanofiltration are highly effective in removing hazardous metals, but their cost makes them mainly used in developed countries. In developing regions, where the pollution problem is often greatest, but financial and technological resources are limited, the implementation of modern treatment technologies lags behind. Similarly, in the case of pollutant recovery technologies, although the recovery of metals from e-waste can bring significant economic benefits, the operating costs, especially when the metal content of the waste is low, often exceed the value of the recovered raw materials. For this reason, many countries are still not implementing recovery technologies on a large scale, despite their potential environmental and economic benefits [55].

Regulatory and legal issues are another major barrier to the development of pollutant remediation and pollutant recovery processes. The lack of consistent regulations at the global level hinders the widespread implementation of advanced technologies and the cross-border movement of waste and raw materials. An example of this is the complex regulations on the export and import of electronic waste, which often lead to illegal treatment practices in countries with less stringent regulations, which in turn contribute to serious health and environmental problems. In addition, the diversity of regulations at national and regional levels often leads to fragmented waste management systems and inefficiencies in recovery processes. In the case of the European Union, although there is a framework of regulations promoting a closed-loop economy, such as Directive 2008/98/EC on waste, the lack of uniform technical standards for the recovery of critical metals limits their widespread implementation at the international level [64].

An additional technological challenge is the threat of secondary pollution that can occur during the implementation of remediation and recovery processes. For example, thermal processes such as waste incineration or pyrolysis, while effective in waste volume reduction and energy recovery, can lead to emissions of hazardous substances such as dioxins, furans, and nitrogen oxides, which are highly toxic. Cleaning flue gases of these substances requires advanced filtration systems, which in turn increases the cost of the process. Similar challenges arise in hydrometallurgical processes, where liquid residues from metal leaching can contain toxic substances that must be properly neutralized before they are released into the environment, placing an additional burden on recovery technologies [65].

Finally, global cooperation and technology transfer are key to overcoming the barriers associated with pollutant remediation and pollutant recovery. Developing countries, where the pollution problem is most acute, lack both advanced technologies and financial resources, making it difficult to implement modern solutions. International initiatives, such as the research cooperation between the European Union and China on rare earth recovery, show that a global approach to the problem of pollution and raw materials can lead to significant environmental progress. However, for these efforts to be effective, more research and development are needed, as well as harmonization of international regulations on waste and raw materials management [66].

In conclusion, both pollutant remediation and pollutant recovery processes offer tremendous opportunities for environmental improvement and resource management. However, their full potential can only be achieved if numerous technological, economic, and regulatory challenges are overcome. Achieving this goal requires an integrated approach, combining the development of new and green technologies, effective resource management, and coordinated international cooperation. Only then will it be possible to build a globally sustainable system that not only minimizes pollution but also promotes the rational use of natural resources [62].

REFERENCES

[1] S. Tanaka, Environmental pollution and remediation, in: S. Tanaka, M. Kurasaki, M. Morikawa, Y. Kamiya (Eds.), *Design of Materials and Technologies for Environmental Remediation*, Springer Nature, Singapore, 2023: pp. 3–32. https://doi.org/10.1007/698_2021_819.

[2] S. Singh, S.K. Paswan, P. Kumar, R.K. Singh, L. Kumar, Chapter 14: Heavy metal water pollution: an overview about remediation, removal and recovery of metals from contaminated water, in: S. K. Shukla, S. Kumar, S. Madhav, P. K. Mishra (Eds.), *Metals in Water*, Elsevier, 2023: pp. 263–284. https://doi.org/10.1016/B978-0-323-95919-3.00018-5.

[3] A. Abidli, Y. Huang, Z. Ben Rejeb, A. Zaoui, C.B. Park, Sustainable and efficient technologies for removal and recovery of toxic and valuable metals from wastewater: Recent progress, challenges, and future perspectives, *Chemosphere* 292 (2022) 133102. https://doi.org/10.1016/j.chemosphere.2021.133102.

[4] A. Vaseashta, M. Vaclavikova, S. Vaseashta, G. Gallios, P. Roy, O. Pummakarnchana, Nanostructures in environmental pollution detection, monitoring, and remediation, *Sci. Technol. Adv. Mater.* 8 (2007) 47. https://doi.org/10.1016/j.stam.2006.11.003.

[5] E.O. Dada, K.I. Njoku, A.A. Osuntoki, M.O. Akinola, A review of current techniques of physico-chemical and biological remediation of heavy metals polluted soil, *Ethiop. J. Environ. Stud. Manage.* 8 (2015) 606–615. https://doi.org/10.4314/ejesm.v8i5.13.

[6] M. Biloshytskyi, H. Tatarchenko, N. Biloshytska, P. Uvarov, Operational lifetime increase of the pumping equipment when pumping-out contaminated groundwater, *Min. Miner. Depos.* 15 (2021) 42–49. https://doi.org/10.33271/mining15.01.042.

[7] G. Saxena, R. Chandra, R.N. Bharagava, Environmental pollution, toxicity profile and treatment approaches for tannery wastewater and its chemical pollutants, in: P. de Voogt (Ed.), *Reviews of Environmental Contamination and Toxicology Volume 240*, Springer International Publishing, Cham, 2017: pp. 31–69. https://doi.org/10.1007/398_2015_5009.

[8] M.U. Saeed, N. Hussain, A. Sumrin, A. Shahbaz, S. Noor, M. Bilal, L. Aleya, H.M.N. Iqbal, Microbial bioremediation strategies with wastewater treatment potentialities – A review, *Sci. Total Environ.* 818 (2022) 151754. https://doi.org/10.1016/j.scitotenv.2021.151754.

[9] V. Gunarathne, A. Ashiq, S. Ramanayaka, P. Wijekoon, M. Vithanage, Biochar from municipal solid waste for resource recovery and pollution remediation, *Environ. Chem. Lett.* 17 (2019) 1225–1235. https://doi.org/10.1007/s10311-019-00866-0.

[10] U.U. Jadhav, H. Hocheng, A review of recovery of metals from industrial waste, *J. Achiev. Mater. Manuf. Eng.* 54 (2012) 159–167.

[11] P. Wilfert, P.S. Kumar, L. Korving, G.-J. Witkamp, M.C.M. van Loosdrecht, The relevance of phosphorus and iron chemistry to the recovery of phosphorus from wastewater: a review, *Environ. Sci. Technol.* 49 (2015) 9400–9414. https://doi.org/10.1021/acs.est.5b00150.

[12] M. Ocak, Z. Ocak, S. Bilgen, S. Keleş, K. Kaygusuz, Energy utilization, environmental pollution and renewable energy sources in Turkey, *Energy Convers. Manag.* 45 (2004) 845–864. https://doi.org/10.1016/S0196-8904(03)00192-4.

[13] D.C. Adriano, W.W. Wenzel, J. Vangronsveld, N.S. Bolan, Role of assisted natural remediation in environmental cleanup, *Geoderma* 122 (2004) 121–142. https://doi.org/10.1016/j.geoderma.2004.01.003.

[14] A. Gungor, S.M. Gupta, Issues in environmentally conscious manufacturing and product recovery: a survey, *Comput. Ind. Eng.* 36 (1999) 811–853. https://doi.org/10.1016/S0360-8352(99)00167-9.

[15] H. Saleem, S.J. Zaidi, A.F. Ismail, P.S. Goh, Advances of nanomaterials for air pollution remediation and their impacts on the environment, *Chemosphere* 287 (2022) 132083. https://doi.org/10.1016/j.chemosphere.2021.132083.

[16] L. He, Y. Xu, M. Zhang, S. Gul, X. Zhang, H. Zhong, Y. Tang, D. Dong, Y. Xu, D. Liu, N. Bolan, H. Wang, Effect of remediation technologies on soil fertility in heavy metal(loid)-contaminated soils: A critical review, *Crit. Rev. Environ. Sci. Technol.* 54 (2024) 1417–1435. https://doi.org/10.1080/10643389.2024.2315009.

[17] M. Simon, H. Joshi, A review on green technologies for the rejuvenation of polluted surface water bodies: Field-scale feasibility, challenges, and future perspectives, *J. Environ. Chem. Eng.* 9 (2021) 105763. https://doi.org/10.1016/j.jece.2021.105763.

[18] C. Yudao, C. Yaping, J. Yaping, Y. Yani, L. Zhiteng, T. Qiling, S. Zisen, Pump-and-treat method to remove nitrate from groundwater with liquor as the carbon source, *Environ. Earth Sci.* 76 (2017) 392. https://doi.org/10.1007/s12665-017-6729-z.

[19] L. Liu, W. Li, W. Song, M. Guo, Remediation techniques for heavy metal-contaminated soils: principles and applicability, *Sci. Total Environ.* 633 (2018) 206–219. https://doi.org/10.1016/j.scitotenv.2018.03.161.

[20] G.R. Najem, J.L. Cappadona, Health effects of hazardous chemical waste disposal sites in New Jersey and in the United States: a review, *Am. J. Prev. Med.* 7 (1991) 352–362. https://doi.org/10.1016/S0749-3797(18)30872-9.

[21] R. Thiruvenkatachari, S. Vigneswaran, R. Naidu, Permeable reactive barrier for groundwater remediation, *J. Ind. Eng. Chem.* 14 (2008) 145–156. https://doi.org/10.1016/j.jiec.2007.10.001.

[22] R.M. Kurakalva, Chapter 8: In situ chemical oxidation (ISCO) remediation: a focus on activated persulfate oxidation of pesticide-contaminated soil and groundwater, in: S. Kathi, S. Devipriya, K. Thamaraiselvi (Eds.), *Cost Effective Technologies for Solid Waste and Wastewater Treatment*, Elsevier, 2022: pp. 75–86. https://doi.org/10.1016/B978-0-12-822933-0.00011-5.

[23] T.T. Tsai, C.M. Kao, T.Y. Yeh, M.S. Lee, Chemical Oxidation of Chlorinated Solvents in Contaminated Groundwater: Review, *J. Hazard. Toxic Radioact. Waste Manage.* 12 (2008) 116–126. https://doi.org/10.1061/(ASCE)1090025X(2008)12:2(116).

[24] J.R. Conner, S.L. Hoeffner, A critical review of stabilization/solidification technology, *Crit. Rev. Environ. Sci. Technol.* 28 (1998) 397–462. https://doi.org/10.1080/10643389891254250.

[25] S. Paria, P.K. Yuet, Solidification–stabilization of organic and inorganic contaminants using portland cement: a literature review, *Environ. Rev.* 14 (2006) 217–255. https://doi.org/10.1139/a06-004.

[26] T. Iwamoto, M. Nasu, Current bioremediation practice and perspective, *J. Biosci. Bioeng.* 92 (2001) 1–8. https://doi.org/10.1016/S1389-1723(01)80190-0.

[27] M. Alexander, *Biodegradation and Bioremediation*, Gulf Professional Publishing, Houston, TX, 1999.

[28] G. Omokhagbor Adams, P. Tawari Fufeyin, S. Eruke Okoro, I. Ehinomen, Bioremediation, biostimulation and bioaugmentation: a review, *IJEBB* 3 (2020) 28–39. https://doi.org/10.12691/ijebb-3-1-5.

[29] D.C. Youvan, *Bioremediation in Oil Spill Response: Case Studies from Exxon Valdez to Modern Environmental Remediation*, Springer, 2024. https://doi.org/10.13140/RG.2.2.34067.49445.

[30] M. Herrero, D.C. Stuckey, Bioaugmentation and its application in wastewater treatment: a review, *Chemosphere* 140 (2015) 119–128. https://doi.org/10.1016/j.chemosphere.2014.10.033.

[31] A.A. Ansari, S.S. Gill, R. Gill, G.R. Lanza, L. Newman, eds., *Phytoremediation: Management of Environmental Contaminants*, Volume 2, Springer International Publishing, Cham, 2015. https://doi.org/10.1007/978-3-319-10969-5.

[32] M.A. Hashim, S. Mukhopadhyay, J.N. Sahu, B. Sengupta, Remediation technologies for heavy metal contaminated groundwater, *J. Environ. Manage.* 92 (2011) 2355–2388. https://doi.org/10.1016/j.jenvman.2011.06.009.

[33] M.N.V. Prasad, K. Shih, *Environmental Materials and Waste: Resource Recovery and Pollution Prevention*, Academic Press, 2016.

[34] J. Singh, B.-K. Lee, Pollution control and metal resource recovery for low grade automobile shredder residue: a mechanism, bioavailability and risk assessment, *Waste Management* 38 (2015) 271–283. https://doi.org/10.1016/j.wasman.2015.01.035.

[35] G. Yadav, A. Mishra, P. Ghosh, R. Sindhu, V. Vinayak, A. Pugazhendhi, Technical, economic and environmental feasibility of resource recovery technologies from wastewater, *Sci. Total Environ.* 796 (2021) 149022. https://doi.org/10.1016/j.scitotenv.2021.149022.

[36] S. Krishnan, N.S. Zulkapli, H. Kamyab, S.M. Taib, M.F.B.M. Din, Z.A. Majid, S. Chaiprapat, I. Kenzo, Y. Ichikawa, M. Nasrullah, S. Chelliapan, N. Othman, Current technologies for recovery of metals from industrial wastes: an overview, *Environ. Technol. Innov.* 22 (2021) 101525. https://doi.org/10.1016/j.eti.2021.101525.

[37] G.R. de Freitas, M.G.C. da Silva, M.G.A. Vieira, Biosorption technology for removal of toxic metals: a review of commercial biosorbents and patents, *Environ. Sci. Pollut. Res.* 26 (2019) 19097–19118. https://doi.org/10.1007/s11356-019-05330-8.

[38] E. Kavitha, E. Poonguzhali, D. Nanditha, A. Kapoor, G. Arthanareeswaran, S. Prabhakar, Current status and future prospects of membrane separation processes for value recovery from wastewater, *Chemosphere* 291 (2022) 132690. https://doi.org/10.1016/j.chemosphere.2021.132690.

[39] P.R. Robinson, Sulfur Removal and Recovery, in: C. S. Hsu, P. R. Robinson (Eds.), Springer *Handbook of Petroleum Technology*, Springer International Publishing, Cham, 2017: pp. 649–673. https://doi.org/10.1007/978-3-319-49347-3_20.

[40] S.K. Gangwal, W.J. McMichael, T.P. Dorchak, The direct sulfur recovery process, *Environ. Prog.* 10 (1991) 186–191. https://doi.org/10.1002/ep.670100317.

[41] S. Brückner, S. Liu, L. Miró, M. Radspieler, L.F. Cabeza, E. Lävemann, Industrial waste heat recovery technologies: an economic analysis of heat transformation technologies, *Appl. Energy* 151 (2015) 157–167. https://doi.org/10.1016/j.apenergy.2015.01.147.

[42] S. Lecompte, H. Huisseune, M. van den Broek, B. Vanslambrouck, M. De Paepe, Review of organic Rankine cycle (ORC) architectures for waste heat recovery, *Renew. Sustain. Energy Rev.* 47 (2015) 448–461. https://doi.org/10.1016/j.rser.2015.03.089.

[43] E. Macchi, M. Astolfi, *Organic Rankine Cycle (ORC) Power Systems: Technologies and Applications*, Woodhead Publishing, 2016.

[44] Z. Yuan, S. Pratt, D.J. Batstone, Phosphorus recovery from wastewater through microbial processes, *Curr. Opin. Biotechnol.* 23 (2012) 878–883. https://doi.org/10.1016/j.copbio.2012.08.001.

[45] B. Tansel, G. Lunn, O. Monje, Struvite formation and decomposition characteristics for ammonia and phosphorus recovery: a review of magnesium-ammonia-phosphate interactions, *Chemosphere* 194 (2018) 504–514. https://doi.org/10.1016/j.chemosphere.2017.12.004.

[46] F.-C. Mihai, *E-Waste in Transition: From Pollution to Resource*, BoD – Books on Demand, 2016.

[47] A. Ashiq, J. Kulkarni, M. Vithanage, Chapter 10: Hydrometallurgical recovery of metals from e-waste, in: M. N. V. Prasad, M. Vithanage (Eds.), *Electronic Waste Management and Treatment Technology*, Butterworth-Heinemann, 2019: pp. 225–246. https://doi.org/10.1016/B978-0-12-816190-6.00010-8.

[48] H. Insam, M.S.A. Seewald, Volatile organic compounds (VOCs) in soils, *Biol. Fertil. Soils* 46 (2010) 199–213. https://doi.org/10.1007/s00374-010-0442-3.

[49] B. Mishra, S. Varjani, G. Kumar, M.K. Awasthi, S.K. Awasthi, R. Sindhu, P. Binod, E.R. Rene, Z. Zhang, Microbial approaches for remediation of pollutants: Innovations, future outlook, and challenges, *Energy Environ.* 32 (2021) 1029–1058. https://doi.org/10.1177/0958305X19896781.

[50] A. Selvi, A. Rajasekar, J. Theerthagiri, A. Ananthaselvam, K. Sathishkumar, J. Madhavan, P.K.S.M. Rahman, Integrated remediation processes toward heavy metal removal/recovery from various environments – a review, *Front. Environ. Sci.* 7 (2019) 66. https://doi.org/10.3389/fenvs.2019.00066.

[51] A.I. Shah, M.U. Din Dar, R.A. Bhat, J.P. Singh, K. Singh, S.A. Bhat, Prospectives and challenges of wastewater treatment technologies to combat contaminants of emerging concerns, *Ecol. Eng.* 152 (2020) 105882. https://doi.org/10.1016/j.ecoleng.2020.105882.

[52] A. Mahar, P. Wang, A. Ali, M.K. Awasthi, A.H. Lahori, Q. Wang, R. Li, Z. Zhang, Challenges and opportunities in the phytoremediation of heavy metals contaminated soils: a review, *Ecotoxicol. Environ. Saf.* 126 (2016) 111–121. https://doi.org/10.1016/j.ecoenv.2015.12.023.

[53] G.R. Parmar, N.N. Rao, Emerging control technologies for volatile organic compounds, *Crit. Rev. Environ. Sci. Technol.* 39 (2008) 41–78. https://doi.org/10.1080/10643380701413658.

[54] X. Zhang, B. Gao, A.E. Creamer, C. Cao, Y. Li, Adsorption of VOCs onto engineered carbon materials: a review, *J. Hazard. Mater.* 338 (2017) 102–123. https://doi.org/10.1016/j.jhazmat.2017.05.013.

[55] R. Goutham, P. Rohit, S.S. Vigneshwar, A. Swetha, J. Arun, K.P. Gopinath, A. Pugazhendhi, Ionic liquids in wastewater treatment: a review on pollutant removal and degradation, recovery of ionic liquids, economics and future perspectives, *J. Mol. Liq.* 349 (2022) 118150. https://doi.org/10.1016/j.molliq.2021.118150.

[56] V. Singh, C.P.S. Chandel, Analytical Study of heavy Metals of Industrial Effluents at Jaipur, Rajasthan (India), *J. Environ. Sci. Eng.* 48 (2006) 103–108.

[57] S.A. Khan, E-products, E-waste and the Basel convention: regulatory challenges and impossibilities of international environmental law, *Rev. Eur. Comp. Int. Environ. Law* 25 (2016) 248–260. https://doi.org/10.1111/reel.12163.

[58] M. Gharfalkar, R. Court, C. Campbell, Z. Ali, G. Hillier, Analysis of waste hierarchy in the European waste directive 2008/98/EC, *Waste Manage.* 39 (2015) 305–313. https://doi.org/10.1016/j.wasman.2015.02.007.

[59] M. Grifoni, E. Franchi, D. Fusini, M. Vocciante, M. Barbafieri, F. Pedron, I. Rosellini, G. Petruzzelli, Soil remediation: Towards a resilient and adaptive approach to deal with the ever-changing environmental challenges, *Environments* 9 (2022) 18. https://doi.org/10.3390/environments9020018.

[60] D.B. Audretsch, E.E. Lehmann, M. Wright, Technology transfer in a global economy, *J. Technol. Transf.* 39 (2014) 301–312. https://doi.org/10.1007/s10961-012-9283-6.

[61] J. Howells, International coordination of technology flows and knowledge activity in innovation, *Int. J. Technol. Manage.* 19 (2000) 806–819. https://doi.org/10.1504/IJTM.2000.002845.

[62] G. Naidu, S. Ryu, R. Thiruvenkatachari, Y. Choi, S. Jeong, S. Vigneswaran, A critical review on remediation, reuse, and resource recovery from acid mine drainage, *Environ. Pollut.* 247 (2019) 1110–1124. https://doi.org/10.1016/j.envpol.2019.01.085.

[63] A. Roy, A. Sharma, S. Yadav, L.T. Jule, R. Krishnaraj, Nanomaterials for remediation of environmental pollutants, *Bioinorg. Chem. Appl.* 2021 (2021) 1764647. https://doi.org/10.1155/2021/1764647.

[64] A. Saravanan, P. Senthil Kumar, S. Jeevanantham, S. Karishma, B. Tajsabreen, P.R. Yaashikaa, B. Reshma, Effective water/wastewater treatment methodologies for toxic pollutants removal: processes and applications towards sustainable development, *Chemosphere* 280 (2021) 130595. https://doi.org/10.1016/j.chemosphere.2021.130595.

[65] O.B. Akpor, D.A. Otohinoyi, D.T. Olaolu, B.I. Aderiye, Pollutants in wastewater effluents: impacts and remediation processes, *Int. J. Environ. Res. Earth Sci.* 3 (2014) 050–059.

[66] D.G. Ockwell, J. Watson, G. MacKerron, P. Pal, F. Yamin, Key policy considerations for facilitating low carbon technology transfer to developing countries, *Energy Policy* 36 (2008) 4104–4115. https://doi.org/10.1016/j.enpol.2008.06.019.

2 Green Strategies for Removal of Emerging Contaminants from Aqueous System

A Promising Approach for Sustainable Future

Riti Thapar Kapoor, Mohd Rafatullah, and Jakub Zdarta

2.1 INTRODUCTION

Water pollution has become a serious problem worldwide due to the upsurge in industrial and anthropogenic activities. Extensive agricultural practices, industrial operations, urbanization, improper waste disposal, and insufficient sewage systems contaminate natural water resources by jeopardizing their safety. In the last decade, emerging contaminants (ECs) have become a serious issue as they are widely distributed throughout the environmental matrices such as water, soil, air, food, and human/animal tissues. They can negatively affect the environment and health of human beings even present at a very less amount, thus also considered contaminants of emerging concern (Rizzo et al. 2019). The disposal of pollutants in water due to rise in anthropogenic activities can contaminate water resources and is categorized as emerging pollutants (Ouda et al. 2021). Emerging pollutants are organic compounds, pseudo-persistent, unregulated new contaminants, and present in a trace amount from parts per trillion to parts per billion (Rout et al. 2021). The United States Geological Survey reported ECs as "any natural or synthetic chemical or microbe which are not generally examined in environment but has capacity to reflect negative effects on environment or human health". As per the US Environmental Protection Agency, ECs are novel compounds that show an adverse impact on living organisms and ecosystems (Besha et al. 2020). The existence of ECs in the environment is not a new phenomenon but was reported back to 2000 years ago because of the exploitation of lead mines by Romans (Sauvé and Desrosiers 2014). The first report on awareness of ECs was highlighted in the book *Silent Spring* by Rachel Carson in 1962, which reflected the link between detrimental impacts of dichloro-diphenyl-trichloroethane (DDT) on the environment (Sauve and Desrosiers 2014). Most ECs

 DOI: 10.1201/9781003536277-2

have been detected in the last two decades, however, because of less availability of identification procedures for the detection of less ECs concentration in the environment (Noguera-Oviedo and Aga 2016). Approximately 700 compounds are classified into 20 groups of ECs (www.norman-network.net) such as surfactants, antibiotics, pharmaceuticals, endocrine-disrupting compounds, sunscreen, pesticides, and fire retardants (Vasilachi et al. 2021).

Mishra et al. (2023) observed that >9 million people die every year due to exposure to contaminated water, food products, microplastics (MPs), drugs, endocrine disruptors, and other chemicals. In 1995, endocrine disruptors were considered ECs as they reflected adverse effects on the function of hormones and the reproductive system (Seralini and Jungers 2021). The application of artificial sweeteners as sugar substitutes in the diet of human beings and the feed of animals has been extensively used. In 1970, US Food and Drug Administration banned the consumption of artificial sweeteners due to their carcinogenic effects on experimental animals. However, the presence of artificial sweeteners in water and soil is due to their extensive use and has shown significant toxicity on ecosystems (Buerge et al. 2024). Among different ECs, approximately 70% are pharmaceutical and personal care products, whereas 30% are agro-wastes and industrial products. ECs can easily reach into the environment via effluents released from septic tanks, municipal wastewater treatment plants, households, hospitals, industries, etc. (Figure 2.1). Antibiotics like amoxicillin, ciprofloxacin, and azithromycin have been reported in wastewater obtained from different countries of Asia, Europe, and America (Tran et al. 2018). Lipophilic substances with less water solubility like triclosan, carbamazepine, and musk xylene reflected high elimination efficacy via adsorption process. The hydrophilic substances with more water solubility like caffeine, atenolol, and glyphosate cannot be

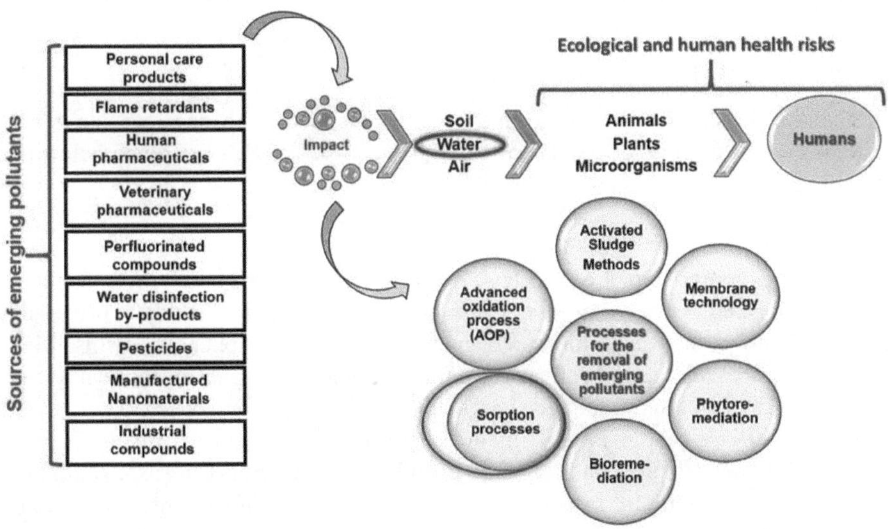

FIGURE 2.1 Various kinds of emerging contaminants and their removal methods. (Reprinted from Vasilachi et al. 2021.)

easily removed through adsorption process. Mohapatra and Kirpalani (2019) reported deleterious effects of ECs on macro-invertebrate diversity in rivers, population of fishes and freshwater mussels, human beings, and wildlife. The long-duration persistence of ECs may lead to the extinction of various flora and fauna from ecosystems (de Oliveira et al. 2019). Thus, wastewater released from industries must be treated prior to its transfer into rivers and lakes to prevent groundwater pollution. Thus, there is an urgent need to maintain strict discard limitations for ECs and application of current technologies for the elimination of ECs from industrial effluents or wastewater before its release into the environment. Traditional treatment procedures are unable to completely eliminate ECs from environment. Several efforts have been made by researchers for enhancement in ECs remediation and novel methods have been applied for the detection and treatment of wastewater. The aim of the present investigation is to compile information on different types of ECs, their detection technologies, environmental and health risks associated with ECs, and procedures used for the removal of ECs for environmental protection.

2.2 SOURCES OF EMERGING CONTAMINANTS

Emerging pollutants are hazardous, non-degradable, easily accumulate, and transmit to different locations via air, water, and migratory species and accumulate in soil and aquatic ecosystems. Pharmaceutical products, pesticides, surfactants, flame retardants, artificial sweeteners, hormones, drugs, ointments, cosmetics, sun protection, and personal care products are major sources of ECs (Richardson and Kimura 2017). Different types of ECs are disposed into natural environment without proper treatment, either intentionally or unintentionally, which poses health-related disorders (Rathi et al. 2021a). The ubiquitous presence of ECs and their adverse repercussions on the environment and human health have become a serious issue (Khan et al. 2021).

2.2.1 Types of Emerging Contaminants

ECs can be classified as per the sources, types, uses, consequences, and fate in ecosystems (Kristanti et al. 2023). Details of various types of ECs are given below and in Table 2.1.

2.2.1.1 Phenolic Compounds

The phenolic compounds are colorless, crystalline compounds with aroma and soluble in water and organic solvents. These are widely used in chemicals, processing of coal, disinfectants, metallurgical, generation of oil, and phenolic resins. Many industrial processes like oil refineries, paint, pesticide, and pharmaceutical sectors use phenolic compounds and discharge toxic substances into wastewater (Couto et al. 2019). Bisphenol A is used as an important component for resins, plastics, and non-polymer supplements. Phenols are also present in mouthwash, ointments, varnish removers, and oral sprays applied for the treatment of sore throat or used for anesthesia (Anku et al. 2017). Phenolic compounds are considered pollutants as they increase oxygen demand and modify water texture. Phenolic compounds also act as endocrine-disrupting compounds like 2,4,6-trichlorophenol and pentachlorophenol

TABLE 2.1
Different Categories and Sources of Emerging Contaminants

Categories of ECs	Sub-types of ECs	Major Pollutants	Sources	References
Biocides	Herbicides, fungicides, molluscicide	Butachlor and epoxiconazole	Surface water, aquafarming, and agricultural runoff	Rathi et al. (2021), Shahid et al. (2021)
Industrial chemicals	Plasticizers and fire retardants	Dimethyl adipate and tris (1-chloro-2-propyl) phosphate	Domestic and industrial wastewater	Lempart et al. (2020)
Personal care products	Insect repellents, synthetic musk, sunscreen UV filters	Diethyltoluamide, 4-benzophenone, galaxolide, tonalide	WWTPs effluent, landfill leachate, surface water	Azaroff et al. (2021)
Pharmaceutically active complexes	Antibiotics, antidepressant, lipid regulators, β-blocker, hormones	Diazepam, ciprofloxacin, metoprolol, diclofenac, carbamazepine, testosterone	Drug manufacturing unit, hospitals, and health centres, domestic or livestock farms wastewater	Jacob et al. (2021)
Stimulants and illegal drugs	Stimulant, human indicator, hallucinogen	Amphetamine, cocaine, caffeine, nicotine, 3,4-methylenedioxymethamphetamine, 3,4-methylenedioxyamphetamine	Industrial wastewater	
Endocrine disruptors	Bisphenol, xenohormone, phthalates	Bisphenol A, xenoestrogen, dioctyl phthalate	Drinking water, surface water, sediments, soil	Shahid et al. (2021)
Regulated compounds	Pesticides and poly aromatic hydrocarbon	Chlorpyrifos and phenanthrene	Agricultural runoffs, sewage treatment plants, soil, surface water	Jacob et al. (2020)
Surfactants	Ionic and non-ionic surfactants	Tweens and sodium lauryl sulfate	Domestic and industrial wastewater	Jagini et al. (2021), Shahid et al. (2021)
Perfluorinated alkylated compounds	Perfluorocarboxylic acid and perfluorosulfonic acid	Perfluorooctanoic acid and perfluorooctane sulfonate	Sediments, groundwater, wastewater	Cheng et al. (2020)

have been recognized as toxic contaminants by WHO. Due to its high toxicity and carcinogenic nature, phenol content should be <1 ppb in drinkable water as per guidelines of the European Union (Cao et al. 2021).

2.2.1.2 Plasticizers

Phthalate esters (PAEs) are dialkyl or alkyl aryl esters of 1,2-benzenedicarboxylic acid and are widely utilized as plasticizers (Huang et al. 2021). Plasticizers such as bisphenol A (BPA) production are around 2 million metric tons annually (Wilkinson et al. 2017). It is predicted that around 500 million tons of PAEs will be generated per year by 2050 at the global level (Sardon and Dove 2018). Phthalates or PAEs are generated through esterification of phthalic anhydride and alcohol. Phthalates and Bisphenol A are semi-volatiles that can quickly transmit into the environment and act as plasticizers and show their potential for endocrine disruption. Low molecular weight phthalates like dimethyl or diethyl phthalates are present in personal care products, varnishes, and coatings, and high molecular weight phthalates such as di-isononyl and di-isodecyl phthalates are present in construction materials (Pang et al. 2021). Phthalates are widely applied in the production of pesticides, lubricants, plastic production, cosmetics, packaging, adhesives and paint additives, etc. (Eichler et al. 2019).

2.2.1.3 Pharmaceutical Sector

Pharmaceutical compounds are complicated polar molecules with various structures, sizes, shapes, and molecular weights (89–900 g/mol) (Almeida-Naranjo et al. 2023). Pharmaceutical compounds are used in diagnostics, treatment, and prevention of diseases (Patel et al. 2019). Due to their complex structure and high durability, pharmaceutical compounds cannot be digested thoroughly and excreted through urine and feces or biliary system; similarly, application of pharmaceuticals in agriculture finally enters in surface water and wastewater treatment plants (Matthee et al. 2023). Pharmaceutical compounds have become troublesome contaminants as they are widely present in water resources worldwide and adversely affect the environment and human health (Patel et al. 2019). Around 3000 substances are utilized as pharmaceutical components such as antibiotics, antidiabetics, beta-blockers, antidepressants, contraceptives, and lipid regulators (Quesada et al. 2019). Triclosan is an antibiotic present in toothpaste, soap, etc. The manufacturing sites of pharmaceutical compounds also release effluents into sewage treatment plants (Ehrhart and Granek, 2023). The presence of sulfamethoxazole, ibuprofen, diclofenac, and carbamazepine has been reported in drinking water (Reyes et al. 2021). In China, ketoprofen and ibuprofen were reported in high concentrations in tissues of living organisms, whereas triclosan was found in the liver of fishes from Japan (Reyes et al. 2021). Traditional treatment plant depends on microbes for degradation; however, due to complex structure and composition, recalcitrant chemicals cannot be degraded and their elimination rate is only 10%.

2.2.1.4 Personal Care Product-Based ECs

Personal care products like cosmetics, perfumes, and skin or hair products are used for nutrition, cleansing, etc. Different ECs are present in personal care products like preservatives, disinfectants, fragrances, insect repellents, shampoos, and sunscreen (Pal et al. 2014). Parabens (20–30 μg/L) are used as preservatives in medicines and cosmetics. Triclosan is present in toothpaste, shampoo, soap, skin care products, cosmetics, and deodorants, and N,N-diethyl-meta-toluamide (DEET) is present in insect's repellent (Mojiri et al. 2022). Soaps, detergents, and deodorants contain a significant amount of fragrance like nitro musks. Polycyclic musks are affordable fragrances and are widely used in cosmetics, washing powder, and detergents (Abedi et al. 2018). Many UV filters such as benzophenones are used in the preparation of sunscreen lotions, which protect skin from sunlight (Campanha et al. 2015). Due to their everyday consumption, personal care products are constantly released into the environment in significant quantities. These substances can easily accumulate, posing a threat to the health of human beings and environment.

2.2.1.5 Pesticides

Pesticides play a pivotal role in agricultural production by preventing diseases, weeds, and insect infestation and ensuring food yield. Pesticides cover a wide variety of substances such as herbicides, fungicides, insecticides, bactericides, rodenticides, and plant growth regulators. In 2020, approximately 3 million tons of pesticides were used in agriculture globally, which is expected to rise in future to meet global food demand (Reddy et al. 2024). They enter in environment from the application site, and depends on chemicals solubility, these pollutants can attach to soil, dissolve in groundwater, absorb by roots, or can be transmitted to surface water and accumulate in plant tissues (Gani and Kazmi 2017). Endosulfan, hexachlorocyclohexane, chlorpyriphos, DDT, atrazine, and alachlor are commonly used pesticides (Gani and Kazmi 2017).

2.2.1.6 Polycyclic Aromatic Hydrocarbon-Based ECs

Polycyclic aromatic hydrocarbons are generated by combustion of fossil fuel and forest fire and affect human beings via ingestion, skin contact, and inhalation (Berthiaume et al. 2021). With an increase in the number of rings, its hydrophobic nature is enhanced but the volatility of contaminants is reduced. The aromatic contaminants can be mixed with water, soil, air, and food because of their excellent volatile characteristics (Berthiaume et al. 2021). Most of the PAHs are carcinogenic and reflect different health-related disorders such as infertility, improper fetal development, gene mutations, and cardiovascular diseases (Ferguson et al. 2017).

2.2.1.7 Dioxins and Polychlorinated Biphenyl-Based ECs

Different ECs like polychlorodibenzofurans, polychlorinated biphenyls, and polychloro-dibenzo-p-dioxins are released into the environment through metal smelting, fossil fuel combustion, burning of pharmaceutical materials, and automobiles. Due to less water solubility and vapor pressure and more lipophilicity, these contaminants are present as suspended solids in environment and can easily enter in food chain (Reddy et al. 2019). These are very harmful and incite serious disorders like cardiovascular diseases, behavioral changes, tumors, and disruption of immune, neurological, and endocrine systems. These are highly durable and accumulate easily as they have a half-life from a few weeks to months.

2.2.1.8 Dyes and Dye-Containing Toxic Pollutants

Dye-contaminated wastewater is generated from textile, cosmetics, printing, paper, and rubber industries. The discharge of dyes in water bodies without proper treatment poses significant hazards to the health of human beings and ecosystems (Khan et al. 2020). Dye-containing wastewater affects the environment by disturbing aquatic life, inhibiting photosynthesis in aquatic plants, and affecting human health by causing skin diseases, respiratory disorders, cancer, etc. Different methods have been used for the removal of dye from wastewater such as coagulation and flocculation, reverse osmosis, and adsorption. However, these methods are expensive, time-consuming, and may generate sludge. As compared to physico-chemical methods, biological treatment methods such as the application of microbes or their consortia and microbial enzymes are better cost-effective alternatives that can eliminate dyes from effluents of industries.

2.2.1.9 Additional Sources of ECs

The surfactants are widely used in the production of pesticides, cleansing agents, cosmetics, paints, etc. (Papaevangelou et al. 2016). Surfactants generate foam on the water surface and adversely affect the aqueous system. The global production of surfactants was 24 million tons in the year 2022 (Palmer and Hatley 2018). Caffeine and nicotine have been reported in groundwater polluted with sewage wastewater. Approximately 159,000 metric tons of synthetic sweeteners are annually consumed (Luo et al. 2019). In marine and groundwater ecosystems, escalated levels of sucralose, saccharin, and cyclamate have been identified due to their excessive use and durability in water resources. Additionally, endocrine disruptors have been found to interfere with endocrine system of body and reflect hazardous effects on reproductive and immune systems of animals and human beings (Gomes et al. 2018).

2.3 ENVIRONMENTAL AND HEALTH RISKS ASSOCIATED WITH EMERGING CONTAMINANTS

ECs can readily enter into food chain and negatively impact non-target organisms and the health of human beings and ecosystems (Khan et al. 2021). Many abnormalities such as male and female sexual disorders, less male fertility, more conception time, high rate of miscarriage, and prostate and breast cancers have been reported after exposure to ECs; however, it depends on contact duration and dosage of ECs (de Oliveira et al. 2019). Sex alteration in male frogs and negative effects on fertility of fishes, mice, and human beings have been reported with atrazine (de Souza et al. 2020). Glyphosate negatively affected the food chain by affecting periphytic colonization and decreasing the population of *Pseudokirchneriella subcapitata* and *Lemma minor*. Paraquat showed neurotoxicity and pulmonary inflammation in rats after inhalation for up to 16 days (Franco et al. 2022). Atrazine and glyphosate show estrogenic impact and cytotoxicity in human beings (Morin-Crini et al. 2022). Various health-related disorders have been observed like cardiovascular diseases, degeneration of retina and muscles, and cancer in humans by atrazine (Saleh et al. 2020). Endocrine disruptors demonstrated a reduction in male sperm count with an increase in chances of cancer and reproductive ailments (Ahmed et al. 2021). Phenolic substances have been associated with disruption in the endocrine system (Singh and Chandra 2019). In *Oncorhynchus mykiss* and mosquito fish, phenolic compounds reduced gonad size, changed sexual properties, and inhibited sex hormones (Sharma and Singh 2021). ECs hepatotoxicity, and mutagenic effects on fish endocrine and reproductive systems have also been reported (Singh and Chandra 2019). Several PAEs demonstrated hazardous impacts on the development and reproduction of animals (Huang et al. 2021). Phthalate exposure is responsible for eczema, allergy, and asthma in children (Jøhnk et al. 2020).

The presence of antibiotics in water adversely affects microbial activity and wastewater treatment efficiency, e.g., ciprofloxacin reduces the partial nitrification process (Gonzalez-Martinez et al. 2014). Exposure to antibiotics promotes the development of antibiotic-resistant bacterial strains, whereas pesticide exposure causes cancer (Grenni et al. 2018). Tetracycline (10–100 μg/L concentration) has declined population of algae, bacteria, nematodes, etc. The reduced population of vultures in Asia was because of diclofenac's presence in the food, which enhanced failure of renal system and death (Tiwari et al. 2017).

Propranolol and fluoxetine reflected toxicity in Benthos and zooplanktons (Jacob et al. 2016). Animals absorb a huge amount of veterinary antibiotics; however, 30%–90% antibiotics are released via feces and are mixed in water and soil (Fawell and Ong 2012). Perfluorinated compounds can induce infertility, decrease sperm production, thyroid disorders and induce cancer, and damage liver tissues. Reproductive instability and lung cancer are health hazards associated with nanoparticles containing ECs (Enyoh et al. 2020). The surfactants and nonylphenol ethoxylates showed hazardous impacts on tadpoles and frogs (Praveena et al. 2019). Artificial sweeteners such as aspartame have been reported to cause headache, nausea, and vomiting in human beings and cancer in rats. Meanwhile, 5% sucralose reflected shrinkage of thymus gland and migraine in rats and human beings, respectively (Praveena et al. 2019). MPs produce digestive disorders and alterations in metabolic profiles of planktones, crustaceans, fish, turtles, and whales (Li et al. 2018). *Lumbricus terrestris* treated with 450 and 600 g/kg polyethylene enhanced 8% and 25% mortality rate, respectively. Shia et al. (2022) eliminated polyethylene, polypropylene, polystyrene, and reflected 63%–87% removal efficiencies.

2.4 DETECTION AND QUANTIFICATION OF ECs

The assessment of the presence of ECs in the aqueous system requires fast and responsive analytical techniques such as liquid or gas chromatography and mass spectroscopy. Mass spectroscopy procedures can be applied in the precision analysis of ECs at ng/L level in wastewater (Aguera et al. 2013). Some recent procedures like triple quadrupole, linear ion traps quadrupole, and quadrupole-time of flight have been used for morphological assessment of modified substances. Application of capillary electrophoresis and immunoanalytical procedures has been observed as an important method for EC assessment.

An immunoanalytical process depends on the antibody applied and is unable to assess different samples at the same time, and microbiological estimation depends on the sample nature (Buchberger 2011). Sample pretreatment is important for ECs analysis in samples. Different pre-concentration methods have been used such as solid phase extraction, magnetic online or automated solid phase extraction, or molecularly grafted polymers prior to determination of final low identification or quantification limits (Moliner-Martínez et al. 2011). Novel analytical methods like liquid chromatography, mass spectrometry, and tandem MS have been recently applied for the detection of extremely low concentrations of ECs in liquid and solid matrices. The detection and quantification of around 3000 biologically active chemical compounds are possible in the environment (Vasilachi et al. 2021).

2.5 REMEDIATION APPROACHES USED FOR ECs REMOVAL

In recent years, a lot of effort has been taken for the development of treatment technologies for the removal of ECs from aqueous systems. ECs removal by applying one treatment procedure is not suitable as it requires a combination of different treatments to overcome the inadequacies of single treatment (Table 2.2).

TABLE 2.2
Advantages and Challenges of Different Methods Used for Removal of ECs

Treatment Methods	Advantages	Challenges
	Physical Process	
Micro- or ultra-filtration	Removal of ECs and pathogens	Unable to remove large-size ECs
Nanofiltration	Treatment of saline water and wastewater treatment plants Dye and pesticides contaminated water can be treated	Limited application in pharmaceuticals removal, energy-intensive, membrane fouling, and disposal problem
Reverse osmosis	Elimination of endocrine-disrupting compounds and pharmaceuticals	Corrosive texture of treated water
	Chemical Process	
Coagulation	Less turbidity Enhanced sedimentation rate via generation of suspended particles	Ineffective ECs removal, sludge production
Ozonation	High affinity to ECs in the presence of H_2O_2 Major ancillary impact on ECs removal	Energy-intensive process, oxidative by-products generation high operational and maintenance cost
Advanced oxidation process	Less degradation rate	Generation of toxic disinfection by-products Interference of radical scavengers
Fenton and photo-Fenton	Degeneration and mineralization of ECs Sunlight can be used in place of UV light	Decrease in hydroxyl ions because of scavenging of OH^- generating chloride and sulfate ions
Photocatalysis	Persistent organic compounds degeneration by catalysts Less cost and stability of TiO_2 catalyst with easy recovery	Unable to treat a large volume of wastewater Artificial UV lamps and electricity expenditure are costly

(Continued)

TABLE 2.2 (*Continued*)
Advantages and Challenges of Different Methods Used for Removal of ECs

Treatment Methods	Advantages	Challenges
		Separation and reutilization of photocatalytic particles is problematic from slurry suspension
	Conventional Methods	
Biological activated carbon	Various ECs removal from wastewater Elimination of residual disinfection/oxidation products Toxic products are not generated	Maintenance and operational cost are high Sludge disposal increases total cost up to 50%–60%
Microalgae reactor	Recovery of algal biomass can be used as a fertilizer Good quality effluent and no toxicity risk with ECs Less operational costs than AOPs	Seasonal effect on removal efficiencies ECs cannot be degenerated Less efficacy for pharmaceuticals and beta-blockers
Activated sludge	More eco-friendly than chlorination	Large amount of sludge containing ECs Unsuitable as COD level is higher than 4000 mg/L
	Non-conventional	
Constructed wetland	Low energy consumption and low operational and maintenance costs Significant removal of estrogens and pathogens	Clogging, solids entrapment, and sediments generation Biofilm growth, chemical precipitation, and seasonal dependent Requires more space and retention time
Membrane biological reactors	Effective for biorecalcitrant ECs removal Small footprint	More energy consumption, fouling More cost of aeration and membrane roughness Pharmaceutical compounds cannot be properly treated

2.5.1 Physico-Chemical Methods Used for Remediation of ECs

Physical remediation can be used for the elimination of ECs. Membrane filtration is frequently used; however, the type of membranes being used depends on the impurities to be removed (Saleh et al. 2020). Pesticides were removed by applying polypiperazine amide thin-film composite (TFC) nanofiltration membranes (Karimi et al. 2016). Atrazine, diuron, and simazine were eliminated in the presence of humic acid by applying organic nanofiltration membrane sheets (Musbah et al. 2018). The membrane-based processes are very costly and tend to fouling. Steam distillation procedure has been used to remove phenol from wastewater due to more volatility of phenol. The degradation of dimethyl phthalate was enhanced to 98% in 2 hours by six lamps when they used sonophotolysis, i.e., application of both UV light and ultrasound (Xu et al. 2013). Yadav et al. (2019) generated ZnO/SnO_2 nanocomposite which degraded triclopyr.

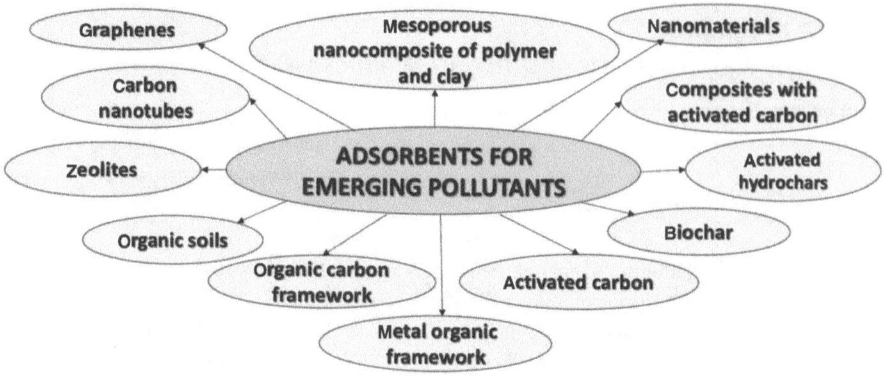

FIGURE 2.2 Adsorbents used for the removal of ECs. (Reprinted from Vasilachi et al. 2021.)

Three coagulants such as aluminum sulfate, aluminum, or ferric chloride were analyzed for their coagulation capacity and PAEs elimination from landfill leachate (Zhang and Wang, 2009). The 30% elimination of PAEs from landfill leachate was reported by aluminum chloride coagulants (Zhang et al. 2019). Advanced oxidation products act as oxidizing agents and convert pollutants into CO_2, H_2O, and acidic mineral and generate OH^- by enhancing mineralization (Pang et al. 2021; Babuponnusami et al. 2023). Activated carbon showed 55 mg/g phthalates sorption; however, after treatment with TBA and Cu, phthalate adsorption was recorded 92 and 117 mg/g, respectively (Ghosh and Sahu 2022). The sodium persulfate acts as an oxidizing species and removes 17 different pesticides (Vela et al. 2019).

Waste biomass serves as an excellent adsorbent due to its huge availability and cost-effectiveness, which can aid in waste management. Adsorption is widely used due to its high pollutants removal efficiency, economical, cost-effective, low maintenance, easy operation, and no sludge generation (Kapoor and Zdarta et al. 2024). Conventional treatment technologies such as membrane filtration, coagulation/electrocoagulation, photocatalysis, ozonation, adsorption, and advanced oxidation processes (Figure 2.2) exhibited an effective removal of ECs (Pesqueira et al. 2020). However, most of these methods are expensive, are time-consuming, and generate toxic by-products, and their application at a large scale is not practically feasible. Different countries are planning to upgrade wastewater treatment plants by using recent treatment procedures to comply with ECs elimination goals for environmental protection (Pesqueira et al. 2020).

2.6 APPLICATION OF ACTIVATED CARBON AND BIOCHAR FOR ECs REMOVAL

Agricultural wastes like fruit seed and peels, husk, legume shells, and cereal crops can be used for the production of biochar. Biochar can be produced from waste biomass through pyrolysis at high temperatures under oxygen-limited conditions. Lignocellulosic biomass is widely available and chemically stable due to the presence of lignin, cellulose, and hemicellulose (Mo et al. 2018). Biochar shows many benefits as compared to

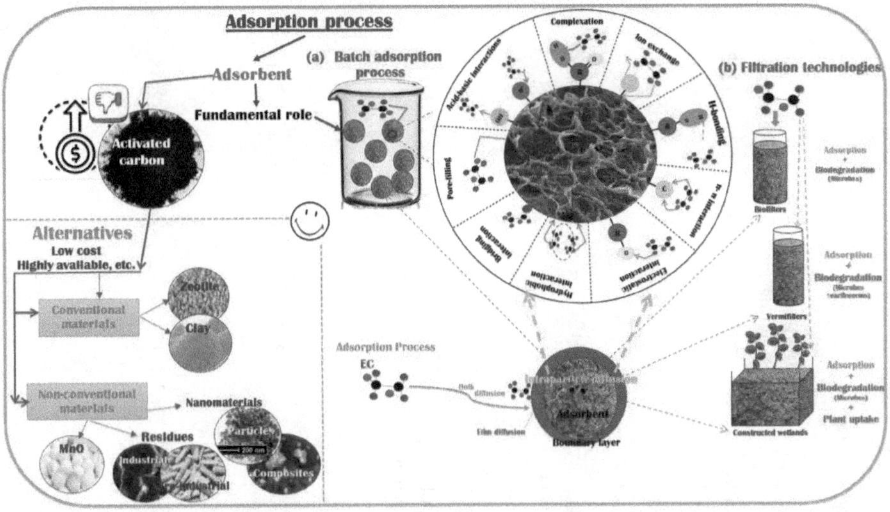

FIGURE 2.3 Different procedures used for the treatment of ECs. (Reprinted from Almeida-Naranjo et al. 2023.)

commercially available activated carbon like huge availability, inexpensive, produced from renewable raw substances, the presence of functional groups, and condensed structure (Thompson et al. 2016). Biochar is applied for ECs elimination at a laboratory scale as it showed high efficiency around 95% and sorption >640 mg/g as compared to activated carbon (Bedia et al. 2018). Sulfamethazine, ibuprofen, triclosan, diclofenac, atrazine, dimethyl phthalate, benzophene, metolachlor, phenol, and pesticides can be removed by biochar (Choi et al. 2020). Figure 2.3 indicates different methods used for the treatment of ECs. Activated carbon removed EC compounds >90% from water due to its large surface area and porosity (Khan et al. 2021). Acetaminophen showed >90% adsorption with activated carbon produced from wood; however, 60%–87% adsorption was reported when activated carbon was generated from various sources (Cabrita et al. 2010). Diclofenac elimination was recorded >90% with activated carbon and olive waste cake (Baccar et al. 2012). Biomass wastes generated activated carbon reduced ECs like antibiotics, paracetamol, and anti-inflammatory drugs (Mirasole et al. 2016). The peanut hulls and sugar beet pulp removed tetracycline from water >90%; however, wood and coconut shell eliminated 75% and 30% (Torres-Pérez et al. 2012). A 99% removal of ciprofloxacin was reported by using commercially available activated carbon (Robles-Molina et al. 2014). Ganie et al. (2021) prepared bagasse biochar at different temperatures of 350°C, 550°C, and 750°C for MPs removal from aqueous systems, and they observed that 750°C biochar showed the maximum removal rate of MPs because of porous texture and large surface area. Biochar production at high temperatures reflected enhanced MPs elimination and immobilization ability (Wang et al. 2020). Wang et al. (2021) showed that magnetic biochar removed 94% polystyrene microspheres from an aqueous solution. The oxidized corncob biochar reflected high polystyrene (>90%) removal in comparison to raw biochar due to the presence of more hydroxyl groups (Singh et al. 2021). Table 2.3 shows the application of materials for EC removal from the aqueous system in batch experiments.

TABLE 2.3
Application of Materials for EC Removal from Aqueous System in Batch Experiments

Removal of EPs	Adsorbent	Operational Conditions	Adsorption Mechanism	Adsorption Capacity (mg/g)	References
Activated carbon from argan seed shell	Paracetamol	pH = 3–11 AD = 0.1 EC = 100 CT = 1–12 T = 25°C	van der Waals force, H-bonding, dipole–dipole interactions, ion exchange, covalent bonding, cation, and water bridging	512	Benjedim et al. (2020)
Activated carbon from argan seed shell	Amoxicillin	pH = 3–11 AD = 0.1 EC = 100 CT = 1–12 T = 25°C	van der Waals forces, H-bonding, dipole–dipole interactions, ion exchange, covalent linkage	319	Benjedim et al. (2020)
Commercial activated carbon	Ibuprofen	pH = 3–8 AD = 0.1 EC = 1.13 CT = 4 T = 30°C	van der Waals force, H-bond, dipole–dipole interactions, ion exchange, covalent bonding, water and cation bridging	72	Kaur et al. (2017)
Activated carbon from olive pomace	Niesulide	pH = 8–11 AD = 0.1–0.5 EC = 10–30 CT = 6	van der Waals force, H-bonding, dipole–dipole interactions, covalent bonding, cation bridging, and water bridging	353	Raupp et al. (2021)
Cu^{2+} zeolite 4A	Glyphosate	pH = 6–8 AD = 2 EC = 100 CT = 2 T = 25°C	van der Waals and acid-base forces, donor–acceptor interaction	113	Zavareh et al. (2018)

(Continued)

TABLE 2.3 (*Continued*)
Application of Materials for EC Removal from Aqueous System in Batch Experiments

Removal of EPs	Adsorbent	Operational Conditions	Adsorption Mechanism	Adsorption Capacity (mg/g)	References
Zeolite Y	Paraquat	pH = 6–8 AD = 2 CT = 2 EC = 20–250 T = 25°C	van der Waals and acid-base forces, donor–acceptor interaction	71.4–92.6	Pukcothanung et al. (2018)
Natural clay	Methomyl	pH = 6.6 AD = 10 EC = 20–44 CT = 3 T = 25°C	Ion exchange, dipole–dipole interactions, van der Waals forces, H-bond, acid-base forces, donor–acceptor interaction, acid-base forces	0.3–0.5	Nassar et al. (2012)
Raw alumina modified with HCl	Acrylic acid	pH = 4.5 AD = 67 EC = 0.08 CT = 0.5–120 T = 20°C–50°C	Electrostatic interactions and hydrogen bonding	0.29–0.31	Chen et al. (2015)
Modified biochar + *Spiruline* sp.	Tetracycline	pH = 3–9 AD = 0.1 EC = 100 CT = 120 T = 20°C–40°C	Ion exchange, electrostatic, hydrophobic, pore-filling, and bridging interactions	132	Choi et al. (2020)
Chitosan	Phenol	pH = 7 AD = 1 EC = 150 CT = 3 T = 20°C	Electrostatic, hydrophobic interactions and hydrogen bonding	60	Zhou et al. (2014)

(*Continued*)

TABLE 2.3 (*Continued*)
Application of Materials for EC Removal from Aqueous System in Batch Experiments

Removal of EPs	Adsorbent	Operational Conditions	Adsorption Mechanism	Adsorption Capacity (mg/g)	References
Rice husk	Norfloxacin	pH = 6.2 AD = 1 EC = 5 CT = 1 T = 25°C	π–π or electrostatic interactions, carbonyl and carboxyl bond formation	20	Paredes-Laverde et al. (2018)
Graphene oxide	Metformin	pH = 4.5–8.5 AD = 0.05–0.15 EC = 300–700 CT = 1–3 *T* = 1545°C	Hydrophobic effect, π–π interactions, hydrogen and covalent bonding, and electrostatic interactions	123	Balasubramani et al. (2020)
Biochar + Chitosan	Ciprofloxacin	pH = 3 CT = 48 AD = 5 EC = 160 *T* = 20°C	Depends on composite components, hydrogen bonding, π–π interaction, amidation reaction, electrostatic and hydrophobic interaction ligand exchange chemisorption	76	Zaheer et al. (2018)
Palm-shell waste AC + magnesium silicate	BPA	pH = 4.5 CT = 24 AD = 0.1 EC = 10–100 *T* = 24°C	Hydrogen bonding, π–π interaction, cation–π bond, electrostatic interaction, hydrophobic interaction, ligand exchange, cation–π bonding, chemisorption	168	Choong et al. (2018)

Here, AD, adsorbent dose (g/L); CT, contact time (hours); EC, concentration of emerging contaminants (mg/L); *T*, temperature [°C].

Poddar et al. (2024) found that coco peat biochar exhibited 99% adsorption capability for norfoxacin. Lu et al. (2017) stated that *Ulva prolifera* biochar removed 90% of endocrine disruptors within 4 hours. Ahmad et al. (2019) observed that nano zero-valent iron-modified biochar removed 98% chlortetracycline from water, which was more than zeolite or silica or non-modified biochar. Hickory chip biochar (produced at 700°C) adsorbed ECs from water (Huang et al. 2020). Arenas et al. (2021) observed PS particles elimination by coconut shell-activated carbon and reported 2.2 mg/g adsorption capacity of PS at pH 7.4 after 240 minutes. Zhu et al. (2022) prepared corncob biochar for PS removal, which showed 56 mg/g adsorption capacity and after five regeneration cycles, sorption ability was maintained at 45 mg/g which showed cost-effectiveness and reproducibility of process. Wang et al. (2021) fabricated magnesium or zinc-modified magnetic biochar which exhibited 95%, 98%, and 99% PS removal efficiencies, respectively, as compared to 26% removal efficiency with unmodified biochar. After five adsorption–desorption cycles, removal efficiencies maintained their high levels of 95%, 94%, and 96%, respectively. Shi et al. (2022) reported 98% removal of polyethylene with magnetic seafoam and due to recyclability of seafoam, removal efficiency of polyethylene was more than 90% under magnetic field after five regeneration cycles. The pyrolytic calcium-enriched biochar prepared from crab shells significantly removed chlortetracycline. At high concentration, adsorption capacity increased up to 5048 mg/g, which was retained even after five regeneration cycles (Xu et al. 2020). The 1 g of wheat straw biochar removed 63 mg 17β-estradiol (Tong et al. 2020). Industrial wastes such as fly ash, red mud, steel slag, and slurry have been used for elimination of tetracycline, sulfamethoxazole, trimethoprim, phenols, carbamazepine, BPA, sulfathiazole, ibuprofen, carbofuran, nimesulide, etc. with removal efficiencies 2%–90%, and sorption capacity was between 0.6 and 212 mg/g (Devi and Saroha 2016).

2.7 APPLICATION OF CLAY MINERALS AND METAL ORGANIC FRAMEWORK FOR ADSORPTION OF ECs

Clay materials are applied in pollutants removal because of their huge availability and lower price as these are 20 times cheaper than activated carbon. Clay minerals are composed of silica, alumina or magnesium, potassium, iron, sodium, calcium, water etc. Surface-modified clay reflected high adsorption capacity for ECs as compared to raw clay. Clay can efficiently remove ECs such as atenolol, metformin, ciprofloxacin, ranitidine, diazepam, naproxen, phenol, and atrazine (Thiebault 2019). However, clay has 11 times less surface area in comparison to activated carbon (Sophia and Lima 2018).

Metal organic frameworks are used for the removal of ECs because of more surface area, large pore structure, multiple topologies, easy regeneration, and high adsorption capacity (Rojas and Horcajada 2020). Metal organic framework can be synthesized on a large scale and can be easily transformed into membranes, beads, and pellets, for columns that are suitable for decontamination devices. Metal organic framework removed ECs as MIL-101 showed adsorption of tetracycline and MOF-5 reflected ciprofloxacin sorption (Gadipelly et al. 2018). However, MOFs are unstable in aqueous solution, thus MOFs cannot adsorb various contaminants simultaneously (Jeong et al. 2023). Lv et al. (2025) synthesized coal gasification slag-based adsorbent for MPs removal and showed 99% removal efficiency. Bele et al. (2016) reported graphene oxide production from hydrazine hydrate, which showed 94 mg/L adsorption capacity for bisphenol A. Less stability of MOFs in aqueous system is the main problem for their applications in industries (Rojas and Horcajada 2020).

2.8 APPLICATION OF NANOMATERIALS FOR ADSORPTION OF ECs

Nanomaterials are also known as novel sorbents because of their high surface area (up to 3200 m^2/g), catalytic ability, high surface energy and reactivity, reactive sites availability, dispersibility, and free surface energy. Carbon nanotubes exhibit sorption that depends on physico-chemical properties, size, and preparation procedure (Rakhi, 2019). They may be of two types single-walled nanotubes (SWNTs) and multi-walled nanotubes (MWNTs), made up of many tubes or graphene layers. The single walled carbon nanotubes (SWCNTs) can be used for the preparation of MWCNTs through chemical processes, which promotes elimination efficacy by rise in active sites. Tang et al. (2021) generated magnetic carbon nanotubes, which showed maximum adsorption capacities of 1650, 1400, and 1100 mg/g, respectively, for polyethylene terephthalate, polyethylene, and polyamide. Ji et al. (2010b) removed 92% tetracycline with SWNT, but only 16% by using MWNT under the same operational conditions. MWNT showed the removal of ECs such as amoxicillin (Teixido et al. 2011), ciprofloxacin (Mitchell et al. 2015), and triclosan or ibuprofen (Zheng et al. 2013) with removal efficiency 7%, 90%, and 100%, respectively. The SWNT and MWCTs showed 100% and 35% removal of norfloxacin (Yang et al. 2012). PET enzyme immobilized on Fe_3O_4 nanoparticles degenerated PET. It reflected a rise in degeneration efficacy up to 80% in comparison to PET enzyme and maintained 50% initial activity after ten cycles (Schwaminger et al. 2021). Nanomaterials are promising alternatives for ECs elimination, and their large-scale production and separation from aqueous medium is needed for their application at a large scale.

2.9 CHITOSAN

Chitosan is a non-porous deacetylated derivative of chitin. Chitosan is a relatively low-cost sorbent that is non-hazardous, biodegradable, reactive, and can be functionalized via physical and chemical procedures. Chitosan has been used for the removal of caffeine, bezafibrate, tannic acid, alkylbenzene sulfonate, BPA, and phenols and it has shown adsorption efficiencies from 11% to 96% (Tarasi et al. 2018). Sun et al. (2021) prepared chitin-based sponges with graphene oxide and chitosan to generate sponges with high elastic and mechanical features. The sponges showed 72%–92% elimination efficacy for PS, carboxyl, and amine functionalized PS at pH 6–8.

2.10 PEAT AND GRAPHENE

Peat is generated through decomposition of organic matter and soil and is easily available worldwide. Peat can compete with commercially available sorbents, like activated carbon and zeolite, because of its high cation exchange potential. The peat removed sulfamethoxazole, metolachlor, phenol, BPA, p-nitrophenol, tris 2-butoxyethyl, or tris 2-chloroethyl phosphate (Zheng et al. 2016).

Graphene is considered the thinnest material and good adsorbent for the removal of ECs (sorption capacity 19–3710 mg/g) due to its more delocalized π electron system and flexible chemical features. However, hydrophobicity and restricted dispersibility of graphene reduce its capacity for adsorption. The removal of bisphenol A, phenol, diclofenac, metformin, nimesulide, sulfonamide, acetaminophen, and nicotine has been reported by graphene (Baig et al. 2019).

2.11 POLYMERIC ADSORBENTS

Polymeric adsorbents can be regenerated and their surface can be functionalized for the removal of different contaminants. Polymeric adsorbents are four times less expensive as compared to activated carbon but reflect 15–2000 times more adsorption capacity than activated carbon. ECs such as penicillin, ibuprofen, cephalexin, caffeine, phenol, erythromycin, alachlor, diclofenac, and BPA were significantly eliminated by using polymeric adsorbents (Li et al. 2018). In wastewater treatment, inorganic–organic porous hybrid crystalline materials have significantly removed pharmaceuticals and personal care products, dyes, benzene, toluene, pesticides, industrial products, and plasticizers (Rojas and Horcajada 2020). However, it is at the initial stages of development and need a thorough analysis of variables like environmental safety and reusability.

2.12 FILTRATION TECHNOLOGIES

Filtration technologies are cost-effective and easy to use with less energy consumption and can be easily used at a large scale. Arenas et al. (2022) reported that the sand filtration process showed 54% removal of polystyrene particles. However, the addition of activated carbon filtration step by following sand filtration enhanced elimination efficacy of PS up to 74%.

Bisphenol A was filtered from water by ceramic membranes, which was earlier gone through biotreatment and recorded 98% endocrine disruptors elimination from wastewater (Zielinska et al. 2017). Si et al. (2018) observed the removal of endocrine disruptors by using ultrafiltration from 8% to 46%, 70% by ozonation alone, and reached up to 99% with the combined application of ultrafiltration and ozonation methods. Filtration procedures have been applied to treat secondary/tertiary effluents (Tejedor et al. 2020). Biofiltration is a biological filtration method, which may be of three different types such as biofilters, constructed wetlands, and vermifilters (Georgin et al. 2022).

2.13 BIOFILTERS

Biofilters are used as filters for ECs removal. These filters may be of two types: rapid or slow based on the retention period or hydraulic load rate applied in the biofilters. Reduction in hydraulic retention time decreases removal efficiency of various ECs present in surface water and groundwater. Biofilters can be continuous or intermittent depending on the feeding type. Rapid filtration is widely used in drinking water treatment plants for water purification processes due to less energy consumption (Wang et al. 2021). Rapid filters can partially degrade >50% of ECs. The removal efficiencies in rapid filters were around 99% (Brunner et al. 2018). The mechanism used for ECs removal is sorption on sand, oxidation, and sorption by metal oxides, and biodegradation. Biofiltration process has high efficacy, less cost, and ease of operation, thus applied in decentralized wastewater treatment systems in developing nations.

Slow filtration has a less filtration rate as compared to the fast filtration process with no need of chemical coagulation. Slow filtration has less operational cost and is simple in operation and maintenance. Slow filtration combined with microbes promotes biodegradation, thus material not only absorbs ECs but also assists microbes by maintaining food till consumption (Verma et al. 2017). Slow filtration was used for ECs elimination like paracetamol, naproxen, methylparaben, diclofenac, tebuconazole, and propiconazole and achieved 15%–98% removal rate (Pompei et al. 2017). In biofilters, wastewater percolates via filter bed material, microbes adhere to the surface of material, colonize, and form stable biofilms, which can degrade hazardous pollutants (Thuptimdang et al. 2020). Clay, sand, and activated carbon can also be applied as sorbents for the removal of ECs in biofilters. Biochar prepared from rice husk, peanut shells, fruit peels, and sawdust has shown 100% removal of ECs (Zhang et al. 2018). Adsorption and biodegradation are the main removal mechanisms of ECs in biofilters (Thuptimdang et al. 2020).

2.14 CONSTRUCTED WETLAND

Application of constructed wetland for the removal of inorganic and organic contaminants from wastewater is a sustainable treatment process. Earlier constructed wetlands were used for secondary and tertiary treatment of domestic and sewage wastewater, and further research has expanded their capacity to remove a wide spectrum of pollutants. Constructed wetlands are multifunctional systems, which can enhance water quality and function as hydrological buffers which conserve natural habitats and recreational sites.

Constructed wetlands are composed of plants, support materials, native microbes, and water (Avila et al. 2021). Conventional materials used are soil, sand, red soil, zeolite, brick, and gravel for the removal of ECs (Chen et al. 2019). Various kinds of constructed wetlands (surface free water, horizontal or vertical groundwater flow, and hybrid) have been used for the removal of ECs. It has been used in the treatment of landfill leachate and municipal and hospital wastewater. However, physico-chemical (adsorption, oxidation, volatilization, and hydrolysis) and biological methods (biodegradation, absorption by plants, etc.) are mainly used for pollutants removal in constructed wetlands (Ji et al. 2022). This technology does not require any additional energy source and has less operational cost. Sabri et al. (2021) applied constructed wetlands for treatment of sewage wastewater and antibiotics, which showed 28%–100% removal rate. Elimination of various ECs (>99%) like phenol, diclofenac, atrazine, endosulfan, erythromycin, azithromycin, sulfamethoxazole, ibuprofen, acetaminophen, triclosan, and BPA was reported by constructed wetlands (Chen et al. 2019).

2.15 VERMIFILTERS

Vermifilters are composed of earthworms, microbes, and filter material, which reflect symbiotic association. Both slow and fast biofilters apply sand as a filter bed for the removal of ECs present in surface water or groundwater. Earthworms maintain microbial activities and biomass and microbes degrade pollutants (Arora et al. 2021). Earthworm species such as *Eisenia foetida, Lumbricus rubellus*, and *Eudrilus eugeniae* are kept on filter bed, which can be soil where degradation of contaminants takes place. Agro-waste materials, sand, and gravel can be applied as filter bed. The removal of ECs such as ciprofloxacin, sulfamethoxazole, tetracycline, amoxicillin, streptomycin, tetracycline, and chloramphenicol showed 40%–98% removal efficiencies by vermifilters (Shokouhi et al. 2019). Mechanism involved in ECs elimination was degradation by earthworms, sorption via biofilm, degradation by microbes, and adsorption in filter material (Arora et al. 2021). However, vermifilters are the least used biofilters for ECs removal in comparison to other filters and constructed wetlands.

2.16 MEMBRANE FILTRATION TECHNOLOGY

As compared to traditional membrane treatment procedures, electrochemical membrane filtration significantly degraded ECs in wastewater. Traditional membrane restricts flow, whereas electrochemical membrane generates free radicals for degradation and mineralization of contaminants. Membrane bioreactor (MBR) can eliminate >95% MPs; however, in real water treatment plant, 79% removal efficiency was observed (Bayo et al. 2020). After combination with ultrafiltration technology, removal efficiency of MPs can exceed 99% (Lares et al. 2018). High conductivity of carbon nanotubes assists their integration into fibre membranes for membrane filtration and electrochemical methods (Yang et al. 2019). In comparison to membrane filtration, electrochemically assisted carbon nanotubes/Al_2O_3 membrane filtration permeation flux was enhanced up to two times, with 300% rise in organic matter elimination efficacy (Yang et al. 2019).

2.17 BIOLOGICAL METHODS USED FOR ECs REMOVAL

Physico-chemical methods are expensive, time-consuming, and generate toxic by-products; hence their application at a large scale is not practically feasible while the use of biological methods such as application of bacteria, fungi, and plants can remove ECs. Mustard and sunflower plants significantly removed phthalates (Zavoda et al. 2001). Dimethyl phthalate, diethyl phthalate, and di-(2-ethylhexyl) phthalate were eliminated from wastewater by *Arundo donax* and *Thalia dealbata* (Xiaoyan et al., 2015). Phthalates were eliminated from municipal wastewater by *Wolffia arrhiza* (Das et al. 2021).

2.18 REMEDIATION OF ECs WITH FUNGI AND ALGAE

Aspergillus flavus degraded low- and high-density polyethylene (Zhang et al. 2020). The *Fusarium culmorum, Pleurotus ostreatus*, and *Aspergillus japonicas* also degraded di-(2-ethylhexyl) phthalate. *Saccharomyces cerevisiae* and *Rhodotorula rubra* degraded phthalates. Five South African fungal species were used for the removal of carbamazepine, diclofenac, and ibuprofen by batch reactor process with 89%, 96%, and 91% removal efficiencies for carbamazepine (CBZ), diclofenac (DFC), and ibuprofen (IBP), respectively, after 24 hours of exposure period (Kasonga et al. 2019). Křesinova et al. (2018) found BPA elimination and triclosan by *Pleurotus ostreatus* and efficiency was >90% within 12 days.

Phycoremediation is a cost-effective, efficient, and eco-friendly treatment of toxic contaminants present in industrial effluents (Nguyen et al. 2023). Microalgae can effectively treat pollutants in cosmetics, pharma, and nutraceutical industries due to their high resistance under harsh environmental conditions, high degradability, economical, and end products are non-toxic, which can generate valuable products. *Scenedesmus* and *Chlorella* species effectively removed ECs as both can survive under harsh environmental conditions (Sutherland and Ralph 2019). Two types of tanks either open or closed, or their combination can be used for increasing micro-algal bioremediation (García-Galan et al. 2018). Open tanks require less energy, economical, and promote microalgal growth but they need large space with more changes of contamination in open tanks and maintenance of temperature is difficult (Molina et al. 2020). The closed reactors have a better distribution of sunlight, which produces more biomass (Rempel et al. 2021). Photobioreactors are widely used for biodegradation of ECs at a large scale, which occurs with the help of microalgae and their enzymes, which convert complex molecules into simple ones (Hena et al. 2021).

Hom-Diaz et al. (2017) used a photobioreactor for the treatment of effluent from drug-dispensing toilets. They removed 33%, 36%, 99%, 98%, 69%, and 33% naproxen, ketoprofen, paracetamol, ibuprofen, and salicylic acid, respectively, from effluent. As per Wang et al. (2013), *Chlorella pyrenoidosa* removed 50% triclosan from water within 60 minutes of exposure period. *Chlorella sorokiniana* eliminated 34% pharmaceuticals mixture by sorption (de Wilt et al. 2016). *Scenedesmus obliquus, Chlorella, Nannochloropsis,* and *Anabaena variabilis* significantly eliminated cresol and nitrophenol from wastewater (Das et al. 2015). *Cyclindrotheca, Closterium* species, and *Chaetoceros muelleri* showed 90% degradation of phthalates within 4 days (Ghosh and Sahu 2022). Bai and Acharya (2019) used *Nannochloris* species for the removal of endocrine disrupters from ultrafiltration and ozonated effluents. In ultrafiltration effluent cultivated with *Nannochloris* sp. for 7 days, the removal rate of 17β-estradiol, 17α-ethinylestradiol, and salicylic acid was 60%. Shi et al. (2021) reported quinolone and indole degradation up to 100% when *Chlorella pyrenoidosa* powder was applied. Phycoremediation faces many hurdles such as a tedious harvesting process, more nutritional requirements, contamination with microbes, complex monitoring, and control. Microalgal treatment capacity can be adversely affected under harsh environmental conditions.

2.19 REMEDIATION OF ECs WITH BACTERIA

Bacillus megaterium YJB3, *B. thuringiensis*, *Providencia*, and *Pseudomonas* species degraded dibutyl phthalates (Wang et al. 2020). *Microbacterium, Pseudomonas, Rhodococcus erythropolis, Rhizobium, Gordonia, and Achromobacter* species significantly degraded phthalates (Lin et al. 2022). *Ochrobacterium* species, *Pseudomonas aeruginosa,* and *Achromobacter xylosoxidans* used endosulfan as a sulfur source (Seralathan et al. 2015). *Esdencode dehalogenases, dehydrochlorinases, hydrolases,* and *monooxygenases* genes were involved in sulfur mineralization and can degenerate organochlorine pesticides (Hussain et al. 2009). However, long-chain PAEs have complicated structures, which create difficulty for microbes for bond breakage within molecules (Ghosh and Sahu 2022).

2.20 GENETICALLY ENGINEERED BACTERIA

The degradation of ECs depends on microbial species or their enzymes (Janssen and Stucki 2020). Different environmental factors such as pH, dissolved oxygen concentration, and temperature can influence the activity of microbes. Application of genetically engineered bacteria is green strategy specifically when wild microbes are unable to degenerate persistent contaminants (Liu et al. 2019). Gong et al. (2018) compared the degradation capacity of wild and engineered strains and proved that the engineered strain has a higher degradation capacity. Mardani et al. (2017) applied genetically modified *Pseudomonas putida* for oil degeneration from spiked soil and the rate of oil degradation was higher in soil supplemented with genetically modified *P. putida* as compared to wild species (Mardani et al. 2017). *Geobacillus thermodenitrifican* was genetically modified and applied for nitro-alkanes degradation (Zhang et al. 2021). *Geobacillus thermodenitrifican* degraded 2-nitropropane twofold more in comparison to wild bacteria under anaerobic conditions; however, threefold higher as compared to wild bacteria under aerobic conditions (Zhang et al. 2021).

2.21 CONCLUSION

The release of ECs into the environment due to anthropogenic or industrial activities and changes in lifestyle have led to health issues. Several treatment procedures have been used for the removal of pollutants, but most of them are unsustainable as generate detrimental by-products. Agro-waste residues and industrial wastes are promising options to replace activated carbon. They are available at zero cost, with high adsorption capacities. Agro-industrial wastes utilization can resolve the waste disposal problem as it aligns with the circular economy model to achieve sustainable development goals. Hence, agro-industrial residues can be applied with biofiltration technologies. A major challenge is to scale up the laboratory investigations for the feasibility of process at a large scale for its practical applications. Future research should focus on novel wastewater treatment process development or a combination of technologies which are efficient, economical, and eco-friendly for the removal of ECs. Furthermore, emphasis should be given on multi-disciplinary coordination to increase the practical deployment and sustainability of available technologies. Further researches are needed for the degradation/removal of ECs by novel materials following green strategies for achieving sustainable development goals.

ACKNOWLEDGEMENTS

The authors extend their appreciation to Amity University, Uttar Pradesh, Noida, India; Universiti Sains Malaysia, Malaysia; and Poznan University of Technology, Poland for providing all necessary facilities and valuable support.

REFERENCES

Abedi G, Talebpour Z, Jamechenarboo F. 2018. The survey of analytical methods for sample preparation and analysis of fragrances in cosmetics and personal care products. *TrAC Trend Anal Chem.* 102: 41–59.

Aguera A, Martínez Bueno MJ, Fernández-Alba AR. 2013. New trends in the analytical determination of emerging contaminants and their transformation products in environmental waters. *Environ Sci Pollut Res.* 20: 3496–3515.

Ahmad M, Usman ARA, Rafique MI, Al-Wabel MI. 2019. Engineered biochar composites with zeolite, silica, and nano-zerovalent iron for the efficient scavenging of chlortetracycline from aqueous solutions. *Environ Sci Pollut Res.* 26: 15136–15152.

Ahmed SF, Mofijur M, Nuzhat S, Chowdhury AT, Rafa N, Uddin MA, Inayat A, Mahlia TMI, Ong HC, Chia WY, Show PL. 2021. Recent developments in physical, biological, chemical, and hybrid treatment techniques for removing emerging contaminants from wastewater. *J Hazard Mater.* 416: 125912.

Almeida-Naranjo CE, Morillo B, Aldás MB, Garcés N, Debut A, Guerrero VH. 2023. Zinc removal from synthetic waters using magnetite/graphene oxide composites. *Remediation.* 33: 135–150.

Anku WW, Mamo MA, Govender PP. 2017. Phenolic compounds in water: sources, reactivity, toxicity and treatment methods. In: *Phenolic Compounds-Natural Sources, Importance and Applications* (Eds.) Marcos Soto-Hernandez, Mariana Palma-Tenango and Maria del Rosario Garcia-Mateos). Intech Open. pp. 419–443.

Arenas LR, Gentile SR, Zimmermann S, Stoll S. 2021. Nanoplastics adsorption and removal efficiency by granular activated carbon used in drinking water treatment process. *Sci Total Environ.* 791: 148175.

Arenas LR, Gentile SR, Zimmermann S, Stoll S. 2022. Fate and removal efficiency of polystyrene nanoplastics in a pilot drinking water treatment plant. *Sci Total Environ.* 813: 152623.

Arora S, Saraswat S, Rajpal A, Shringi H, Mishra R, Sethi J, Rajvanshi J, Nag A, Saxena S, Kazmi AA. 2021. Effect of earthworms in reduction and fate of antibiotic resistant bacteria and antibiotic resistant genes during clinical laboratory wastewater treatment by vermifiltration. *Sci Total Environ.* 773: 145152.

Ávila C, García-Galán MJ, Borrego CM, Rodríguez-Mozaz S, García J, Barceló D. 2021. New insights on the combined removal of antibiotics and ARGs in urban wastewater through the use of two configurations of vertical subsurface flow constructed wetlands. *Sci Total Environ.* 755: 142554.

Azaroff A, Monperrus M, Miossec C, Gassie C, Guyoneaud R. 2021. Microbial degradation of hydrophobic emerging contaminants from marine sediment slurries (Capbreton Canyon) to pure bacterial strain. *J Hazard Mater.* 402: 123477.

Babuponnusami A, Sinha S, Ashokan H, Paul MV, Hariharan SP, Arun J, Gopinath KP, Le QH, Pugazhendhi A. 2023. Advanced oxidation process combined biological process for wastewater treatment: a review on advancements, feasibility and practicability of combined techniques. *Environ Res.* 22: 116944.

Baccar R, Sarrà M, Bouzid J, Feki M, Blánquez P. 2012. Removal of pharmaceutical compounds by activated carbon prepared from agricultural by-product. *Chem Eng J.* 211: 310–317.

Bai X, Acharya K. 2017. Algae-mediated removal of selected pharmaceutical and personal care products from Lake Mead water. *Sci Total Environ.* 581–582: 734–740.

Baig N, Sajid M, Saleh A. 2019. Graphene-based adsorbents for the removal of toxic organic pollutants: a review. *J Environ Manag.* 244: 370–382.

Balasubramani K, Sivarajasekar N, Naushad M. 2020. Effective adsorption of antidiabetic pharmaceutical (metformin) from aqueous medium using graphene oxide nanoparticles: equilibrium and statistical modelling. *J Mol Liq.* 301: 112426.

Bayo J, López-Castellanos J, Olmos S. 2020. Membrane bioreactor and rapid sand filtration for the removal of microplastics in an urban wastewater treatment plant. *Mar Pollut Bull.* 156: 111211.

Bedia J, Peñas-garz M, Almudena G, Rodriguez JJ. 2018. A review on the synthesis and characterization of biomass-derived carbons for adsorption of emerging contaminants from water. *J Carbon Res.* 4: 63.

Bele S, Samanidou V, Deliyanni E. 2016. Effect of the reduction degree of graphene oxide on the adsorption of bisphenol A. *Chem Eng Res Des.* 109: 573–585.

Benjedim S, Romero-Cano LA, Pérez-Cadenas AF, Bautista-Toledo MI, Carrasco-Marín F. 2020. Removal of emerging pollutants present in water using an E-coli biofilm supported onto activated carbons prepared from argan wastes: adsorption studies in batch and fixed bed. *Sci Total Environ.* 720: 137491.

Berthiaume A, Galarneau E, Marson G. 2021. Polycyclic aromatic compounds in the Canadian environment: sources and emissions. *Environ Pollut.* 269: 116008.

Besha AT, Liu Y, Fang C, Bekele DN, Naidu R. 2020. Assessing the interactions between micropollutants and nanoparticles in engineered and natural aquatic environments. *Crit Rev Environ Sci Technol.* 50: 135–215.

Buchberger WW. 2011. Current approaches to trace analysis of pharmaceuticals and personal care products in the environment. *J Chromatogr A.*, 1218 (4): 603–618.

Buerge I J, Kasteel R, Poiger T. 2024. Leaching of herbicides and their metabolites in lysimeters filled with soils from railway tracks. *Sci Total Environ.* 909: 168396.

Cabrita I, Ruiz B, Mestre AS, Fonseca IM, Carvalho AP, Ania CO. 2010. Removal of an analgesic using activated carbons prepared from urban and industrial residues. *Chem Eng J.* 163: 249–255.

Campanha MB, Awan AT, de Sousa DN, Grosseli GM, Mozeto AA, Fadini PS. 2015. A 3-year study on occurrence of emerging contaminants in an urban stream of São Paulo State of Southeast Brazil. *Environ Sci Pollut Res.* 22 (10): 7936–7947.

Cao X, Wang K, Feng X. 2021. Removal of phenolic contaminants from water by pervaporation. *J Membr Sci.* 623: 119043.

Chen J, Liu Y, Deng W, Ying G. 2019. Removal of steroid hormones and biocides from rural wastewater by an integrated constructed wetland. *Sci Total Environ.* 660: 358–365.

Chen W, Xue M, Xue F, Mu X, Xu Z, Meng Z, Zhu G, Shea KJ. 2015. Molecularly imprinted hollow spheres for the solid phase extraction of estrogens. *Talanta* 140: 68–72.

Cheng D, Ngo HH, Guo W, Chang SW, Nguyen DD, Liu Y, Shan X, Nghiem LD, Nguyen LN. 2020. Removal process of antibiotics during anaerobic treatment of swine wastewater. *Bioresour Technol.* 300: 122707.

Choi YK, Choi TR, Gurav R, Bhatia SK, Park YL, Kim HJ, Kan E, Yang YH. 2020. Adsorption behavior of tetracycline onto *Spirulina* sp. derived biochars produced at different temperatures. *Sci Total Environ.* 710: 136282.

Choong CE, Ibrahim S, Yoon Y, Jang M. 2018. Removal of lead and bisphenol A using magnesium silicate impregnated palm-shell waste powdered activated carbon: comparative studies on single and binary pollutant adsorption. *Ecotoxicol Environ Saf.* 148: 142–151.

Couto CF, Lange LC, Amaral MC. 2019. Occurrence, fate and removal of pharmaceutically active compounds (PhACs) in water and wastewater treatment plants: a review. *J Water Process Eng.* 32: 100927.

Das MT, Kumar SS, Ghosh P, Shah G, Malyan SK, Bajar S, Thakur IS, Singh L. 2021. Remediation strategies for mitigation of phthalate pollution: challenges and future perspectives. *J Hazard Mater.* 409: 124496.

de Oliveira M, Frihling BEF, Velasques J, Magalhães Filho FJC, Cavalheri PS, Migliolo L. 2019. Pharmaceuticals residues and xenobiotics contaminants: occurrence, analytical techniques and sustainable alternatives for wastewater treatment. *Sci Total Environ.* 705: 135568.

de Souza RM, Seibert D, Quesada HB, de Jesus Bassetti F, Fagundes-Klen MR, Bergamasco R. 2020. Occurrence, impacts and general aspects of pesticides in surface water: a review. *Process Saf Environ Prot.* 135: 22–37.

Devi P, Saroha AK. 2016. Utilization of sludge based adsorbents for the removal of various pollutants: a review. *Sci Total Environ.* 578: 16–33.

de Wilt A, Butkovskyi A, Tuantet K, Leal LH, Fernandes TV, Langenhoff A, Zeeman G. 2016. Micropollutant removal in an algal treatment system fed with source separated wastewater streams. *J Hazard Mater.* 304: 84–92.

Ehrhart AL, Granek EF. 2023. PPCPs in coastal wastewater treatment plant effluent and uptake by Pacific oysters (*Crassostrea gigas*): findings from a laboratory experiment. *Sci Total Environ.* 900: 165728.

Eichler CM, Cohen Hubal EA, Little JC. 2019. Assessing human exposure to chemicals in materials, products and articles: the international risk management landscape for phthalates. *Environ Sci Technol.* 53 (23): 13583–13597.

Enyoh CE, Verla AW, Qingyue W, Ohiagu FO, Chowdhury AH, Enyoh EC, Chowdhury T, Verla EN, Chinwendu UP. 2020. An overview of emerging pollutants in air: method of analysis and potential public health concern from human environmental exposure. *Trends Environ Anal Chem.* 28: e00107.

Fawell J, Ong CN. 2012. Emerging contaminants and the implications for drinking water. *Int J Water Resour Dev.* 28 (2): 247–263.

Ferguson KK, McElrath TF, Chen YH, Loch-Caruso R, Mukherjee B, Meeker JD. 2015. Repeated measures of urinary oxidative stress biomarkers during pregnancy and preterm birth. *Amer J Obstet Gynecol.* 212 (2): 208.e1–e8.

Franco DSP, Georgin J, Lima EC, Silva LFO. 2022. Advances made in removing paraquat herbicide by adsorption technology: a review. *J Water Process Eng.* 49: 102988.

Gadipelly CR, Marathe KV, Rathod VK. 2018. Effective adsorption of ciprofloxacin hydrochloride from aqueous solutions using metal-organic framework. *Sep Sci and Technol.* 53: 2826–2832.

Gani KM, Kazmi AA. 2017. Contamination of emerging contaminants in Indian aquatic sources: first overview of the situation. *J Hazard Toxic Radioact Waste.* 21: 04016026.

Ganie ZA, Khandelwal N, Tiwari E, Singh N, Darbha GK. 2021. Biochar-facilitated remediation of nanoplastic contaminated water: effect of pyrolysis temperature induced surface modifications. *J Hazard Mater.* 417: 126096.

Ghosh S, Sahu M. 2022. Phthalate pollution and remediation strategies: a review. *J Hazard Mater Advan.* 6: 100065.

Gomes IB, Simoes LC, Simoes M. 2018. The effects of emerging environmental contaminants on Stenotrophomonas maltophilia isolated from drinking water in planktonic and sessile states. *Sci Total Environ.* 643: 1348–1356.

Gong J, Kong T, Li Y, Li Q, Li Z, Zhang J. 2018. Biodegradation of microplastic derived from poly (ethylene terephthalate) with bacterial whole-cell biocatalysts. *Polymers* 10 (12): 1326.

Gonzalez-Martinez A, Rodriguez-Sanchez A, Martinez-Toledo MV, Garcia-Ruiz MJ, Hontoria E, Osorio-Robles F, Gonzalez–Lopez J. 2014. Effect of ciprofloxacin antibiotic on the partial-nitritation process and bacterial community structure of a submerged biofilter. *Sci Total Environ.* 476–477: 276–287.

Grenni P, Ancona V, Barra Caracciolo A. 2018. Ecological effects of antibiotics on natural ecosystems: a review. *Microchem J.* 136: 25–39.

Hena S, Gutierrez L, Croué JP. 2021. Removal of pharmaceutical and personal care products from wastewater using microalgae: a review. *J Hazard Mater.* 403: 124041.

Hom-Diaz A, Jaen-Gil A, Bello-Laserna I, Rodríguez-Mozaz S, Vicent T, Barcelo D, Blanquez P. 2017. Performance of a microalgal photobioreactor treating toilet wastewater: pharmaceutically active compound removal and biomass harvesting. *Sci Total Environ.* 592: 1–11.

Huang L, Zhu X, Zhou S, Cheng Z, Shi K, Zhang C, Shao H. 2021. Phthalic acid esters: natural sources and biological activities. *Toxins* 13 (7): 495.

Hung CM, Huang CP, Chen CW, Wu C H, Lin YL, Dong C D. 2020. Activation of percarbonate by water treatment sludge-derived biochar for the remediation of PAH-contaminated sediments. *Environ Pollut.* 265 (Pt B): 114914.

Hussain S, Siddique T, Arshad M, Saleem M. 2009. Bioremediation and phytoremediation of pesticides: recent advances. *Crit Rev Environ Sci Technol.* 39 (10): 843–907.

Jacob MM, Ponnuchamy M, Kapoor A, Sivaraman P. 2020. Bagasse based biochar for the adsorptive removal of chlorpyrifos from contaminated water. *J Environ Chem Eng.* 8: 103904.

Jacob RS, Araújo CV, de Souza Santos LV, Moreira VR, Lebron YAR, Lange L C. 2021. The environmental risks of pharmaceuticals beyond traditional toxic effects: chemical differences that can repel or entrap aquatic organisms. *Environ Pollut.* 268: 115902.

Jagini S, Thaduri S, Konda S, Saranga VK, Dheeravath B, Vurimindi H. 2021. Emerging contaminant (Triclosan) removal by adsorption and oxidation process: comparative study. *Model Earth Syst Environ.* 7: 2431–2438.

Jeong C, Ansari M Z, Hakeem Anwer A, Kim SH, Nasar A, Shoeb M. et al. 2023. A review on metal-organic frameworks for the removal of hazardous environmental contaminants. *Sep Purif Technol.* 305: 122416.

Ji L, Shao Y, Xu Z, Zheng S, Zhu D. 2010. Adsorption of monoaromatic compounds and pharmaceutical antibiotics on carbon nanotubes activated by KOH etching. *Environ Sci Technol.* 44 (16): 6429–6436.

Ji Z, Tang W, Pei Y, 2022. Constructed wetland substrates: a review on development, function mechanisms, and application in contaminants removal. *Chemosphere* 286 (1): 131564.

Kapoor RT, Zdarta J. 2024. Fabrication of engineered biochar for remediation of toxic contaminants in soil matrices and soil valorization. *Chemosphere* 358: 142101.

Karimi H, Rahimpour A, Shirzad Kebria MR. 2016. Pesticides removal from water using modified piperazine-based nanofiltration membranes. *Desalin Water Treat.* 57 (52): 24844–24854.

Kasonga TK, Coetzee MA, Van Zijl C, Momba MNB. 2019. Removal of pharmaceutical' estrogenic activity of sequencing batch reactor effluents assessed in the T47D-KBluc reporter gene assay. *J Environ Manag.* 240: 209–218.

Kaur Y, Bhatia Y, Chaudhary S, Chaudhary GR. 2017. Comparative performance of bare and functionalize ZnO nanoadsorbents for pesticide removal from aqueous solution. *J Mol Liq.* 234: 94–103.

Khan NA, Khan SU, Ahmed S, Farooqi IH, Yousef M, Mohammadi AA, Changani F. 2020. Recent trends in disposal and treatment technologies of emerging-pollutants – a critical review. *Trends Anal Chem.* 122: 115744.

Khan S, Naushad M, Govarthanan M, Iqbal J, Alfadul SM. 2022. Emerging contaminants of high concern for the environment: current trends and future research. *Environ Res.* 207: 112609.

Křesinová Z, Linhartová L, Filipová A, Ezechiáš M, Mašín P, Cajthaml T. 2018. Biodegradation of endocrine disruptors in urban wastewater using Pleurotus ostreatus bioreactor. *New Biotechnol.* 43: 53–61.

Kristanti RA, Tirtalistyani R, Tang YY, Thao NT, Kasongo J, Wijayanti Y. 2023. Phytoremediation mechanism for emerging pollutants: a review. *Trop Aquat Soil Pollut.* 3 (1): 88–108.

Lares M, Ncibi MC, Sillanpää M, Sillanpää, M. 2018. Occurrence, identification and removal of microplastic particles and fibers in conventional activated sludge process and advanced MBR technology. *Water Res.* 133: 236–246.

Lempart A, Kudlek E, Dudziak M. 2020. The potential of the organic micropollutants emission from swimming accessories into pool water. *Environ Int.* 136: 105442.

Li J, Zhou Q, Campos LC. 2018. The application of GAC sandwich slow sand filtration to remove pharmaceutical and personal care products. *Sci Total Environ.* 635: 1182–1190.

Lin C, Cheruiyot NK, Bui XT, Ngo HH. 2022. Composting and its application in bioremediation of organic contaminants. *Bioengineered* 13 (1): 1073–1089.

Liu Y, Blowes DW, Ptacek CJ, Groza LG. 2019. Removal of pharmaceutical compounds, artificial sweeteners, and perfluoroalkyl substances from water using a passive treatment system containing zero-valent iron and biochar. *Sci Total Environ.* 691: 165–177.

Lu J, Zhang C, Wu J et al. 2017. Adsorptive removal of bisphenol A using n-doped biochar made of *Ulva prolifera. Water Air Soil Pollut.* 228: 327.

Luo J, Zhang Q, Cao M, Wu L, Cao J, Fang F, Li C, Xue Z, Feng Q. 2019. Ecotoxicity and environmental fates of newly recognized contaminants-artificial sweeteners: a review. *Sci Total Environ.* 653: 1149–1160.

Lv B, Jiao Y, Deng X, Fan W, Xing B. 2025. Adsorptive removal of microplastics from aquatic environments using coal gasification slag-based adsorbent in a liquid–solid fluidized bed. *Sep Purif Technol.* 354: 128935.

Mardani G, Mahvi AH, Hashemzadeh-Chaleshtori M, Naseri S, Dehghani MH, et al. 2017. Application of genetically engineered dioxygenase producing *Pseudomonas putida* on decomposition of oil from spiked soil. *Jundishapur J Nat Pharm Prod.* 12 (3): e64313.

Matthee C, Brown AR, Lange A, Tyler CR. 2023. Factors determining the susceptibility of fish to effects of human pharmaceuticals. *Environ Sci Technol.* 57 (24): 8845–8862.

Mirasole C, Emond Di Carro M, Tanwar S, Magi E. 2016. Liquid chromatography–tandem mass spectrometry and passive sampling: powerful tools for the determination of emerging pollutants in water for human consumption. *J Mass Spectrom.* 51 (9): 814–820.

Molina MC, Bautista LF, Catala´ M, de las Heras MR, Martínez-Hidalgo P, SanSebastian J, Gonz´ alez-Benítez N. 2020. From laboratory tests to the ecoremedial system: the importance of microorganisms in the recovery of PPCPs-disturbed ecosystems. *Appl Sci.* 10 (10): 3391.

Morin-Crini N, Lichtfouse E, Liu G, Balaram V, Ribeiro AR, Lu Z, Stock F, Carmona E, Teixeira MR, Picos-Corrales LA, Moreno-Pirajan JC. 2022. Worldwide cases of water pollution by emerging contaminants: a review. *Environ Chem Lett.* 20 (4): 2311–2338.

Mishra RK, Mentha SS, Misra Y, Dwivedi N. 2023. Emerging pollutants of severe environmental concern in water and wastewater: a comprehensive review on current developments and future research. *Water-Energy Nexus.* 6: 74–95.

Mitchell SM, Subbiah M, Ullman JL, Frear C, Call DR. 2015. Evaluation of 27 different biochars for potential sequestration of antibiotic residues in food animal production environments. *J Environ Chem Eng.* 3 (1): 162–169.

Mo J, Yang Q, Zhang N, Zhang W, Zheng Y, Zhang Z. 2018. A review on agro-industrial waste (AIW) derived adsorbents for water and wastewater treatment. *J Environ Manag.* 227: 395–405.

Mohapatra DP, Kirpalani DM. 2019. Advancement in treatment of wastewater: fate of emerging contaminants. *Can J Chem Eng.* 97 (10): 2621–2631.

Mojiri A, Zhou JL, Ratnaweera H, Rezania S, Nazari M. 2022. Pharmaceuticals and personal care products in aquatic environments and their removal by algaebased systems. *Chemosphere* 288: 132580.

Moliner-Martínez Y, Ribera A, Coronado E, Campíns-Falcó P. 2011. Preconcentration of emerging contaminants in environmental water samples by using silica supported Fe_3O_4 magnetic nanoparticles for improving mass detection in capillary liquid chromatography. *J Chromatogr A.* 1218 (16): 2276–2283

Musbah I, Ciceron D, Saboni A, Alexandrova S. 2018. Removal of pesticides and desethylatrazine (DEA) by nanofiltration: effects of organic and inorganic solutes on solute rejection. *J Chem Technol Metall.* 53 (4): 657–663.

Nassar MM, Farrag TE, Ahmed MH. 2012. Removal of an insecticide (methomyl) from aqueous solutions using natural clay. *Alexandria Eng J.* 51: 11–18.

Ouda M, Pandey RP, Banat F, Hasan SW. 2023. Advancing water treatment sustainability: investigating electrified $Ti_3C_2T_x$ composite membranes for minimizing microplastic fouling. *Chemosphere* 343: 140224.

Pal A, He Y, Jekel M, Reinhard M, Gin KYH. 2014. Emerging contaminants of public health significance as water quality indicator compounds in the urban water cycle. *Environ Int.* 71: 46–62.

Palmer M, Hatley H. 2018. The role of surfactants in wastewater treatment: impact, removal and future techniques: a critical review. *Water Res* 147: 60–72.

Pang X, Skillen N, Gunaratne N, Rooney DW, Robertson PK. 2021. Removal of phthalates from aqueous solution by semiconductor photocatalysis: a review. *J Hazard Mater.* 402: 123461.

Paredes-Laverde M, Silva-agredo J, Torres-palma RA. 2018. Removal of norfloxacin in deionized, municipal water and urine using rice (Oryza sativa) and coffee (Coffee angmu) husk wastes as natural adsorbents. *J Environ Manag.* 213: 98–108.

Pesqueira JF, Pereira MFR, Silva A. 2020. Environmental impact assessment of advanced urban wastewater treatment technologies for the removal of priority substances and contaminants of emerging concern: a review. *J Clean Prod.* 261: 121078.

Poddar K, Sarkar D, Sarkar A. 2024. Norfloxacin adsorption by torrefied coco peat biochar as a novel adsorbent in a circular economy framework. *Environ Res.* 251 (Part 2): 118711.

Pompei CME, Ciric L, Canales M, Karu K, Vieira EM, Campos LC. 2017. Influence of PPCPs on the performance of intermittently operated slow sand filters for household water purification. *Sci Total Environ.* 581–582: 174–185.

Praveena SM, Cheema MS, Guo HR. 2019. Non-nutritive artificial sweeteners as an emerging contaminant in environment: a global review and risks perspectives. *Ecotoxicol Environ Saf.* 170: 699–707.

Pukcothanung Y, Siritanon T, Rangsriwatananon K. 2018. The efficiency of zeolite Y and surfactant-modified zeolite Y for removal of 2,4-dichlorophenoxyacetic acid and 1,1"-dimethyl-4,4"-bipyridinium ion. *Microporous Mesoporous Mater.* 258: 131–140.

Quesada HB, Alves Baptista AT, Cusioli LF, Seibert D, de Oliveira Bezerra C, Bergamasco R. 2019. Surface water pollution by pharmaceuticals and an alternative of removal by low-cost adsorbents: a review. *Chemosphere* 222: 766–780.

Rathi BS, Kumar PS, Show PL. 2021. A review on effective removal of emerging contaminants from aquatic systems: current trends and scope for further research. *J Hazard Mater.* 409: 124413.

Raupp ÍN, Filho AV, Arim AL, Muniz ARC, da Rosa GS. 2021. Development and characterization of activated carbon from olive pomace: experimental design, kinetic and equilibrium studies in nimesulide adsorption. *Materials.* 14: 6820.

Reddy AA, Reddy M, Mathur V. 2024. Pesticide use, regulation, and policies in Indian agriculture. *Sustainability.* 16: 7839.

Reddy AVB, Moniruzzaman M, Aminabhavi TM. 2019. Polychlorinated biphenyls in the environment: recent updates on sampling, pretreatment, cleanup technologies and their analysis. *Chem Eng J.* 358: 1186–1207.

Reyes NJ, Geronimo FK, Yano KA, Guerra HB, Kim LH. 2021. Pharmaceutical and personal care products in different matrices: occurrence, pathways, and treatment processes. *Water* 13 (9): 1159.

Richardson SD, Kimura SY. 2017. Emerging environmental contaminants: challenges facing our next generation and potential engineering solutions. *Environ Technol In*nov. 8: 40–56.

Rizzo L, Malato S, Antakyali D, Beretsou VG, Đolić MB, Gernjak W, Heath E, Ivancev-Tumbas I, Karaolia P, Ribeiro ARL, Mascolo G. 2019. Consolidated vs new advanced treatment methods for the removal of contaminants of emerging concern from urban wastewater. *Sci Total Environ.* 655: 986–1008.

Robles-Molina J, Lara-Ortega FJ, Gilbert-López B, García-Reyes JF, Molina-Díaz A. 2014. Multi-residue method for the determination of over 400 priority and emerging pollutants in water and wastewater by solid-phase extraction and liquid chromatography-time-of-flight mass spectrometry. *J Chromatogr A.* 1350: 30–43.

Rojas S, Horcajada P. 2020. Metal–organic frameworks for the removal of emerging organic contaminants in water. *Chem Rev.* 120 (16): 8378–8415.

Rout PR, Zhang TC, Bhunia P, Surampalli RY. 2021. Treatment technologies for emerging contaminants in wastewater treatment plants: a review. *Sci Total Environ.* 753: 141990.

Sabri NA, Schmitt H, van der Zaan BM, Gerritsen HW, Rijnaarts HHM, Langenhoff AAM. 2021. Performance of full scale constructed wetlands in removing antibiotics and antibiotic resistance genes. *Sci Total Environ.* 786: 147368.

Saleh IA, Zouari N, Al-Ghouti MA. 2020. Removal of pesticides from water and wastewater: chemical, physical and biological treatment approaches. *Environ Technol Innovat.* 19: 101026.

Sardon H, Dove A.P. 2018. Plastics recycling with a difference. *Science* 360 (6387): 380–381.

Sauvé S, Desrosiers M. 2014. A review of what is an emerging contaminant. *Chem Central J.* 8 (1): 1–7.

Schwaminger SP, Fehn S, Steegmüller T, Rauwolf S, Löwe H, Pflüger-Grau K, Berensmeier S. 2021. Immobilization of PETase enzymes on magnetic iron oxide nanoparticles for the decomposition of microplastic PET. *Nanoscale Adv.* 3: 4395–4399.

Seralini GE, Jungers G. 2021. Endocrine disruptors also function as nervous disruptors and can be renamed endocrine and nervous disruptors (ENDs). *Toxicol Rep.* 8: 1538–1557.

Sharma P, Singh SP. 2021. Pollutants characterization and toxicity assessment of pulp and paper industry sludge for safe environmental disposal. In: *Emerging Treatment Technologies for Waste Management* (Eds. I Haq, AS Kalamdhad). Springer Nature. pp. 207–223.

Shahid MK, Kashif A, Fuwad A, Choi Y. 2021. Current advances in treatment technologies for removal of emerging contaminants from water – a critical review. *Coord Chem Rev.* 442: 213993.

Shi C, Zhang S, Zhao J, Ma J, Wu H, Sun H, Cheng S. 2022. Experimental study on removal of microplastics from aqueous solution by magnetic force effect on the magnetic sepiolite. *Sep Purif Technol.* 288: 120564.

Shi J, Zheng M, Zhang Z, Han H, Xu C. 2021. Enhanced anaerobic degradation of quinoline and indole with dried Chlorella pyrenoidosa powder as co-metabolic substance. *J Clean Prod.* 285: 124909.

Shokouhi R, Ghobadi N, Godini K, Hadi M, Atashzaban Z. 2019. Antibiotic detection in a hospital wastewater and comparison of their removal rate by activated sludge and earthworm-based vermifiltration: environmental risk assessment. *Process Saf Environ Prot.* 134: 169–177.

Si X, Hu Z, Huang S. 2018. Combined process of ozone oxidation and ultrafiltration as an effective treatment technology for the removal of endocrine-disrupting chemicals. *Appl Sci.* 8: 1240.

Singh AK, Chandra R. 2019. Pollutants released from the pulp paper industry: aquatic toxicity and their health hazards. *Aquat Toxicol.* 211: 202–216.

Singh N, Khandelwal N, Ganie ZA, Tiwari E, Darbha GK. 2021. Eco-friendly magnetic biochar: an effective trap for nanoplastics of varying surface functionality and size in the aqueous environment. *Chem Eng J.* 418: 129405.

Sophia AC, Lima EC. 2018. Removal of emerging contaminants from the environment by adsorption. *Ecotoxicol Environ Saf.* 150: 1–17.

Sun Y, Lu G, Li J, Dang T, Xue C, Liu J. 2021. Multimedia distribution and trophic transfer of PPCPs in the middle and lower reaches of the Yarlung Zangbo river. *Environ Pollut.* 271: 116408.

Tang J, Liu Z, Lu W, Wang L, Zhang C, Su P. 2021. Electrochemical degradation of perfluorinated compounds by Ag coated Ti (Ti/Ag) anode: electrode preparation, characterization and application. *Environ Sci Water Res Technol.* 7: 455–467.

Tarasi R, Alipour M, Gorgannezhad L, Imanparast S, Yousefi-Ahmadipour A, Ramezani A. 2018. Laccase immobilization onto magnetic β-cyclodextrin-modified chitosan: improved enzyme stability and efficient performance for phenolic compounds elimination. *Macromol Res.* 26: 755–762.

Teixidó M, Pignatello JJ, Beltrán JL, Granados M, Peccia J. 2011. Speciation of the ionizable antibiotic sulfamethazine on black carbon (biochar). *Environ Sci Technol.* 45 (23): 10020–10027.

Tejedor J, Cóndor V, Almeida-Naranjo CE, Guerrero VH, Villamar CA. 2020. Performance of wood chips/peanut shells biofilters used to remove organic matter from domestic wastewater. *Sci Total Environ.* 738: 139589.

Thiebault T. 2019. Raw and modified clays and clay minerals for the removal of pharmaceutical products from aqueous solutions: state of the art and future perspectives. *Crit Rev Environ Sci Technol.* 50: 1451–1514.

Thuptimdang P, Siripattanakul-Ratpukdi STR, Youngwilai A, Khan E. 2020. Biofiltration for treatment of recent emerging contaminants in water: current and future perspectives. *Water Environ Res.* 93: 972–992.

Tiwari B, Sellamuthu B, Ouarda Y, Drogui P, Tyagi RD, Buelna G. 2017. Review on fate and mechanism of removal of pharmaceutical pollutants from wastewater using biological approach. *Bioresour Technol.* 224: 1–12.

Tong X, Jiang L, Li Y, Chen X, Zhao Y, Hu B, Zhang F. 2020. Function of agricultural waste montmorillonite-biochars for sorptive removal of 17β-estradiol. *Bioresour Technol.* 296: 122368.

Torres-Pérez J, Gérente C, Andrès Y. 2012. Sustainable activated carbons from agricultural residues dedicated to antibiotic removal by adsorption. *Chin J Chem Eng.* 20 (3): 524–529.

Tran NH, Reinhard M, Gin KYH. 2018. Occurrence and fate of emerging contaminants in municipal wastewater treatment plants from different geographical regions: a review. *Water Res.* 133: 182–207.

Vasilachi IC, Asiminicesei DM, Fertu DI, Gavrilescu M. 2021. Occurrence and fate of emerging pollutants in water environment and options for their removal. *Water.* 13: 181.

Vela N, Fenoll J, Garrido I, P´erez-Lucas G, Flores P, Hellín P, Navarro S. 2019. Reclamation of agro-wastewater polluted with pesticide residues using sunlight activated persulfate for agricultural reuse. *Sci Total Environ.* 66: 923–930.

Wang S, Wang X, Poon K, Wang Y, Li S, Liu H, Lin S, Cai Z. 2013. Removal and reductive dechlorination of triclosan by Chlorella pyrenoidosa. *Chemosphere* 92 (11): 1498–1505.

Wang X, Li J, Zhang X, Chen Z, Shen J, Kang J. 2021. The performance of aerobic granular sludge for simulated swine wastewater treatment and the removal mechanism of tetracycline. *J Hazard Mater J.* 408: 124762.

Wang Z, Sedighi M, Lea-Langton A. 2020. Filtration of microplastic spheres by biochar: removal efficiency and immobilisation mechanisms. *Water Res.* 184: 116165.

Wilkinson J, Hooda PS, Barker J, Barton S, Swinden J. 2017. Occurrence, fate and transformation of emerging contaminants in water: an overarching review of the field. *Environ Pollut.* 231: 954–970.

Xiaoyan T, Suyu W, Yang Y, Ran T, Yunv D, Dan A, Li L. 2015. Removal of six phthalic acid esters (PAEs) from domestic sewage by constructed wetlands. *Chem Eng J.* 275: 198–205.

Xu LJ, Chu W, Graham N. 2013. A systematic study of the degradation of dimethyl phthalate using a high-frequency ultrasonic process. *Ultrason Sonochem.* 20 (3): 892–899.

Xu Q, Zhou Q, Pan M, Dai L. 2020. Interaction between chlortetracycline and calcium-rich biochar: enhanced removal by adsorption coupled with flocculation. *Chem Eng J.* 382: 122705.

Yadav S, Kumar N, Kumari V, Mittal A, Sharma S. 2019. Photocatalytic degradation of Triclopyt, a persistent pesticide by ZnO/SnO_2 nano-composities. *Mater Today Proc.* 19 (2): 642–645.

Yang W, Lu Y, Zheng F, Xue X, Li N, Liu D. 2012. Adsorption behavior and mechanisms of norfloxacin onto porous resins and carbon nanotube. *Chem Eng J.* 179: 112–118.

Yang Y, Qiao S, Zheng M, Zhou J, Quan X. 2019. Enhanced permeability, contaminants removal and antifouling ability of CNTs-based hollow fiber membranes under electrochemical assistance. *J Membr Sci.* 582: 335–341.

Zaheer M, Sun X, Liu J, Song C, Wang S, Javed A. 2018. Enhancement of ciprofloxacin sorption on chitosan/biochar hydrogel beads. *Sci Total Environ.* 639: 560–569.

Zavareh S, Farrokhzad Z, Darvishi F. 2018. Modification of zeolite 4A for use as an adsorbent for glyphosate and as an antibacterial agent for water. *Ecotoxicol Environ Saf.* 155: 1–8.

Zavoda J, Cutright T, Szpak J, Fallon E. 2001. Uptake, selectivity, and inhibition of hydroponic treatment of contaminants. *J Environ Eng.* 127: 502–508.

Zhang C, Lu J, Wu J. 2021. Enhanced removal of phenolic endocrine disrupting chemicals from coastal waters by intertidal macroalgae. *J Hazard Mater.* 411: 125105.

Zhang J, Gao D, Li Q, Zhao Y, Li L, Lin H, Bi Q, Zhao Y. 2020. Biodegradation of polyethylene microplastic particles by the fungus Aspergillus flavus from the guts of wax moth Galleria mellonella. *Sci Total Environ.* 704: 135931.

Zhang M, Wang S, Ji B, Liu Y. 2019. Towards mainstream deammonification of municipal wastewater: partial nitrification-anammox versus partial denitrification-anammox. *Sci Total Environ.* 692: 393–401.

Zhang SS, Yang N, Ni SQ, Natarajan V, Ahmad HA, Xu S, Fang X, Zhan J. 2018. One-pot synthesis of highly active Ni/Fe nano-bimetal by simultaneous ball milling and in situ chemical deposition. *RSC Adv.* 8: 26469–26475.

Zheng C, Feng S, Liu P, Fries E, Wang Q, Shen Z, Liu H, Zhang T. 2016. Sorption of organophosphate flame retardants on Pahokee peat soil. *Clean Soil Air Water* 44 (9): 1163–1173.
Zheng H, Wang Z, Zhao J, Herbert S, Xing B. 2013. Sorption of antibiotic sulfamethoxazole varies with biochars produced at different temperatures. *Environ Pollut.* 181: 60–67.
Zielińska M, Cydzik-Kwiatkowska A, Bułkowska K, Bernat K, Wojnowska-Baryła I. 2017. Treatment of bisphenol A-containing effluents from aerobic granular sludge reactors with the use of microfiltration and ultrafiltration ceramic membranes. *Water Air Soil Pollut.* 228: 282.
Zhou LC, Meng X, Fu J, Yang Y. 2014. Highly efficient adsorption of chlorophenols onto chemically modified chitosan. *Appl Surf Sci.* 292: 735–741.
Zhu N, Yan Q, He Y, Wang X, Wei Z, Liang D, Yue H, Yun Y, Li G, Sang. 2022. Insights into the removal of polystyrene nanoplastics using the contaminated corncob-derived mesoporous biochar from mining area. *J Hazard Mater.* 433: 128756.

3 Green Solutions for the Removal of Pharmaceutics from Wastewater

Hanna Sieracka, Aleksandra Rybak, Michał Kapczyński, Wojciech Smułek, and Ewa Kaczorek

3.1 INTRODUCTION

In recent decades, global consumption of pharmaceuticals has significantly increased, particularly regarding over-the-counter medications such as diclofenac, ibuprofen, and naproxen, and also antibiotics and antidepressants. This trend has been driven by the rapid development of the pharmaceutical industry and a substantial rise in drug production. Contemporary societies spend billions of dollars every year on pharmaceuticals [1]. However, increased drug sales, despite the low doses of active substances in tablets, result in significant concentrations of pollutants being introduced into the ecosystem [2]. What is important? Most of them are purchased unnecessarily and end up in landfills or sewage after their expiration date. In addition, pharmaceuticals are used on a large scale in animal production, which also contributes significantly to the spread of pharmaceuticals in the environment [3].

The properties of pharmaceutical compounds make traditional wastewater treatment plants incapable of removing them, resulting in trace concentrations of drugs reaching the environment along with treated wastewater [4]. Pharmaceuticals biodegrade at varying rates or may not decompose at all, leading to their accumulation in surface water or infiltration into groundwater, and even entering drinking water supplies, posing risks to the health and life of living organisms [5]. This has heightened interest in finding effective methods for removing pharmaceuticals from the biosphere. What is especially important from the perspective of the One Health approach recommended by the European Commission?

The choice of wastewater treatment technique for pharmaceutics removal is based on the premise that the method should be economical, simple, quick, reusable, and environment- friendly [6]. However, various procedures differ in their effectiveness in removing and biodegrading pharmaceuticals, which means that the choice of the appropriate methodology primarily relies on the requirements that treated wastewater must meet. The removal of drugs from wastewater is a very challenging process

 DOI: 10.1201/9781003536277-3

due to the diversity of the active substances and the numerous toxic metabolites of pharmaceuticals [7]. Many factors influence the effectiveness of wastewater treatment, including the properties, type, concentration, and biological activity of pharmaceutical preparations, making it difficult to find a universal technique.

3.2 PHARMACEUTICALS IN THE ENVIRONMENT

Drugs are organic, biologically active chemical compounds used for their specific effects on living organisms. They have therapeutic and medicinal properties for both humans and animals. They are characterized by high biological activity, hydrophilicity, and challenging biodegradability. Their low molecular weight and susceptibility to metabolic transformations make drugs mobile in aqueous solutions. Pharmaceuticals can be divided into numerous groups of synthetic and natural compounds. Table 3.1 presents the classification of drugs according to Ellis [8].

The pharmaceuticals mentioned above differ in their degree of chemical and biological transformation within the ecosystem. They may undergo complete elimination, partial degradation, or permanently contaminate the natural environment [9]. Active ingredients from pharmaceuticals can be found in surface water, groundwater, drinking water, soil, sewage sludge, manure, flora, and fauna as well as in the food chain of animals.

Pharmaceuticals and their metabolites are found globally in the natural environment, in both developing and industrialized countries. They have been detected in 71 countries, spanning all continents. A total of 631 different active substances have been identified above the detection limits of the methods used [10].

As noted by Wilkinson et al. [11] who compared the pharmaceutical content of rivers across the globe, pharmaceutical contamination of surface water is most relevant to countries in Western Europe, North America, India, and Australia.

TABLE 3.1
Most Commonly Used Pharmaceuticals According to Ellis [8]

Group Name	Active Substance
Antibiotics	Trimethoprim, erythromycin, lincomycin, sulfamethoxazole, chloramphenicol, amoxicillin
β-Blockers	Metoprolol, propranolol, timolol, sotalol, atenolol
X-ray contrast agents	Iopromide, iopamidol, diatrizoate
Bactericidal drugs	Nitrofurantoin, furazidine, furazolidone, nifuroxazide, nifurzide, nitrofurazone
Pain relievers, anti-inflammatory drugs	Ibuprofen, diclofenac, fenoprofen, acetaminophen, naproxen, acetylsalicylic acid, fluoxetine, ketoprofen, indomethacin, paracetamol
Psychotropic drugs	Diazepam, carbamazepine, primidone, salbutamol, fluoxetine
Lipid metabolism regulators	Clofibric acid, bezafibrate, fenofibrate, etofibrate, gemfibrozil
Steroids and hormones	Estradiol, estrone, estriol, diethylstilbestrol, ethinylestradiol

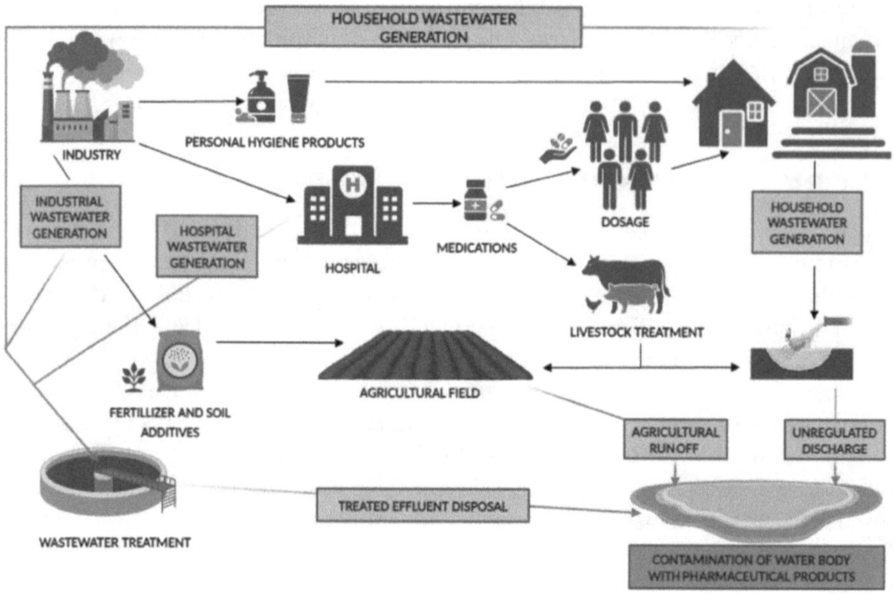

FIGURE 3.1 Sources and transformations of pharmaceuticals in the aquatic environment according to Khan and Barros [12].

However, countries in South America, Africa, and Asia are also affected, as pharmaceutical concentrations also pose an environmental risk [11].

The occurrence and metabolic transformations of most active substances in the ecosystem are poorly understood and not legally regulated. A major issue is the lack of knowledge regarding which pharmaceuticals are harmful to living organisms in trace amounts [12].

There are many pathways through which drugs and their metabolites enter the environment (Figure 3.1). The main sources are municipal and medical wastewater containing pharmaceuticals excreted from the body in unchanged form or as metabolites. Expired medications are often poorly disposed of and also end up in sewage or landfills. Wastewater treatment plants are not legally required to monitor the effectiveness of pharmaceutical removal. Pharmaceutical manufacturing facilities significantly contribute to environmental pollution as well [13]. Wastewater discharged from pharmaceutical factories contains much higher levels of drugs than municipal wastewater. This is due to the fact that many facilities have not implemented routine processes for treating wastewater to remove drug residues. Another source of contamination is veterinary medications used in livestock, pets, or aquaculture. Pharmaceuticals can also enter the ecosystem through the agricultural use of contaminated sewage sludge or manure. The unique mechanisms of action of pharmaceuticals and their design for mammals mean that their residues pose a greater threat to human life and health than pesticides. The negative effects of drugs are not limited to higher organisms but impact all levels of the food chain [14].

The sources of pharmaceuticals entering the environment can be classified as follows [15]:

- **continuous:** originating from wastewater treatment plants,
- **seasonal:** dependent on increased drug usage due to allergies, flu season, or periods of intensified agriculture, as well as lower temperatures, which reduce the biological activity of the activated sludge in treatment plants,
- **periodic:** resulting from rainfall, which affects the dilution of drug concentrations and their leaching from waste disposal sites.

Pharmaceuticals entering the environment undergo various transformations due to abiotic or biotic factors (Figure 3.1) [12]. These transformations can lead to the formation of more toxic and persistent substances, a decrease in the concentration of the parent compounds, or a reduction in the active substance's impact on the environment [16]. Biological degradation plays a lesser role in the elimination of drugs from the environment [16]. Photodegradation due to light exposure and biodegradation is a result of the activity of specific microbial strains. Hydrophobic pharmaceuticals may adsorb onto sediment particles found on the bottoms of aquatic areas. This represents a temporary and apparent immobilization of pollutants due to the possibility of occurrence of reverse processes, which significantly minimize the bioavailability of the pharmaceuticals. However, this sorption process can be reversed as a result of changes in the pH of the environment, the presence of surfactants, temperature, etc. [17]. Moreover, lipid-soluble pharmaceuticals can bioaccumulate in the fatty tissues of animals, subsequently entering the food chain as a result of this process [18].

3.3 REMOVAL OF PHARMACEUTICALS FROM WASTEWATER

Municipal wastewater treatment plants are designed to effectively remove many synthetic and natural compounds [19]. However, they are not equipped to efficiently treat wastewater for pharmaceuticals and their metabolites. As a result, the wastewater released into the environment is not sufficiently free of these substances, making it a major source of ecosystem contamination [20]. The efficiency of pharmaceutical removal in standard wastewater treatment plants typically ranges from 0% to 99% for various substances (Table 3.1) [19,21,22]. Table 3.2 provides a comprehensive overview of the removal efficiencies of various pharmaceuticals in standard wastewater treatment plants. The data indicate that while substances such as ibuprofen and thymol demonstrate relatively high removal efficiencies, ranging from 53% to 99% and 93% to 99%, respectively; others, such as clofibric acid and carbamazepine, exhibit significantly lower efficiencies with removal rates ranging from 0% to 91% and 8% to 20%, respectively.

This variability serves to illustrate the inherent complexity and challenges associated with the treatment of pharmaceutical contaminants in wastewater. The observed differences in removal efficiency are the results of a number of contributing factors. The effectiveness of wastewater treatment technologies is significantly influenced by the difference in treatment processes, environmental conditions, and the inherent properties of the pharmaceutical compounds themselves. For example, the structural

TABLE 3.2
Efficiency of Wastewater Treatment for Selected Pharmaceuticals

Compound	Removal Efficiency (%)	References
Ibuprofen	53–99	[19,21]
Aspirin	67–99	[21]
Fenoprofen	65–96	[19,21]
Naproxen	15–83	[21]
Ketoprofen	8–98	[21]
Thymol	93–99	[21]
Triclosan	46–92	[21]
Diclofenac	17–71	[19]
Clofibric acid	0–91	[19]
Carbamazepine	8–20	[21]
Estrone	0–83	[19,21]

properties of compounds, such as polarity and solubility, can affect their biodegradability and adsorption properties, which in turn affect their removal rates.

The varied effectiveness of municipal wastewater treatment plants, along with the high durability and resistance of certain pharmaceuticals to the methods used, has highlighted the need for effective techniques to remove a wide range of pharmaceuticals [23].

Wastewater treatment technologies can be divided into biotic and abiotic methods. Physicochemical methods include adsorption, filtration, and advanced oxidation processes (AOPs). Biological techniques consist of microbial degradation, bioremediation using microalgae, phytoremediation, and biosorption. It is also possible to combine different biotic and abiotic technologies for more efficient removal of pharmaceuticals from the environment to create hybrid ones [24]. When selecting an appropriate technique, considerations should primarily include its cost, simplicity, effectiveness, environmental friendliness, duration, and applicability [25].

3.3.1 Adsorption

Adsorption is an effective method for removing pharmaceuticals from wastewater due to its low cost, simplicity, resistance to the toxicity of pollutants, lack of unwanted by-products, and scalability. The process occurs at the interface between the liquid phase (contaminated wastewater) and the solid phase (the adsorbent used) [26]. An important aspect of conducting this process is the proper selection of the adsorbing substance, which depends on adsorption parameters as well as the concentration and chemical structure (functional groups) of the pharmaceuticals being removed [27]. The chosen adsorbent should possess appropriate properties, shape selectivity, and the ability to be regenerated [28]. Various materials are used for adsorption, including porous polymers, clay minerals, zeolites, silica, activated carbon, graphene, carbon nanotubes, graphene oxides, polymer resins, chitosan, and metal–organic frameworks [26,28]. Adsorbents can undergo various modifications to enhance their adsorption properties. Chitosan materials can be functionalized with anionic groups and crosslinked with glutaraldehyde to increase their resistance to extreme process conditions. Additionally, micro- and mesoporous carbons are synthesized with pore sizes tailored to the size of the pharmaceutical compound being removed. This results in greater selectivity, capacity, and efficiency of adsorption, as well as faster kinetics [29,30]. Iron oxides also serve as effective adsorbents for drug molecules containing highly electron-donating functional groups (e.g., tetracycline), forming very stable metal–ligand complexes [31].

In the process of adsorbing pharmaceuticals, carbon-based adsorbents exhibit the highest efficiency [26]. The removal efficiencies of pharmaceuticals from wastewater for selected compounds using activated carbon, graphene, and magnetic multi-walled carbon nanotubes are presented in Table 3.3.

The table shows that the efficiency of pharmaceutical adsorption from wastewater depends on the type of adsorbent, with advanced materials like graphene and magnetic carbon nanotubes achieving over 90% effectiveness. Activated carbon performs worse, with more variable results. Incorporating advanced adsorbents has the potential to significantly enhance the effectiveness of pharmaceutical contaminant removal in wastewater treatment processes. The difficult, costly, and time-consuming regeneration of carbon adsorbents drives the search for alternative materials based on inexpensive and environmentally friendly sorbents [30]. Biochar composites are

TABLE 3.3
Efficiency of Wastewater Treatment for Selected Pharmaceuticals Using Adsorption Methods

Adsorbent	Compound	Removal Efficiency (%)	References
Activated carbon	Estriol	60	[32]
	Estradiol	80	[32]
	Sulfamethoxazole	60	[32]
	Thrimethoprin	75	[32]
	Ibuprofen	15	[32]
	Diclofenac	40	[32]
	Paracetamol (Acetaminophen)	70	[32]
	Carbamazepine	75	[32]
	Primidone	90	[33]
Graphene	Ibuprofen	96	[34]
	Diclofenac	97	[34]
	Carbamazepine	97	[34]
Magnetic multi-walled carbon nanotubes	Furazolidone	98	[35]
	Nitrofurazone	93	[35]
	Nitrofurantoin	97	[35]
	Furaltadone	94	[35]

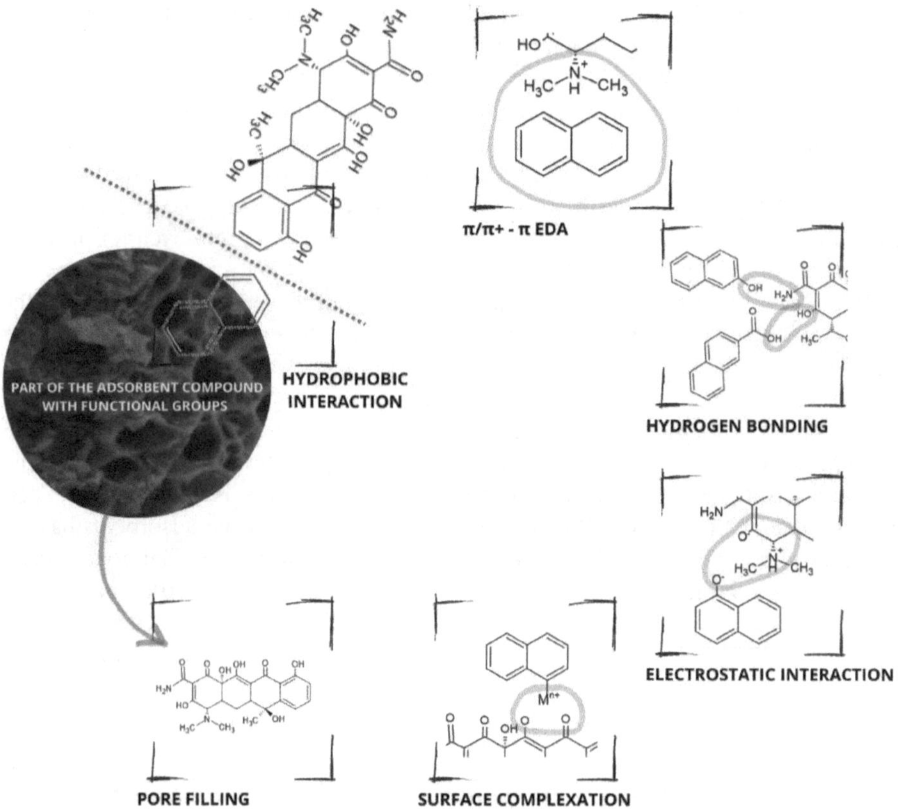

FIGURE 3.2 Interaction between the adsorbent and the pharmaceutical substance based on tetracycline according to Krasucka *et al.* [26].

gaining popularity due to their similar adsorption properties, lower cost, and greater environmental friendliness compared to activated carbons [26].

The adsorption of pharmaceuticals onto biochars is facilitated by various interactions between the adsorbent and the molecules being removed (see Figure 3.2). These interactions can include electrostatic, hydrophobic, dipole (through hydrogen bonding), and π-π interactions. The process may also occur through the filling of pores with the adsorbed substance and as a result of surface complexation reactions [26,36].

Not all of the mechanisms responsible for adsorption occur in every biochar-pharmaceutical system. The type of interaction depends on the structure of the adsorbent and the substance being removed, as well as the process conditions. When removing aromatic pharmaceutical compounds with electron-acceptor or electron-donor functional groups, hydrophobic or π-π interactions with the surface of biochar are primarily occur reactions [26,36].

Adsorption is a process that involves transferring contaminants (pharmaceuticals) to another phase (solid), indicating that it does not solve the problem of the presence of pharmaceutical substances in the environment. It also does not allow for the purification of wastewater from compounds characterized by low adsorption coefficients [19,22].

3.3.2 Advanced Oxidation Processes

AOPs are employed to remove a wide range of pharmaceuticals from wastewater [37]. Their mechanism involves the generation of highly reactive hydroxyl radicals and superoxide anions. These radicals are produced from various oxidizing agents, such as ozone or hydrogen peroxide. Often, catalysts such as UV or visible radiation or ultrasound are additionally involved in the process [24,37]. The standard oxidation potential of radicals, e.g., hydroxyl radical, is significantly higher than that of conventional oxidizers, giving them much stronger reactivity [38]. Radicals can react with pharmaceuticals through hydrogen abstraction (Equation 3.1), radical–radical interactions (Equation 3.2), or by accepting electrons (Equation 3.3).

$$HO^{\bullet} + RH \rightarrow R^{\bullet} + H_2O \tag{3.1}$$

$$R^{\bullet} + O_2 \rightarrow RO_2^{\bullet} \tag{3.2}$$

$$HO^{\bullet} + RH \rightarrow RX^{\bullet +} + HO^{-} \tag{3.3}$$

This results in the formation of oxidized intermediate forms, and potentially, through complete mineralization, carbon dioxide and water [37,39]. However, the oxidized intermediates formed may exhibit lower or higher toxicity than the parent compounds [27].

AOPs can be divided into thermal, photochemical, and hybrid methods, which are combinations of the two previously mentioned types of AOPs. Such integration of methods enhances the removal capabilities of pharmaceuticals from wastewater due to the synergistic effect and the increased number of reactive radicals [37]. The classification of advanced oxidation methods is presented in Figure 3.3.

The aforementioned variants of AOPs are used in the treatment of wastewater containing various pharmaceutical compounds [37]. Table 3.4 provides a brief overview of the most commonly used types of AOPs for the removal of pharmaceuticals.

The removal of nitrofuran pharmaceuticals using AOP methods most commonly employs direct photolysis [43]. This technique relies on the absorption of UV radiation by the degraded pharmaceutical compounds [24]. In the case of nitrofurans, the chromophore is the nitro group, and the process occurs with the use of an acid catalyst. As a result of the degradation of nitrofuran derivatives, a photolabile intermediate product—nitrofuraldehyde—along with an acid is formed [43]. The disadvantages of AOPs include high costs, the formation of toxic by-products, and, in some cases, harsh reaction conditions for the degradation of pharmaceutical compounds [46].

3.3.3 Microorganisms' Degradation of Pharmaceuticals

Some microorganisms, such as bacteria, fungi, and algae, can be used to remove pharmaceuticals from wastewater [47]. Table 3.5 presents examples of drugs removed using various microorganisms.

Microorganisms utilize drugs as a source of carbon and energy by transforming pharmaceutical compounds into easily metabolizable substrates [62]. The

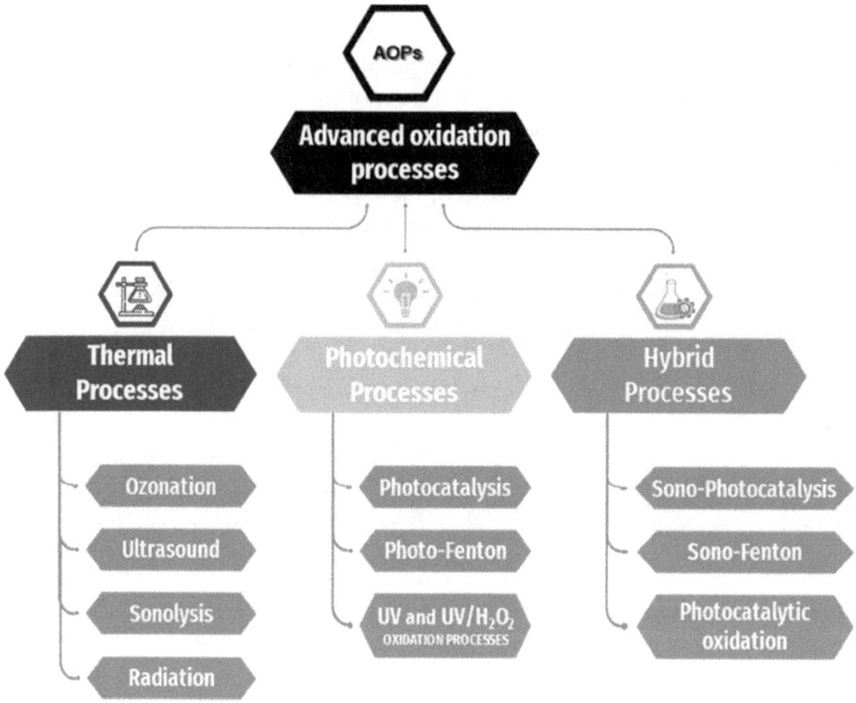

FIGURE 3.3 Types of AOPs used in the removal of pharmaceuticals [37,40,41].

TABLE 3.4
Examples of Drugs Removed by the Mainly Chosen AOP Methods

Method	Removed Pharmaceuticals	References
Photocatalysis	Tetracycline, naproxen, ibuprofen, estriol, estrone	[42]
Photolysis	Furaltadone, furazolidone, nitrofurantoin	[43]
Ozonation	Ibuprofen, ciprofloxacin, sulfamethoxazole, phenazone, aspirin, phenacetin, naproxen, metoprolol	[40, 44]
Non-thermal plasma	Oxacillin, amoxicillin, sulfadiazine, ibuprofen, carbamazepine, primidone, paracetamol, diclofenac	[45]
Fenton	Ofloxacin, trimethoprim, carbamazepine, ibuprofen, sulfamethoxazole, flumequine, tramadol, oxazepam, methadone, citalopram, methamphetamine	[37, 40]
Photo-Fenton	Amoxicillin, antipyrine, oxacillin, ciprofloxacin, carbamazepine, diclofenac, benzotriazole, mecoprop	[37, 44, 46]
UV and UV/H_2O_2 oxidation processes	Sulfamethoxazole, diclofenac, trimethoprim, amoxicillin, ketoprofen, ibuprofen	[37,46]
Sonolysis	Diclofenac, carbamazepine, ibuprofen	[37]
Electrochemical oxidation	Diclofenac, sulfamethoxazole, carbamazepine	[37]

(Continued)

TABLE 3.4 (*Continued*)
Examples of Drugs Removed by the Mainly Chosen AOP Methods

Method	Removed Pharmaceuticals	References
Radiation	Carbamazepine, fluoxetine, piperacillin	[37]
Sono-photocatalysis	Diclofenac, paracetamol, chloramphenicol, ibuprofen	[37]
Photocatalytic oxidation	Ibuprofen, carbamazepine, diclofenac, trimethoprim,	[37]

TABLE 3.5
Selected Microorganisms Used for Wastewater Treatment

Microorganism	Name	Removed Compound	Removal Efficiency (%)	References
Algae	*Chlamydomonas acidophila*	Erythromycin	93	[48]
		Clarithromycin	64	[48]
	Scenedesmus obliquus	Progesterone	95	[49]
	Chlorella pyrenoidosa	Progesterone	95	[49]
Fungi	*Trametes versicolor*	Ibuprofen	100	[50]
		Paracetamol	100	[50]
		Ketoprofen	100	[50]
		Azithromycin	100	[50]
Bacteria	*Pseudomonas extremaustralis* CSW01	Paracetamol	100	[51]
	Stutzerimonas stutzeri CSW02	Paracetamol	100	[51]
	Rhodococcus zopfii	Estrone	100	[52]
		Estriol	100	[52]
		17β-estradiol	100	[52]
	Bacillus thuringiensis B1(2015b)	Naproxene	100	[53]
	Stenotrophomonas maltophilia KB2	Naproxene	100	[54]
	Planococcus sp. S5	Naproxene	100	[55]
	Achromobacter denitrificans PR1	Sulfamethoxazole	100	[56]
	Labrys portucalensis F11	Ofloxacin	91	[57]
		Ciprofloxacin	85	[57]
		Norfloxacin	85	[57]
	Patulibacter sp. I11	Ibuprofen	92	[58]
	Pseudomonas aeruginosa HJ1012	Paracetamol	99	[59]
	Bacillus thuringiensis B1(2015b)	Ibuprophene	100	[60]
	Serratia marcescens ODW152	Nitrofurantoin	97	[61]

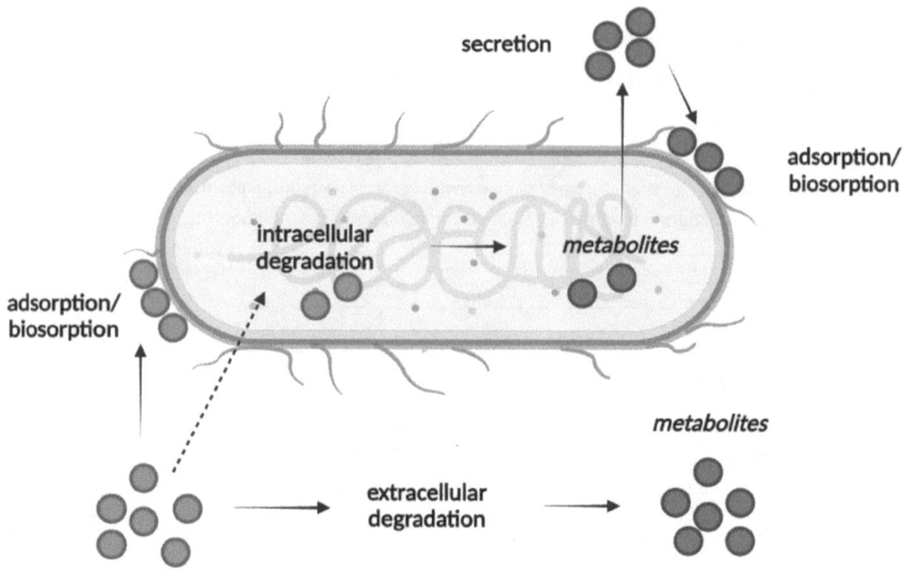

FIGURE 3.4 Main mechanisms of pharmaceutical removal by microorganisms [22].

effectiveness of microorganisms in treating wastewater largely depends on the structure and concentration of the pharmaceutical compounds themselves, as well as the presence of substances that inhibit microbial activity [63].

The processes for removing pharmaceuticals are based on various non-metabolic and metabolic pathways that lead to the sorption of substances or the degradation of compounds into harmless by-products (Figure 3.4).

The most commonly occurring processes for drug removal are biodegradation and biosorption. Biological sorption involves the binding of pharmaceutical compound molecules to the surfaces of microbial cells, while degradation relies on the transformation of drugs into simpler substances using enzymes, potentially even into carbon dioxide [22].

Microbial methods are characterized by their simplicity of use, cost-effectiveness, accessibility, and, above all, environmental friendliness [64]. However, they have certain limitations; they cannot be applied to substances that are toxic to microorganisms. Additionally, biological degradation may lead to the formation of more harmful products [22,61].

3.3.4 Phytoremediation

Methods based on plants and their associated rhizospheric microorganisms are an effective strategy for the sustainable removal of pharmaceuticals from wastewater [65]. The so-called "constructed wetlands" have significant potential. These are terrestrial systems of shallow ditches, settling ponds, or lagoons, planted with specialized vegetation—hydrophytes [66]. They have been designed to mimic the natural processes occurring in wetland environments. Wastewater treatment occurs through the coupling of physical, biological, and chemical interactions between the substrate, plants, and soil [67]. The removal of pharmaceuticals from wastewater using constructed wetlands can occur through various mechanisms, as illustrated in Figure 3.5.

The treatment of wastewater in constructed wetlands is a very complex process, where simpler, less toxic compounds can be obtained. Contaminants can also undergo sorption, meaning they are retained in the substrate. An important mechanism for the removal of pharmaceuticals is phytodegradation. After absorbing drugs, plants transform complex organic substances into less harmful end products, which are then sequestered in plant tissues [68]. Constructed wetlands demonstrate a high capacity for treating wastewater containing a wide range of pharmaceutical compounds [67]. Examples of drugs removed using constructed wetlands are presented in Table 3.6.

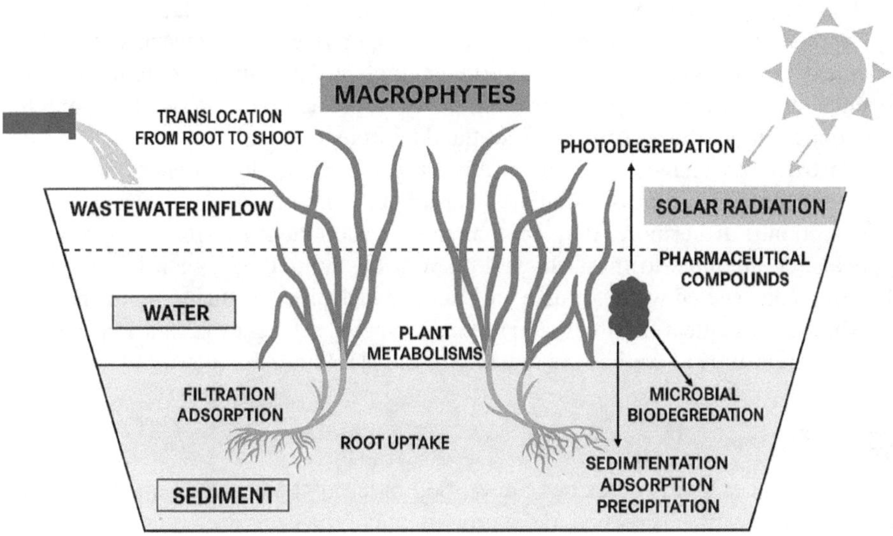

FIGURE 3.5 Mechanism of pharmaceutical removal [68].

TABLE 3.6
Selected Pharmaceuticals and Their Removal Efficiencies in Constructed Wetlands

Compound	Removal Efficiency (%)	References
Ibuprofen	71–96	[67]
Diclofenac	73–96	[67]
Naproxen	52–92	[67]
Ketoprofen	97–99	[67]
Carbamazepine	30–97	[67]
Triclosan	71–99	[68]
Tonalide	59–97	[67]
Flunixin	64	[68]

The data presented in Table 3.6 illustrate the varying removal efficiencies of different pharmaceuticals with ibuprofen and diclofenac exhibiting removal rates of 71%–96% and 73%–96% and ketoprofen demonstrating exceptional removal efficiencies between 97% and 99%. This variation suggests that the effectiveness of constructed wetlands in eliminating pharmaceutical contaminants is not uniform and may be influenced by the chemical properties of each compound, such as solubility and biodegradability.

The data demonstrate that the removal efficiencies of constructed wetlands systems are dependent on the specific compound, indicating that different categories of compounds may exhibit a greater or lesser range of variability in removal rates. Seasonal and spatial factors, such as temperature and insolation, exert a considerable influence on biodegradation and photodegradation processes. The removal efficiency of pharmaceutical compounds also depends on the species of plants present, the type and depth of the substrate, and the operational mode of the constructed wetland treatment system [68]. The removal efficiency of pharmaceutical compounds also depends on the species of plants present, the type and depth of the substrate, and the operational mode of the constructed wetland treatment system [68].

Constructed wetlands have many advantages, such as high cost-effectiveness, environmental friendliness, simple construction, easy operation, and low energy consumption. Furthermore, they demonstrate high efficiency in removing pharmaceuticals, comparable to that achieved by very costly methods, such as ozonation. However, constructed wetlands also have some disadvantages, including the need for large areas, long retention times, and reduced removal efficiency of pharmaceutics at lower temperatures as well as the removal of the plant biomass [65,67,69].

3.3.5 Enzymatic Degradation

The biological methods described above, both microbial and phytoremediation, are relatively slow, which is related to the growth rate of the degrading (micro)organism. In addition, the effectiveness of biological degradation is sometimes limited by the toxicity of the pharmaceutical (or its metabolites) on the (micro)organism capable of biodegrading it [70,71]. Overcoming these drawbacks is possible with the use of enzyme biocatalysts.

Enzyme degradation primarily uses biocatalysts isolated from fungi or bacteria. As reported earlier predominant enzymes used in pharmaceutics removal are oxidoreductases [72] such as laccases [73,74], peroxidases [75], lipases [76,77], and dioxygenases [78].

Enzymes, by their nature, participate in degradation processes as catalysts that are not consumed in the process. However, over time it degrades and its biocatalytic activity declines. Moreover, it is a challenge to recover the enzyme from the biocatalytic system, which makes it difficult to use free enzymes in continuous drug purification processes [79]. Hence, enzymatic degradation is often used with immobilized enzymes. The carrier used for immobilization can be either organic or inorganic in nature, and its purpose is to facilitate the separation of the biocatalyst from pharmaceutical-containing wastewater and, in addition, to provide greater physicochemical stability [76,80], similarly as it is observed for immobilized microorganisms [54].

Another solution to ease the enzyme application is to use enzymatic membrane bioreactors [81]. Nevertheless, the application of enzymatic methods is strongly limited by the cost of the enzymes themselves and their immobilization on membrane or other carrier.

3.3.6 Hybrid Processes

While combinations of methods for removing pharmaceuticals from the environment are not yet a widespread practice, they represent the most powerful approach to wastewater treatment. These combinations can reduce the toxicity of treated wastewater and minimize the drawbacks of individual processes [24]. Physical–chemical methods are considered expensive and able to create toxic degradation products [82]. However, integrating methodologies is particularly useful in biological techniques, as some microorganisms are sensitive to specific contaminants, limiting the effectiveness of biodegradation. To expand its applicability, AOPs can be used as a preliminary step. In this case, the compounds contaminating the wastewater are transformed into by-products that are less toxic to microorganisms and more easily biodegradable [83]. The combination of AOP and biodegradation also minimizes the risk of additional contamination of the wastewater with by-products generated during oxidation [84]. For tetracyclines, neither biodegradation nor ozonation can fully eliminate them from wastewater. However, the combination of both processes shows high effectiveness in removing these compounds from wastewater [85]. A similar situation occurs with the removal of carbamazepine, furosemide, and ranitidine [84]. Another option is ozonation followed by biodegradation, which can result in a significant increase in the biodegradation rate [86]. The photolysis and photocatalytic oxidation can also be an interesting solution, as shown in the experiments with nitrofurazone antibiotic [87].

Combined processes are not commonly used due to their complexity, high costs, and challenges in continuous operation [24]. Nevertheless, research continues to explore their broader application.

3.4 CONCLUSIONS AND PERSPECTIVES

The presence of pharmaceuticals in wastewater is a widespread and pressing problem. Awareness of the negative consequences of this phenomenon for people and the environment as a whole has resulted in the development of a number of methods aimed at removing pharmaceutical pollutants from wastewater. Adsorption methods, which are relatively cheap and versatile, are widely used, although they do not completely solve the problem of contaminants, but only lead to their displacement and concentration on or in the sorbent. Oxidation methods, while effective and fast, are considered expensive and have a high risk of toxic degradation products. Biological methods, bioremediation with microorganisms and phytoremediation, are considered the safest for the environment. Their faster but much more costly development is the use of enzymatic degradation. All of these biological methods currently have very high potential, as new microorganisms and plants capable of degrading and mineralizing pharmaceuticals are still being discovered. To increase the speed and efficiency of biological methods, it seems valuable to support them with the aforementioned physicochemical methods. It is a hybrid method that can contribute to the adequately effective removal of pharmaceuticals from wastewater.

ACKNOWLEDGMENTS

This publication has been prepared as part of a grant (No. 2020/39/B/NZ9/03196) funded by the National Science Centre, Poland.

REFERENCES

1. Mulcahy AW, Schwam D, Lovejoy SL (2024) International prescription drug price comparisons: Estimates using 2022 data. *Rand Health Q* 11:5.
2. Samal K, Mahapatra S, Hibzur Ali M (2022) Pharmaceutical wastewater as emerging contaminants (EC): Treatment technologies, impact on environment and human health. *Energy Nexus* 6:100076. https://doi.org/10.1016/j.nexus.2022.100076
3. Kaczala F, Blum SE (2016) The occurrence of veterinary pharmaceuticals in the environment: A review. *Curr Anal Chem* 12:169–182. https://doi.org/10.2174/1573411012666151009193108
4. Massima Mouele ES, Tijani JO, Badmus KO, et al (2021) Removal of pharmaceutical residues from water and wastewater using dielectric barrier discharge methods—A review. *Int J Environ Res Public Health* 18:1683. https://doi.org/10.3390/ijerph18041683
5. Wee SY, Ismail NAH, Haron DEM, et al (2022) Pharmaceuticals, hormones, plasticizers, and pesticides in drinking water. *J Hazard Mater* 424:127327. https://doi.org/10.1016/j.jhazmat.2021.127327
6. De Oliveira M, Frihling BEF, Velasques J, et al (2020) Pharmaceuticals residues and xenobiotics contaminants: Occurrence, analytical techniques and sustainable alternatives for wastewater treatment. *Sci Total Environ* 705:135568. https://doi.org/10.1016/j.scitotenv.2019.135568
7. Brown AK, Wong CS (2018) Distribution and fate of pharmaceuticals and their metabolite conjugates in a municipal wastewater treatment plant. *Water Res* 144:774–783. https://doi.org/10.1016/j.watres.2018.08.034
8. Ellis JB (2006) Pharmaceutical and personal care products (PPCPs) in urban receiving waters. *Environ Pollut* 144:184–189. https://doi.org/10.1016/j.envpol.2005.12.018
9. Gworek B, Kijeńska M, Zaborowska M, et al (2019) Pharmaceuticals in aquatic environment. Fate and behaviour, ecotoxicology and risk assessment – A review. *Acta Pol Pharm – Drug Res* 76:397–407. https://doi.org/10.32383/appdr/103368
10. Aus Der Beek T, Weber F, Bergmann A, et al (2016) Pharmaceuticals in the environment—Global occurrences and perspectives. *Environ Toxicol Chem* 35:823–835. https://doi.org/10.1002/etc.3339
11. Wilkinson JL, Boxall ABA, Kolpin DW, et al (2022) Pharmaceutical pollution of the world's rivers. *Proc Natl Acad Sci* 119:e2113947119. https://doi.org/10.1073/pnas.2113947119
12. Khan AHA, Barros R (2023) Pharmaceuticals in water: Risks to aquatic life and remediation strategies. *Hydrobiology* 2:395–409. https://doi.org/10.3390/hydrobiology2020026
13. Larsson DGJ (2014) Pollution from drug manufacturing: review and perspectives. *Philos Trans R Soc B Biol Sci* 369:20130571. https://doi.org/10.1098/rstb.2013.0571
14. Cleuvers M (2003) Aquatic ecotoxicity of pharmaceuticals including the assessment of combination effects. *Toxicol Lett* 142:185–194. https://doi.org/10.1016/S0378-4274(03)00068-7

15. Pal A, Gin KYH, Lin AYC, Reinhard M (2010) Impacts of emerging organic contaminants on freshwater resources: Review of recent occurrences, sources, fate and effects. *Sci Total Environ* 408:6062–6069. https://doi.org/10.1016/j.scitotenv.2010.09.026
16. Rzymski P, Drewek A, Klimaszyk P (2017) Pharmaceutical pollution of aquatic environment: an emerging and enormous challenge. *Limnol Rev* 17:97–107. https://doi.org/10.1515/limre-2017-0010
17. Smułek W, Kaczorek E (2022) Factors influencing the bioavailability of organic molecules to bacterial cells—A mini-review. *Molecules* 27:6579. https://doi.org/10.3390/molecules27196579
18. Gouju J, Legeay S (2023) Pharmacokinetics of obese adults: Not only an increase in weight. *Biomed Pharmacother* 166:115281. https://doi.org/10.1016/j.biopha.2023.115281
19. Carballa M, Omil F, Lema JM, et al (2004) Behavior of pharmaceuticals, cosmetics and hormones in a sewage treatment plant. *Water Res* 38:2918–2926. https://doi.org/10.1016/j.watres.2004.03.029
20. Duan L, Zhang Y, Wang B, et al (2022) Wastewater surveillance for 168 pharmaceuticals and metabolites in a WWTP: Occurrence, temporal variations and feasibility of metabolic biomarkers for intake estimation. *Water Res* 216:118321. https://doi.org/10.1016/j.watres.2022.118321
21. Nakada N, Tanishima T, Shinohara H, et al (2006) Pharmaceutical chemicals and endocrine disrupters in municipal wastewater in Tokyo and their removal during activated sludge treatment. *Water Res* 40:3297–3303. https://doi.org/10.1016/j.watres.2006.06.039
22. Costa F, Lago A, Rocha V, et al (2019) A review on biological processes for pharmaceuticals wastes abatement—A growing threat to modern society. *Environ Sci Technol* 53:7185–7202. https://doi.org/10.1021/acs.est.8b06977
23. Dong H, Yuan X, Wang W, Qiang Z (2016) Occurrence and removal of antibiotics in ecological and conventional wastewater treatment processes: A field study. *J Environ Manage* 178:11–19. https://doi.org/10.1016/j.jenvman.2016.04.037
24. Homem V, Santos L (2011) Degradation and removal methods of antibiotics from aqueous matrices – A review. *J Environ Manage* 92:2304–2347. https://doi.org/10.1016/j.jenvman.2011.05.023
25. Amouzgar P, Wong MY, Horri BA, Salamatinia B (2016) Advanced Material for Pharmaceutical Removal from Wastewater. In: Mishra AK (ed) *Smart Materials for Waste Water Applications*, 1st ed. Wiley, pp 179–212.
26. Krasucka P, Pan B, Sik Ok Y, et al (2021) Engineered biochar – A sustainable solution for the removal of antibiotics from water. *Chem Eng J* 405:126926. https://doi.org/10.1016/j.cej.2020.126926
27. Kyzas GZ, Fu J, Lazaridis NK, et al (2015) New approaches on the removal of pharmaceuticals from wastewaters with adsorbent materials. *J Mol Liq* 209:87–93. https://doi.org/10.1016/j.molliq.2015.05.025
28. De Andrade JR, Oliveira MF, Da Silva MGC, Vieira MGA (2018) Adsorption of pharmaceuticals from water and wastewater using nonconventional low-cost materials: A review. *Ind Eng Chem Res* 57:3103–3127. https://doi.org/10.1021/acs.iecr.7b05137
29. Ji L, Liu F, Xu Z, et al (2010) Adsorption of pharmaceutical antibiotics on template-synthesized ordered micro- and mesoporous carbons. *Environ Sci Technol* 44:3116–3122. https://doi.org/10.1021/es903716s
30. Kyzas GZ, Kostoglou M, Lazaridis NK, et al (2013) Environmental friendly technology for the removal of pharmaceutical contaminants from wastewaters using modified chitosan adsorbents. *Chem Eng J* 222:248–258. https://doi.org/10.1016/j.cej.2013.02.048
31. Liu Q, Zhong L-B, Zhao Q-B, et al (2015) Synthesis of Fe_3O_4/polyacrylonitrile composite electrospun nanofiber mat for effective adsorption of tetracycline. *ACS Appl Mater Interfaces* 7:14573–14583. https://doi.org/10.1021/acsami.5b04598

32. Snyder SA, Adham S, Redding AM, et al (2007) Role of membranes and activated carbon in the removal of endocrine disruptors and pharmaceuticals. *Desalination* 202:156–181. https://doi.org/10.1016/j.desal.2005.12.052
33. Altmann J, Ruhl AS, Zietzschmann F, Jekel M (2014) Direct comparison of ozonation and adsorption onto powdered activated carbon for micropollutant removal in advanced wastewater treatment. *Water Res* 55:185–193. https://doi.org/10.1016/j.watres.2014.02.025
34. Rizzo L, Fiorentino A, Grassi M, et al (2015) Advanced treatment of urban wastewater by sand filtration and graphene adsorption for wastewater reuse: Effect on a mixture of pharmaceuticals and toxicity. *J Environ Chem Eng* 3:122–128. https://doi.org/10.1016/j.jece.2014.11.011
35. Zhen-Yuan Z, Zhen-Hu X (2015) Removal of four nitrofurans drugs from aqueous solution by magnetic multi-wall carbon nanotubes. *Fuller Nanotub Carbon Nanostruct* 23:640–648. https://doi.org/10.1080/1536383X.2014.947646
36. Tong Y, McNamara PJ, Mayer BK (2019) Adsorption of organic micropollutants onto biochar: a review of relevant kinetics, mechanisms and equilibrium. *Environ Sci Water Res Technol* 5:821–838. https://doi.org/10.1039/C8EW00938D
37. Kanakaraju D, Glass BD, Oelgemöller M (2018) Advanced oxidation process-mediated removal of pharmaceuticals from water: A review. *J Environ Manage* 219:189–207. https://doi.org/10.1016/j.jenvman.2018.04.103
38. Guerra-Rodríguez S, Rodríguez E, Singh D, Rodríguez-Chueca J (2018) Assessment of sulfate radical-based advanced oxidation processes for water and wastewater treatment: A review. *Water* 10:1828. https://doi.org/10.3390/w10121828
39. Lian L, Yao B, Hou S, et al (2017) Kinetic study of hydroxyl and sulfate radical-mediated oxidation of pharmaceuticals in wastewater effluents. *Environ Sci Technol* 51:2954–2962. https://doi.org/10.1021/acs.est.6b05536
40. Mackuľak T, Birošová L, Bodík I, et al (2016) Zerovalent iron and iron(VI): Effective means for the removal of psychoactive pharmaceuticals and illicit drugs from wastewaters. *Sci Total Environ* 539:420–426. https://doi.org/10.1016/j.scitotenv.2015.08.138
41. Wu S, Zhang L, Chen J (2012) Paracetamol in the environment and its degradation by microorganisms. *Appl Microbiol Biotechnol* 96:875–884. https://doi.org/10.1007/s00253-012-4414-4
42. Dalrymple OK, Yeh DH, Trotz MA (2007) Removing pharmaceuticals and endocrine-disrupting compounds from wastewater by photocatalysis. *J Chem Technol Biotechnol* 82:121–134. https://doi.org/10.1002/jctb.1657
43. Edhlund BL, Arnold WA, McNeill K (2006) Aquatic photochemistry of nitrofuran antibiotics. *Environ Sci Technol* 40:5422–5427. https://doi.org/10.1021/es0606778
44. Wang J, Bai Z (2017) Fe-based catalysts for heterogeneous catalytic ozonation of emerging contaminants in water and wastewater. *Chem Eng J* 312:79–98. https://doi.org/10.1016/j.cej.2016.11.118
45. Magureanu M, Mandache NB, Parvulescu VI (2015) Degradation of pharmaceutical compounds in water by non-thermal plasma treatment. *Water Res* 81:124–136. https://doi.org/10.1016/j.watres.2015.05.037
46. De La Cruz N, Esquius L, Grandjean D, et al (2013) Degradation of emergent contaminants by UV, UV/H_2O_2 and neutral photo-Fenton at pilot scale in a domestic wastewater treatment plant. *Water Res* 47:5836–5845. https://doi.org/10.1016/j.watres.2013.07.005
47. Patel M, Kumar R, Kishor K, et al (2019) Pharmaceuticals of emerging concern in aquatic systems: Chemistry, occurrence, effects, and removal methods. *Chem Rev* 119:3510–3673. https://doi.org/10.1021/acs.chemrev.8b00299
48. Escudero A, Hunter C, Roberts J, et al (2020) Pharmaceuticals removal and nutrient recovery from wastewaters by Chlamydomonas acidophila. *Biochem Eng J* 156:107517. https://doi.org/10.1016/j.bej.2020.107517

49. Peng F-Q, Ying G-G, Yang B, et al (2014) Biotransformation of progesterone and norgestrel by two freshwater microalgae (Scenedesmus obliquus and Chlorella pyrenoidosa): Transformation kinetics and products identification. *Chemosphere* 95:581–588. https://doi.org/10.1016/j.chemosphere.2013.10.013
50. Cruz-Morató C, Ferrando-Climent L, Rodriguez-Mozaz S, et al (2013) Degradation of pharmaceuticals in non-sterile urban wastewater by Trametes versicolor in a fluidized bed bioreactor. *Water Res* 47:5200–5210. https://doi.org/10.1016/j.watres.2013.06.007
51. Vargas-Ordóñez A, Aguilar-Romero I, Villaverde J, et al (2023) Isolation of novel bacterial strains Pseudomonas extremaustralis CSW01 and Stutzerimonas stutzeri CSW02 from sewage sludge for Paracetamol biodegradation. *Microorganisms* 11:196. https://doi.org/10.3390/microorganisms11010196
52. Yoshimoto T, Nagai F, Fujimoto J, et al (2004) Degradation of estrogens by *Rhodococcus zopfii* and *Rhodococcus equi* Isolates from activated sludge in wastewater treatment plants. *Appl Environ Microbiol* 70:5283–5289. https://doi.org/10.1128/AEM.70.9.5283-5289.2004
53. Górny D, Guzik U, Hupert-Kocurek K, Wojcieszyńska D (2019) A new pathway for naproxen utilisation by Bacillus thuringiensis B1(2015b) and its decomposition in the presence of organic and inorganic contaminants. *J Environ Manage* 239:1–7. https://doi.org/10.1016/j.jenvman.2019.03.034
54. Wojcieszyńska D, Klamka J, Marchlewicz A, et al (2022) Immobilized Stenotrophomonas maltophilia KB2 in Naproxen degradation. *Molecules* 27:5795. https://doi.org/10.3390/molecules27185795
55. Wojcieszyńska D, Domaradzka D, Hupert-Kocurek K, Guzik U (2016) Enzymes involved in Naproxen degradation by Planococcus sp. S5. *Pol J Microbiol* 65:177–182
56. Reis PJM, Reis AC, Ricken B, et al (2014) Biodegradation of sulfamethoxazole and other sulfonamides by Achromobacter denitrificans PR1. *J Hazard Mater* 280:741–749. https://doi.org/10.1016/j.jhazmat.2014.08.039
57. Amorim CL, Moreira IS, Maia AS, et al (2014) Biodegradation of ofloxacin, norfloxacin, and ciprofloxacin as single and mixed substrates by Labrys portucalensis F11. *Appl Microbiol Biotechnol* 98:3181–3190. https://doi.org/10.1007/s00253-013-5333-8
58. Almeida B, Kjeldal H, Lolas I, et al (2013) Quantitative proteomic analysis of ibuprofen-degrading Patulibacter sp. strain I11. *Biodegradation* 24:615–630. https://doi.org/10.1007/s10532-012-9610-5
59. Hu J, Zhang LL, Chen JM, Liu Y (2013) Degradation of Paracetamol by *Pseudomonas aeruginosa* strain HJ1012. *J Environ Sci Health Part A* 48:791–799. https://doi.org/10.1080/10934529.2013.744650
60. Marchlewicz A, Guzik U, Smułek W, Wojcieszyńska D (2017) Exploring the degradation of ibuprofen by Bacillus thuringiensis B1(2015b): The new pathway and factors affecting degradation. *Molecules* 22:1676. https://doi.org/10.3390/molecules22101676
61. Pacholak A, Juzwa W, Zgoła-Grześkowiak A, Kaczorek E (2023) Multi-faceted analysis of bacterial transformation of nitrofurantoin. *Sci Total Environ* 874:162422. https://doi.org/10.1016/j.scitotenv.2023.162422
62. Escapa C, Coimbra RN, Paniagua S, et al (2015) Nutrients and pharmaceuticals removal from wastewater by culture and harvesting of Chlorella sorokiniana. *Bioresour Technol* 185:276–284. https://doi.org/10.1016/j.biortech.2015.03.004
63. Muter O, Perkons I, Selga T, et al (2017) Removal of pharmaceuticals from municipal wastewaters at laboratory scale by treatment with activated sludge and biostimulation. *Sci Total Environ* 584–585:402–413. https://doi.org/10.1016/j.scitotenv.2017.01.023
64. Wu Y, Li T, Yang L (2012) Mechanisms of removing pollutants from aqueous solutions by microorganisms and their aggregates: A review. *Bioresour Technol* 107:10–18. https://doi.org/10.1016/j.biortech.2011.12.088

65. Xiong J-Q, Cui P, Ru S, et al (2021) Unravelling metabolism and microbial community of a phytobed co-planted with Typha angustifolia and Ipomoea aquatica for biodegradation of doxylamine from wastewater. *J Hazard Mater* 401:123404. https://doi.org/10.1016/j.jhazmat.2020.123404
66. Matamoros V, Bayona JM (2006) Elimination of pharmaceuticals and personal care products in subsurface flow constructed wetlands. *Environ Sci Technol* 40:5811–5816. https://doi.org/10.1021/es0607741
67. Töre GY, Meriç S, Lofrano G, De Feo G (2012) Removal of Trace Pollutants from Wastewater in Constructed Wetlands. In: Lofrano G (ed) *Emerging Compounds Removal from Wastewater.* Springer Netherlands, Dordrecht, pp. 39–58.
68. Zhang D, Gersberg RM, Ng WJ, Tan SK (2014) Removal of pharmaceuticals and personal care products in aquatic plant-based systems: A review. *Environ Pollut* 184:620–639. https://doi.org/10.1016/j.envpol.2013.09.009
69. Matamoros V, García J, Bayona JM (2008) Organic micropollutant removal in a full-scale surface flow constructed wetland fed with secondary effluent. *Water Res* 42:653–660. https://doi.org/10.1016/j.watres.2007.08.016
70. Dionisi D (2014) Potential and limits of biodegradation processes for the removal of organic xenobiotics from wastewaters. *ChemBioEng Rev* 1:67–82. https://doi.org/10.1002/cben.201300008
71. Hemmami H, Amor IB, Zeghoud S, et al (2023) Phytoremediation of Xenobiotics: Principles and Applications in Environmental Pollution Removal. In: Singh R, Singh P, Tripathi S, et al (eds) *Xenobiotics in Urban Ecosystems.* Springer International Publishing, Cham, pp 261–290.
72. Sá H, Michelin M, Tavares T, Silva B (2022) Current challenges for biological treatment of pharmaceutical-based contaminants with oxidoreductase enzymes: Immobilization processes, real aqueous matrices and hybrid techniques. *Biomolecules* 12:1489. https://doi.org/10.3390/biom12101489
73. Yang J, Li W, Ng TB, et al (2017) Laccases: Production, expression regulation, and applications in pharmaceutical biodegradation. *Front Microbiol* 8:832. https://doi.org/10.3389/fmicb.2017.00832
74. Navada KK, Kulal A (2019) Enzymatic degradation of chloramphenicol by laccase from Trametes hirsuta and comparison among mediators. *Int Biodeterior Biodegrad* 138:63–69. https://doi.org/10.1016/j.ibiod.2018.12.012
75. Sellami K, Couvert A, Nasrallah N, et al (2022) Peroxidase enzymes as green catalysts for bioremediation and biotechnological applications: A review. *Sci Total Environ* 806:150500. https://doi.org/10.1016/j.scitotenv.2021.150500
76. Degórska O, Szada D, Zdarta A, et al (2022) Immobilized lipase in resolution of ketoprofen enantiomers: Examination of biocatalysts properties and process characterization. *Pharmaceutics* 14:1443. https://doi.org/10.3390/pharmaceutics14071443
77. Degórska O, Szada D, Fu Q, et al (2024) Ionic liquid supported hydrogel–lipase biocatalytic systems in asymmetric synthesis of enantiomerically pure S-ibuprofen. *Int J Biol Macromol* 281:136221. https://doi.org/10.1016/j.ijbiomac.2024.136221
78. Quintana J, Weiss S, Reemtsma T (2005) Pathways and metabolites of microbial degradation of selected acidic pharmaceutical and their occurrence in municipal wastewater treated by a membrane bioreactor. *Water Res* 39:2654–2664. https://doi.org/10.1016/j.watres.2005.04.068
79. Wackett LP, Robinson SL (2020) The ever-expanding limits of enzyme catalysis and biodegradation: Polyaromatic, polychlorinated, polyfluorinated, and polymeric compounds. *Biochem J* 477:2875–2891. https://doi.org/10.1042/BCJ20190720

80. Rybarczyk A, Smułek W, Grzywaczyk A, et al (2023) 3D printed polylactide scaffolding for laccase immobilization to improve enzyme stability and estrogen removal from wastewater. *Bioresour Technol* 381:129144. https://doi.org/10.1016/j.biortech.2023.129144
81. Zdarta J, Jesionowski T, Meyer AS, Pinelo M (2022) Removal of tetracycline in enzymatic membrane reactor: Enzymatic conversion as the predominant mechanism over adsorption and membrane rejection. *J Environ Chem Eng* 10:106973. https://doi.org/10.1016/j.jece.2021.106973
82. Mora-Gamboa MPC, Rincón-Gamboa SM, Ardila-Leal LD, et al (2022) Impact of antibiotics as waste, physical, chemical, and enzymatical degradation: Use of laccases. *Molecules* 27:4436. https://doi.org/10.3390/molecules27144436
83. Tekin H, Bilkay O, Ataberk S, et al (2006) Use of Fenton oxidation to improve the biodegradability of a pharmaceutical wastewater. *J Hazard Mater* 136:258–265. https://doi.org/10.1016/j.jhazmat.2005.12.012
84. Wang J, Wang S (2016) Removal of pharmaceuticals and personal care products (PPCPs) from wastewater: A review. *J Environ Manage* 182:620–640. https://doi.org/10.1016/j.jenvman.2016.07.049
85. Gómez-Pacheco CV, Sánchez-Polo M, Rivera-Utrilla J, López-Peñalver J (2011) Tetracycline removal from waters by integrated technologies based on ozonation and biodegradation. *Chem Eng J* 178:115–121. https://doi.org/10.1016/j.cej.2011.10.023
86. Olak-Kucharczyk M, Festinger N, Smułek W (2023) Application of ozonation-biodegradation hybrid system for polycyclic aromatic hydrocarbons degradation. *Int J Environ Res Public Health* 20:5347. https://doi.org/10.3390/ijerph20075347
87. Smułek W, Bielan Z, Pacholak A, et al (2021) Nitrofurazone removal from water enhanced by coupling photocatalysis and biodegradation. *Int J Mol Sci* 22:2186. https://doi.org/10.3390/ijms22042186

4 Advancing Biotechnology Toward Pollutant Recirculation and Recovery in a Changing World

Tazeen Malik, Hussain Ibrabeem, Fatima Azeem, and Nazim Hussain

4.1 INTRODUCTION

Pollutant recirculation and recovery is the process in which harmful pollutants such as gases, liquids, or solid wastes are collected to extract useful products and are returned to the environment by converting them into harmless substances. This is particularly an industrial approach for the minimization of wastes and reusing the pollutants. Pollution is not only impacting the quality of air, water, and soil but can also have detrimental effects on animal and plant life. According to the World Health Organization (WHO), one of the significant environmental risks to human health is air pollution [1]. The harmful gases and particles released into the atmosphere cause major health problems across the world. Exposure to air pollutants during pregnancy leads to premature births, low birth weight, behavioral and cognitive issues, asthma, and other respiratory illnesses [2]. Climatic changes mostly caused by air pollution lead to heat-related disorders, malnutrition, mental health issues, and physical trauma [2]. The aquatic ecosystem is polluted by the introduction of insecticides, pesticides, heavy metals, mill wastes, and crude oil. A significant amount of these pollutants in water bodies results in the rapid death of aquatic life. The effects of water pollution include suppressed immune response, decreased metabolism, damaged liver tissue, ulcers, and damaged gills and epithelial layer in fish. Heavy metals such as lead, mercury, and cadmium can accumulate in the tissues of various aquatic species and result in abnormal growth, poor reproduction, and can cause death. Moreover, predators like birds and animals feeding on them are also affected by these heavy metals [3].

Human activities such as industrial operations, mining, residential and commercial wastes, and human and animal prescription medications are contaminating the soil and disturbing the ecological equilibrium. The soil fertility is reduced due to the presence of contaminants, which in turn decreases the yield. The loss of natural nutrients, increased soil salinity, and hazardous dirt formation are some of the

 DOI: 10.1201/9781003536277-4

impacts of soil pollution. A foul smell caused by the presence of industrial chemicals results in headaches and nausea. Moreover, the crops cultivated on contaminated land are hazardous to one's health when ingested [4]. To control the pests, pesticides are frequently sprayed on crops, which kills not only the pests but also butterflies and bees. The killing of these major pollinators by pesticides greatly impacts the reproductive cycles of plants, which in turn affects the entire food chain [5]. To combat these harmful effects caused by pollutants, several innovative solutions are employed.

Gaseous pollutants including volatile organic compounds (VOCs) and greenhouse gases are controlled by three fundamental techniques absorption, adsorption, and incineration. Carbon sequestration, another technique, is applied to check the level of carbon dioxide. Absorption involves the transfer of gaseous contaminants from air into the water, a contacting liquid. This liquid can capture the pollutant using a chemical reaction and can also serve as a solvent. This task is accomplished by packed scrubbers or wet scrubbers. Chemical factories use scrubbers for the removal of water-soluble solvents like acetone and methyl alcohol, glass manufacturing plants for the removal of hydrogen fluoride, and fertilizer production plants for the removal of ammonia from the airstream [6]. Fossil-fuel power plants release flue gas containing sulfur dioxide. It can be regulated by means of an absorption process known as flue gas desulfurization (FGD). FGD can also utilize dry or wet scrubbing. Adsorption, in contrast to absorption, is a surface phenomenon that involves the attachment of gas molecules on a solid surface. This technique is employed by food processing industries to control odors and recover and manage VOCs. The most common adsorbent material is activated carbon or heated charcoal due to its extremely high ratio of surface area to volume and porous nature [6]. Incineration is a rapid oxidation process for the conversion of VOCs and other gaseous hydrocarbons into carbon dioxide and water. This task is performed in a special incinerator called an afterburner that not only controls odors but also destroys toxic compounds and phytochemically reactive substances [6].

In addition, carbon dioxide is captured and stored underground in forests and oceans to check its level through a process named carbon sequestration. The absorption of carbon dioxide by phytoplankton can be increased by a geoengineering method known as ocean fertilization. Phytoplanktons are tiny plants living close to the ocean surface that would take photosynthesized carbon dioxide into their tissues when they die [7]. Another environmental strategy that aims to reduce air pollution effectively is emission trading. It provides the polluters with a limited number of allowances allowing them to find the most economical way to decrease the number of emissions [8]. However, various systems and procedures are created for safeguarding public health by disposal and supply of water and pollution management in a technique known as environmental engineering. Managing hazardous waste sites; collecting, treating, and disposing of wastewater; controlling air and noise pollution; managing municipal solid waste; managing hazardous waste; treating and distributing drinking water; and creating environmental assessments, audits, and impact studies are some major projects running under this field [9].

However, biotechnology plays a major role in pollutant control and management. An eco-friendly method of bioremediation is defined as the use of microbes for the removal of contaminants from the ecosystem. Researchers employed the use of prokaryotes in bioreactors, artificial wetlands, and active sludge for microbial degradation. Fuzzy white rot fungi are also used in complex systems for this purpose. They belong to the family Basidiomycota and have the potential to degrade a wide variety of organic chemicals [10]. Plants can also absorb, stabilize, metabolize, and detoxify pollutants through a process termed phytoremediation. These pollutants are used as nutrient sources (carbon, nitrogen, and phosphorus) in some plants while other plants absorb and store these contaminants in their vegetable structure [11]. For wastewater treatment, microbial treatment methods such as activated sludge, biofilm, anaerobic treatment, microbial nitrogen removal, and microbial phosphorus removal are employed. Another novel approach for the treatment of pollutants involves the use of nanoparticles. Montmorillonite, a nanoparticle, can remove heavy metals and organic wastes. Similarly, kaolinite plays a role in the fuel treatment and the removal of heavy metals [12].

This chapter focuses on the importance of various biotechnological techniques such as bioremediation, phytoremediation, genetic engineering, microbial pathway engineering, advances in enzyme production, anaerobic treatments, biomonitoring, integration of synthetic biology, capture and recycling techniques, and the production of biofuels for the development of pollution-free world.

4.2 CURRENT STATUS OF BIOTECHNOLOGY IN POLLUTION MANAGEMENT

4.2.1 Traditional Bioremediation Techniques

The environmental pollutants are neutralized, eradicated, detoxified, or removed by the metabolic processes of microbes such as bacteria, fungi, or algae in a technique termed bioremediation. This technique has been employed for various decades to produce an eco-friendly environment and restore contaminated sites. These bioremediation strategies are considered a silver bullet collectively for cleaning polluted environments. Different types of bioremediation techniques are categorized into two types based on the nature of the pollutant, site characteristics, the cost of treatment, the depth of pollution, the degree of pollution, the geographical location, and the geology of the polluted site. These traditional types include ex-situ bioremediation and in-situ bioremediation [13], which are represented in Figure 4.1.

Ex-situ bioremediation involves the digging of pollutants from the contaminated site and transporting them to a different location to convert them into less toxic substances. Some traditional ex-situ techniques that are in practice include land farming, windrows, and biopile-mediated bioremediation [14]. Land farming is one of the effective bioremediation approaches because it requires little investment, less operating equipment, and can treat a significant volume of polluted soil with minimal energy usage. It is the most commonly employed ex-situ bioremediation technique in which a layer support is fixed on a ground surface and excavated polluted soil is placed on it. These pollutants are biodegraded aerobically with the help of autochthonous microorganisms [15]. Native and/or transient hydrocarbonoclastic bacteria found in the contaminated soil have the potential to degrade the pollutants.

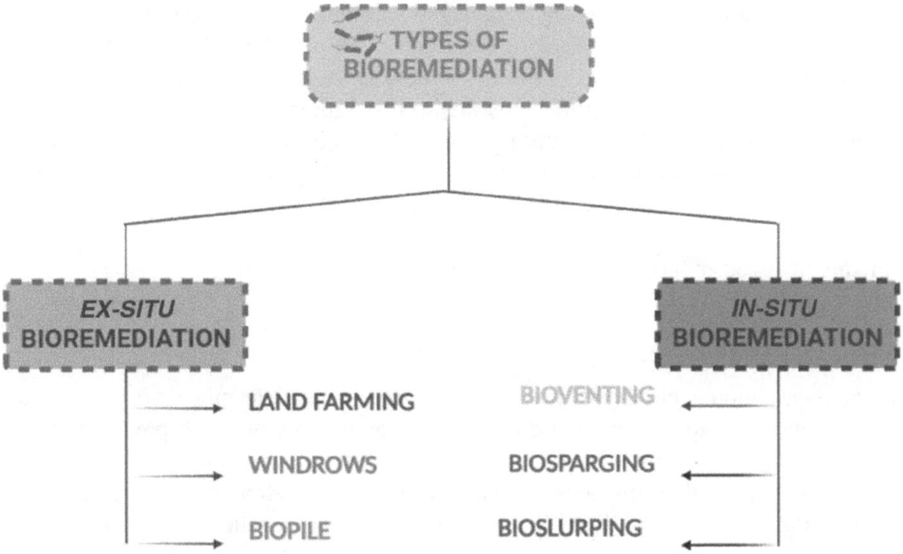

FIGURE 4.1 Representation of different types of bioremediation and their subclasses based on the nature of the pollutant, site characteristics, degree of pollution, and geology of the polluted site. (Created with www.biorender.com.)

This bacterial activity is enhanced in soil by periodically rotating stacked polluted material by an ex-situ bioremediation technology called Windrows. The regular turning of polluted soil and the uniform distribution of water and nutrients accelerate the process of degradation. Windrow treatment shows a greater degree of hydrocarbon degradation due to the friable nature of the soil [16]. Another ex-situ bioremediation technology involves the digging and piling of soil above ground on which nutrients and water are added occasionally to boost the microbial degradation activity. It requires a treatment bed, aeration, irrigation, and the leachate and nutrients collection system. This method has gained popularity due to its cost-effectiveness and its efficient biodegradation under appropriate nutrients, temperature, and aeration. Low molecular weight contaminants are eradicated even in extremely cold polluted climates by this method [17].

Another natural passive bioremediation technology involves the cleaning of contaminated sites without the need for human intervention by stimulating the natural degradation ability of microorganisms or introducing certain microbes to the contaminated site termed in-situ bioremediation. Microbial development is enhanced by improving the physiochemical conditions by introducing genetically engineered microbes [13]. Bioventing is an in-situ microbial transformation process in which the contaminants are converted into harmless substances. In bioventing, oxygen, water, and nutrients are delivered to the unsaturated(vadose) zone, which enhances the degradation ability of native bacteria. This in-situ method is particularly useful for restoring the places that are contaminated by minor petroleum product spills. Bioslurping is another in-situ process to clean up contaminated soil and groundwater that is contaminated with volatile or semi-VOCs. The products are recovered from unsaturated and saturated zones, light nonaqueous phase liquids, and remediating capillaries by indirectly supplying oxygen and stimulating the contaminant degradation [18]. Another in-situ method for the removal of pollutants from contaminated sites involves the enhancement of microbial activity by injecting the air into the soil subsurface is termed biosparging. This method is similar to bioventing in which air is injected into saturated zones that moves the VOCs into unsaturated zones, which improves the degradation process. The biodegradability of pollutants and the soil permeability determine the efficacy of biosparging. This method is applied to treat aquifers that are contaminated with kerosene and diesel [14]. Examples of microorganisms involved in bioremediation are described in Table 4.1.

4.2.2 Barriers to Widespread Adoption

Although traditional bioremediation methods have been determined to produce an eco-friendly environment, there are certain limitations to their widespread adoption. The traditional ex-situ methods discussed above have several limitations. Biopiles are not preferred due to their high cost of maintenance and operation. Moreover, electricity is not available at distant polluted sites to run the air pump for the uniform distribution

TABLE 4.1
Representation of Different Microorganisms Involved in Different Types of Bioremediation, Their Function, and the Sources from Which the Microbes Are Isolated Along with References

Microorganisms	Source of isolation	Function	Type of bioremediation	References
1. Bacillus cereus	Soil contaminated with pesticides	Degradation of organophosphates	Biostimulation	[19]
2. Dehalococcoides mccartyi	Anaerobic sewage sludge	Anaerobic reduction of chlorinated solvents	Biosparging	[20]
3. *Pseudomonas putida*	Soil contaminated with hydrocarbons	Aerobic degradation of petroleum hydrocarbons	Bioventing	[21]
4. *Arthrobacter chlorophenolicus*	Soil contaminated with chlorophenol	Chlorophenol degradation	Biostimulation	[21]
5. *Shewanella oneidensis*	Sediments of lake Oneida	Reduces metals and radionucleotides	Bioaugmentation	[20]
6. *Rhodococcus rhodochrous*	Soil contaminated with oil	Degradation of aliphatic and aromatic hydrocarbons	Bioaugmentation	[21]
7. *Methanosarcina barkeri*	Anaerobic Sewage Sludge	Degradation of aromatic compounds	Biostimulation	[20]
8. *Mycobacterium gilvum*	Soil contaminated with polycyclic aromatic compounds	Aerobic degradation of polycyclic aromatic compounds	Bioventing	[21]

of air. The microbial activity is inhibited and soil is dried up by the action of overheated air. Windrow treatment is not implemented for cleaning the soil that is contaminated with harmful volatiles due to its periodic turning. However, land farming requires a large operating area, it is more time-consuming, and microbial activity is decreased due to unfavorable environmental conditions. Also, excavation requires heavy costs and it is not efficient against inorganic pollutants and harmful volatiles, particularly in hot (tropical) climatic zones. If the transformation stops at an intermediate stage, the resulting substance can be more toxic and hazardous as compared to the parent contaminant [14].

Similarly, traditional in-situ methods described cannot be widely adopted due to various limitations. In bioventing, expected results cannot be obtained as compared to laboratory findings due to environmental variations in the unsaturated zone to which air was introduced. Volatile organic chemicals are moved into the soil vapor phase due to the high airflow that requires the treatment of resultant gases before being released into the atmosphere and increases the treatment time [22]. The main disadvantage of using biosparging is the decreased prediction of the direction of airflow. Native microorganisms are reduced due to the presence of heavy metals and organic chemicals. However, the introduction of nutrients, water, and oxygen results in increased microbial growth to the extent that it blocks the injection wells [14]. The microbial activity is reduced in bioslurping due to a limited rate of oxygen transfer resulting from excessive soil moisture. Also, it requires the creation of a vacuum on a deep, highly permeable site that disturbs the water table [23].

4.3 EMERGING BIOTECHNOLOGY INNOVATIONS

As conventional bioremediation techniques have certain limitations in degrading pollutants, effective ways are required to combat pollution because the world's environmental issues are rising day by day. Researchers are finding new biotechnological innovations to develop a pollution-free environment by degrading the contaminants. Some emerging biotechnological methods are described below:

4.3.1 Cutting-Edge Microbial Technologies

Cutting-edge microbial technologies play a vital role in the degradation of contaminants from polluted sites by enhancing the ability of microbes to degrade, remove, or detoxify them. The microbial consortia for bioremediation and metagenomics approaches are some strategies implemented in cutting-edge microbial technologies. The availability and action of appropriate microbial consortia is one of the key factors that affect the bioremediation efficacy. Microbial consortia refer to the collection of different microbial strains belonging to different genera or species that work together in complex systems whether natural or artificial [24]. Bioremediation of soil contaminants is preferred using microbial consortia because it has the potential to degrade various simple contaminants or complex combinations by the action of various enzymes or metabolic pathways. The degradation potential is improved by collaborating the consortium with the syntrophic interactions, cross-feeding, co-metabolism, or quorum sensing, which enhances detoxification and lowers the accumulation of hazardous intermediates [25].

The action of the microbial consortium is regulated by adopting the changes in the surrounding environment. Improved diversity, microbial community resilience, and stability, as well as higher mineralization efficiency, are some of the advantages due to which microbial consortium is adopted over single cultures. Some bacteria have the natural potential to break down the contaminants including *Salmonella*, *Escherichia coli*, *Lactobacilli*, *Actinobacteria*, *Pseudomonas*, and *Clostridium* [25]. The pollutants such as heavy metals, herbicides, phthalates, polycyclic aromatic hydrocarbon, and crude oil can be degraded by the action of microbial consortia to clean up the polluted soil. Metagenomics is the method of studying DNA without isolating and growing microorganisms. It reveals nonculture microbial communities' genetic information in a natural habitat by using environmental samples [26]. This information is obtained by using function and sequence-based research approaches. Not only novel pollutant-degrading enzymes or encoding gene clusters can be identified, but microbes are also tracked and identified by metagenomics approaches. The application of metagenomics in bioremediation involves the identification of microbes from different habitats and finding pollutant degradation procedures by studying the bacterial genes [27] as described in Figure 4.2.

4.3.2 Advances in Enzymatic Process

Microbial pathway engineering is the method of improving genetic and metabolic pathways that produce bacterial cell hosts with improved enzyme activity, improved

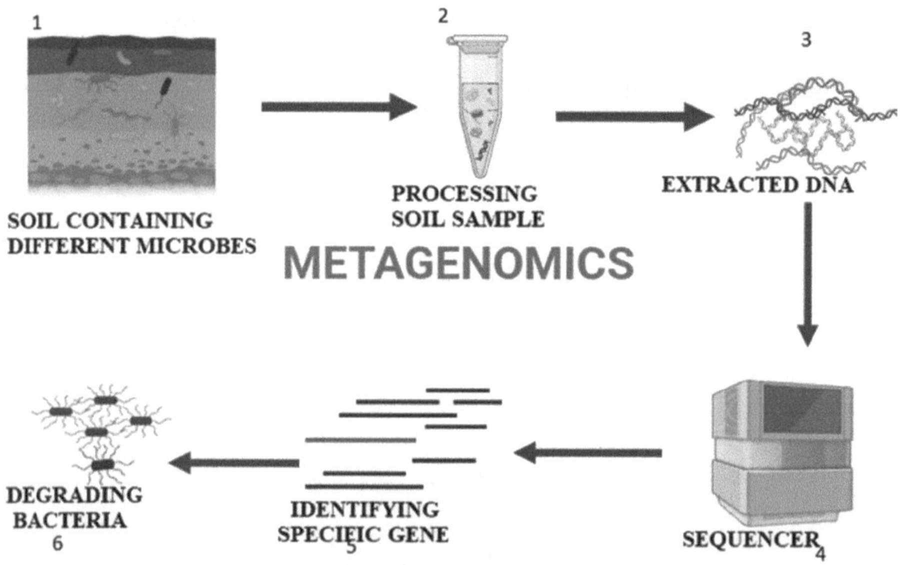

FIGURE 4.2 Representation of how metagenomics helps identify the degrading bacteria.[1] Types of microorganisms present in soil.[2] The soil sample is processed to extract DNA.[3] Extracted DNA.[4] Sequencing.[5] Identifying biodegrading genes.[6] Biodegrading bacteria are isolated by performing genome analysis. (Created with www.biorender.com.)

target pollutants bioavailability, higher synthesis and efficiency of generated biomolecules, and improved biodegradation functions. It was first described by Bailey in 1991 [28]. The different degrading genes are upregulated and downregulated to quickly degrade different kinds of pollutants. Blocking the competing pathways and exogenous enzymes is the primary function of this technology to achieve the desired degradation activity. However, the selection of a suitable host organism, appropriate computational methods, enzyme transcriptional regulation, generating novel synthetic biology techniques, and boosting microbial productivity are some essentials of this technology [29]. With the recent advancement in DNA manipulation techniques and genome sequencing technology, it becomes easy to find the gene clusters that are involved in specific regulatory pathways. These gene clusters can be altered to improve or lose a specific function by inserting or deleting genes using techniques such as gene editing and CRISPR/Cas9. In bioremediation, this technique is applied to provide a precursor or inducer that encourages the degradation of contaminants [30].

Different microbial enzymes, such as oxidases, oxidoreductases, monooxygenases, phenoloxidases, and esterases, are targeted in metabolic engineering due to their degrading activity. Research has been conducted on the enzyme degradation potential of various organic compounds, especially oxidoreductases. The degradation of methyl parathion and hexachlorocyclo hexane is performed by the action of genetically modified microorganisms. Naphthenic acid degradation occurs by the chcpca and pobA gene clusters present in *Rhodococcus opacus* [31]. Enzymes are immobilized on different substrates and their stability is improved by increasing the operating pH and temperature, which works as an economical and recyclable remediation technology [32]. It improves enzyme activity, stability, and half-life. Different enzymes produced by genetic engineering that play a role in pollutant degradation are described in Table 4.2.

TABLE 4.2
Representation of Different Enzymes Produced by Using Biotechnology Measures That Are Involved in Pollution Control Along with the Microbial Source of Enzymes and Their Function

	Enzyme	Microbes	Function in Pollution Control	References
1.	Laccase	*Ganoderma lucidum*	Engineered to breakdown phenolic compounds in wastewater	[33]
2.	Dinitroaniline hydrolase	*Bacillus cereus*	Manufactured to detoxify dinitroaniline compounds in contaminated environments	[34]
3.	Triazine hydrolase	*Pseudomonas* sp.	Effective in degrading triazine herbicides	[35]
4.	PETase	*Ideonella sakaiensis*	Engineered to degrade PET plastics	[36]
5.	Organophosphate hydrolase	*Pseudomonas putida*	Engineered to enhance the degradation of organophosphate pesticides	[37]

4.3.3 Integration of Synthetic Biology

The ability of microbes to degrade environmental pollutants can be enhanced by the generation of various novel biological systems and different devices by the application of synthetic biology. It improves the degradation process by engineering the bioremediation pathways and allows the breakdown of those substances that are not previously metabolized. The resistance of the microbe, *Pseudomonas putida*, against xenobiotics and hydrocarbons allows its survival in the soil due to which it might serve as a host for the breakdown of hydrocarbons. Pangenome analysis of this strain identifies genes that provide the flexible metabolism for the breakdown of both natural and foreign substances. These bacterial strains are provided with a protective mechanism against harmful substances. A set of efflux pumps is attached to this strain that expels the harmful substances from cytoplasm, periplasm, and outer membrane to the surrounding environment [38]. Also, the metabolism of this species allows it to quickly adapt the temperature change, oxidative stress, and sudden osmotic perturbations. These characteristics allow *P. putida* to work best in contaminated surroundings. The addition of this species in the biosensor degrades the hydrocarbons and xenobiotics and converts them into less toxic substances as described in Figure 4.3. Some studies reveal that different catabolic traits are also present in some species, which include the degradation of aromatic hydrocarbons by the action of chromosomally encoded pathways and the breakdown of toluene by the toluene monooxygenase pathway [39]. Moreover, the mutagenesis of this strain allows the degradation of resistant substances like p-ethylbenzoate.

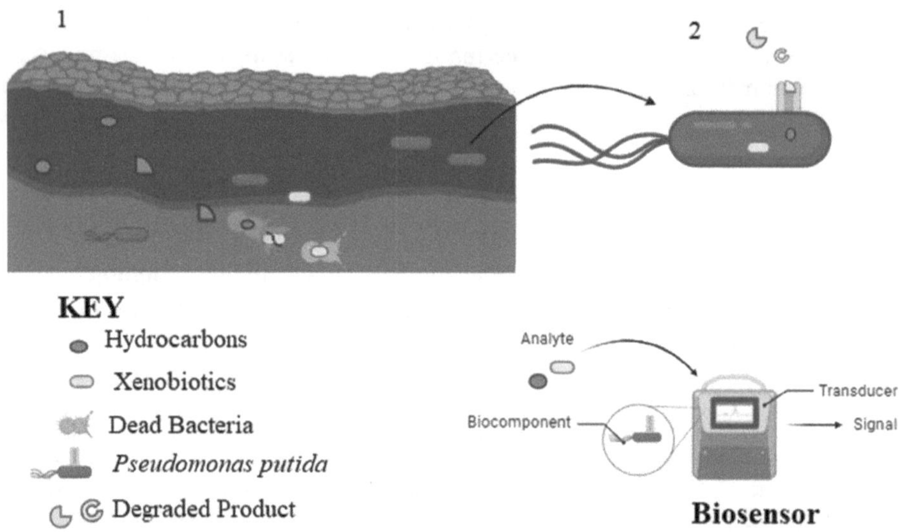

FIGURE 4.3 Degradation of hydrocarbons and xenobiotics by the action of *Pseudomonas putida*.[1] Contaminated soil causes death of microbes except *P. putida*.[2] Efflux pump on its surface pumps out toxins.[3] This strain degrades hydrocarbons and xenobiotics in biosensors. (Created with www.biorender.com.)

One of the *Pseudomonas putida* strains, KT2440 strain, can serve as a model laboratory bacterium for bioremediation [40]. The breakdown of polycyclic organic compounds results in the production of salicylic acid, which helps in the production of artificial degradation modules. A reliable output signal is produced after many dose-dependent experiments to analyze the pollutant concentrations by providing a steady pollutant concentration in the biosensor [40]. The degradation process starts after obtaining biosensor data and the biomass is increased after degradation rates are stimulated as in high-density fermentation. However, it is essential to kill the modified cell after the completion of pollutant breakdown that depends on external inducers or environmental factors. The functions of original component modules must be improved by taking long-term stability and robustness into account in bioremediation. Salicylic acid can be identified by producing red fluorescence after 6 hours with a microbial strain having three modified modules. Salicylic acid is degraded to produce gentisic acid in the early bacterial stationary growth phase. After the completion conversion of salicylic acid, the suicide mechanism is activated by the action of artificial strain [41]. The breakdown of salicylic acid present in industrial wastewater is improved by enhancing the sensitivity of biosensors by controlling the expression of receptors and reporters. Synthetic biology along with certain computational techniques such as transcriptomics, bioinformatics, proteomics, and molecular docking improves the bioremediation efficacy by producing genetically modified microbes [42].

4.4 POLLUTANT RECIRCULATION TECHNOLOGIES

Pollutant recirculation technologies aim to reuse the indoor air which reduces its harmful effects on surroundings by the use of sophisticated filtering techniques. Particulate matter and VOCs can be filtered by using activated carbon and HEPA (high-efficiency particulate air) filters.

4.4.1 Capture and Recycling Techniques

Leachate recirculation is the on-site treatment of leachates, a by-product of the landfill biodegradation process. High amounts of organic and inorganic wastes can be accumulated as a result of landfill leachates whose primary biological, chemical, and physical treatments cause environmental issues. Due to the presence of high concentrations of pollutants, integrated treatment and co-treatment with wastewater are recommended. Leachate recycling in a landfill containing a wide variety of kitchen and food wastes is performed by breaking down the bulk of solid wastes. However, the loss of landfill gases produced as a result of fast decomposition must be taken into consideration. In municipal solid waste landfills, it works as a bioreactor providing environmental and economic advantages. The recovered leachate is moved downwards at the bottom of the landfill due to the action of gravity that results in the loss of leachate. The chances of contact between methanogenic bacteria and dissolved organic matter are improved due to buffering of pH in hydrolysis during leachate recirculation [43].

Another method, carbon capture technology, comes to the rescue to minimize the formation of NOx and reduce the requirement for CO_2 separation by providing oxygen through a solid carrier. Researchers are working on the stability and reactivity of oxygen carriers, which resulted in their reduction and reoxidation process models but still further research is required in the fields of surface chemistry and financial stability [44]. Another efficient postcombustion method is adsorption. However, coal-fired power plants require different conditions such as the type of coal, adsorption methods, and operating conditions [44]. Another advanced thermochemical recycling method, the co-recycling scenario has been presented in which the carbon material system is separated from extracted fossil fuels. Natural and synthetic C-materials are combined in a single recycling system resulting in less waste production. Carbon flow such as CO_2 manufactured or natural and losses can be recovered and superior quality goods can be manufactured by this method. The quality of wood, paper, plastics, and textiles can be enhanced by focusing on carbon recycling because it improves the use of resources [45].

Another modern pollutant degradation technique for the removal of odor and VOCs has gained popularity due to its easy functionality, high reduction efficacy, low requirement of energy, and low cost is termed biofiltration. Its residual particles do not require further treatment. Sustainable machinery requires the immobilization of bacteria on the surface of the medium that breaks down the adsorbed pollutants present in air and water sources. Harmful contaminants such as VOCs present in wastewater or polluted air are removed by the application of fungi and bacteria using biofiltration-based systems recently [46], which is represented in Figure 4.4. To deal with the smell issues associated with open sand filters, a recirculating sand filter (RSF) system has been formed. The oxygen content has been transferred evenly over the filter bed, which removes the foul smells by using biological, physical, and chemical methods for the degradation of impurities present in wastewater. However, the biological processes are more essential to degrade the pollutant. RSF reduces the

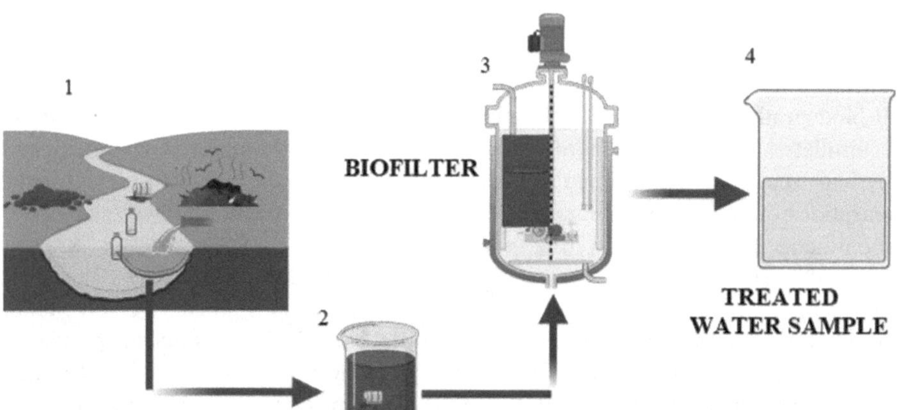

FIGURE 4.4 Representation of wastewater treatment.[1] Contaminated water due to human activities.[2] Water sample is taken.[3] Degradation of contaminants by microorganisms immobilized in biofilter.[4] Sample free of pollutants. (Created with www.biorender.com.)

biochemical oxygen demand (BOD) and total suspended solids up to 85%–95% and also causes denitrification producing a good quality effluent. Up to 50% of the nitrogen applied can be removed by this method. Its efficacy depends on the environmental conditions present inside the filter, the design of the filter, and the biodegradability of the wastewater. The optimum temperature for microbial growth must be taken into consideration to enhance the biodegradation of wastewater [47].

4.4.2 Development of Closed-Loop Systems

For the recycling of used items into raw materials and reuse them in the production of new products, closed-loop systems are created. The quality and integrity of materials are maintained in a closed-loop system as compared to typical recycling in which the quality of substances may deteriorate. Industries are focusing on improving the quality of polymers so they can withstand many cycles. Polyethylene terephthalate (PET) and high-density polyethylene are designed in such a way that they can be recycled indefinitely and are used in various ways. The production of naturally degradable polymers such as bioplastics and biodegradable substances from renewable resources in closed-loop systems is gaining popularity these days [48].

Researchers are working to make these materials more resilient so that they can be recycled again and again. Closed-loop systems are designed with modular components so that materials can be separated easily at the end of the cycle so that the specific materials are preserved after recycling for future use. Moreover, processes like pyrolysis and depolymerization convert recyclable materials into high-quality raw materials by breaking the complex compounds into their component parts or monomers. Even complex materials such as mixed plastics and composite materials can be recycled by using a closed-loop system [49].

4.5 RECOVERY STRATEGIES FOR A CHANGING ENVIRONMENT

The over-dependence of industries on nonrenewable resources produces large amounts of solid waste that results in global climate change. Various recovery strategies are adopted to control the widespread climatic changes.

4.5.1 Circular Economy

The circular economy is an efficient waste management strategy that is implemented in all the industrial sectors to reduce carbon emissions by focusing particularly on waste management, energy, climate change, food production, water, and air quality. The primary goals of the circular economy are restoring ecosystems, eliminating waste, recirculating materials, and restoring ecosystems [50]. The quantity of garbage is increasing day by day, which requires various efficient methods to reduce their production. This put pressure on the government globally to implement various solid waste management methods to recover solid wastes. Environmental engineering applies various strategies such as containment heaps, incinerators, landfills, stockpiling, and biodegradation to dispose off and recycle solid waste that depends on the type, source, and properties of the waste. Landfills, incinerators, containment heaps, stockpiling, and biodegradation are the usual methods used to dispose of solid waste. However, researchers are trying to implement a zero-waste plan that reduces solid waste by recycling and reusing all waste products after being transferred to incinerators, landfills, or other disposal sites [51].

Recycling solid wastes into circulating materials involves the following:

i. Recycling solid waste such as metal, glass, plastic, and paper [52].
ii. The use of coal gangue, red mud, and fly ash in cement production and the use of chrome slag in iron production produce less solid waste [52].
iii. Sulfur iron ore from coal gangue, glass beads from fly ash, and precious metals from metal smelting slag must be recovered [52].
iv. Solid wastes produced as a result of fermentation are crushed in small sizes and are subjected to magnetic separation to recover important substances like rubber, glass, paper, and plastics [52].
v. Biochemical fiberboard, concrete, decorative panels, thermal insulation materials, and waterproofing rolls are produced by recycling building materials such as blast furnace slag, waste plastics, fly ash, coal gangue, sludge, and construction wastes [52].

The circular economy also aims to produce an eco-friendly environment by promoting the harmonious flow of energy, materials, and information between the organisms and the surrounding environment in a process named eco-dynamics. It promotes the maintenance and integration of natural habitats, which results in ecological civilization in human society. A healthy ecosystem can be created by following the ecological rules [53].

4.5.2 Urban Air Pollution Control Strategies

Urban air pollution strategies have been classified into general policies, energy, industry, and transportation that are followed by several countries across the world to control air pollution. For instance, during the last 40 years, $PM_{2.5-}$ and NO_{2-} mortality has decreased to 56% and 44%, respectively, in the United Kingdom. To control air pollution especially greenhouse gas emissions, international agreements must be practiced such as the Paris Agreement [54].

Some of the general policies that are followed to control air pollution include the Third Development Plan of the Economy, the Comprehensive Urban Air Quality Control Strategy, the Road Vehicles Act, the Atmospheric Pollution Prevention Act, the Environmental Protection Law, the Blue-Sky Defense War and Environmental Protection Tax Law, the Air Pollution Control Act (APCA), and the Air Quality Act the 13th Five-Year Plan (13th FYP), Social and Cultural of the Nation, the Clean Air Action, the Clean Air Act (CAA), the National Framework for Air Quality Management in South Africa, the Comprehensive Action issued and Improve Environmental Law, the Establishment of the Basic Environmental Pollution Control and the Establishment of the Air Pollution Control Act, the Air Quality Management Plan, the Integrated Master Plan of Air Pollution Control, and the Smoke and Soot Law [54].

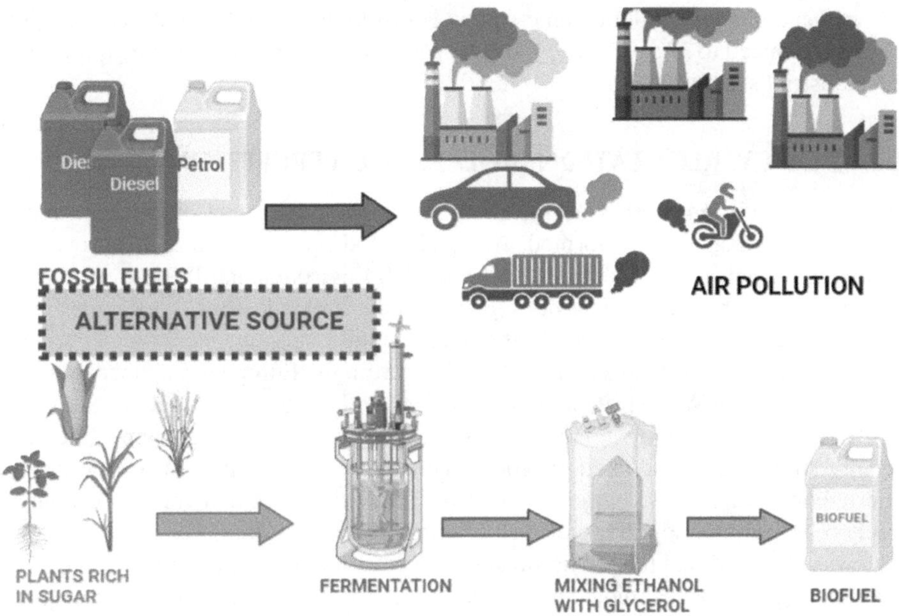

FIGURE 4.5 Representation of increased air pollution by burning of fossil fuels. Biofuel is produced by fermenting sugar-rich plants. Ethanol produced is mixed with glycerol and the resulting product is used in vehicles in place of fuel. (Created with www.biorender.com.)

Air pollution is controlled by focusing on improving the transport and producing biofuels. Biofuel is produced by subjecting the sugar-rich plants to fermentation by the action of microorganisms such as ethanol derived from sugarcane. The process of biofuel production is described in Figure 4.5. The use of diesel with low sulfur content, lead-free petrol, compressed natural gas, liquid natural gas, and cleaner-burning fuel, i.e., methanol in vehicles produces less air pollution. Moreover, the use of electric vehicles powered by battery or fuel cells, biofuel vehicles, hybrid vehicles, and public transport and converting diesel vehicles to liquefied petroleum gas (LPG) vehicles produces an eco-friendly environment [55]. Energy is an essential requirement of human survival. Some of the energy conservation strategies to control air pollution include the use of solar energy, energy alternatives such as coal-to-electricity and coal-to-gas, renewable energy sources, cooking fuels such as electricity and LPG, fuel cells, and energy savers. However, the use of coal in electricity production, bulk coal in civil engineering, domestic heating, bituminous coal distribution, wood heaters, coal and wood-based tandoors in restaurants, and solid fuel must be restricted. Taxes should be implemented on the use of wood and single fuel such as coal for heating and cooking [56].

The number of industries is increasing day by day in urban areas, which produces a lot of waste and pollutes the environment. Modern advances in science claim that improving the industrial structure controls the air pollution caused by industries [57]. The government should implement rules and regulations strictly to manage pollution and should grant licenses for different industrial processes. The industries should be fined for releasing contaminants into the atmosphere and be closed by the government during a pollution emergency. For instance, the release of SO_2 and NO_2 has been reduced to 37.51% and 7.47%, respectively, in China and Jinan by improving the industrial structure [58].

4.6 POLICY, REGULATION, AND PUBLIC PERCEPTION

To combat environmental issues, the evaluation and adjustment of pollutant recirculation technologies is performed. Air quality and public health are improved by following the rules and regulations set by the US Environmental Protection Agency and the CCA by human communities, governments, and industries. These rules must be upgraded by considering the latest environmental issues and technologies so that the environmental problems can be solved accordingly. European member states in collaboration with WHO principles aim to identify the source of pollutants and various reduction methods by monitoring the air quality. Encouraging, supporting, and penalty strategies are adopted by governments globally to reduce urban pollution. The public is forced by the government to follow the rules and regulations either freely or in combination to produce a pollution-free ecosystem [54].

It was observed that the public suggests the government for the implementation of laws and regulations that biotechnological innovations guarantee environmental protection and food safety. They also suggest the researchers to describe the appropriate ways for the best use of biotechnological techniques that improve environmental conditions. Moreover, the federal government should take appropriate actions that enable the public to promote suitable methods of creating different products regarding biotechnology. The government should increase public trust in biotechnological innovations by upholding the laws and regulations because neglecting the biotechnological laws lowers public confidence in these pollution control strategies. Moreover, the public expects that the government should make transparent policies and provide knowledge regarding biotechnological techniques for pollutant degradation [59].

Biotechnological innovations can be adopted by human communities to produce a sustainable ecosystem if they are aware of the latest biotechnological processes and their advantages regarding public health. Transparent information must be provided to the general public in participatory decision-making, public platforms, and educational conferences because people are interested in learning about the processes that are directly targeting their health and environment. Public interest in these techniques can be improved by providing clear knowledge about the advantages and disadvantages of pollutant recirculation methods, listening to their queries regarding these technologies by providing them with local platforms, making sure that the general public understands the concepts regarding biotechnology procedures, and making them believe that biotechnological companies aim to restore the ecosystem and public health [60].

4.7 CONCLUSION

To conclude, biotechnological innovations for pollutant recovery and recirculation are the solutions to deal with environmental issues caused by pollution. Traditional bioremediation techniques were used to degrade the contaminants but heavy metals and organic compounds require complex treatment methods for degradation. Recent advancements in biotechnology focused on improving the metabolic and enzymatic processes of microorganisms so that they can convert complex harmful pollutants into less toxic substances. Moreover, the manufacturing of unique microorganisms with the help of synthetic biology has further improved the pollutant degradation process by making it possible for microbes to withstand extreme environmental conditions such as temperature, pH, and salinity. Recycling and capture technologies not only degrade the pollutants but also recover valuable substances from waste. Industrial waste can be reduced by manufacturing closed-loop systems and practicing a circular economy to create a pollution-free ecosystem. However, improving the infrastructure of industries and urban pollutant management strategies are implemented to reduce the waste and pollutants in cities. The use of bioremediation and pollutant recirculation technologies in combination is an efficient way of creating a sustainable environment. To deal with the growing environmental problems and degrade the pollutants, cutting-edge biotechnological techniques such as synthetic biology and improvements in enzymatic and microbiological procedures came to the rescue. These cost-effective, long-lasting, and efficient biotechnological innovations not only reduce pollution but also improve public health. In summary, these biotechnology advancements open the door to a cleaner and more sustainable world.

REFERENCES

1. Burnett Richard, T., et al., An integrated risk function for estimating the global burden of disease attributable to ambient fine particulate matter exposure. *Environmental Health Perspectives*, 2014. **122**(4): p. 397–403.
2. Salim, S., et al., An Early Warning System for Air Pollution Surveillance: An IoT Based Big Data Framework to Monitor Risks Associated with Air Pollution. *2023 IEEE Global Conference on Artificial Intelligence and Internet of Things (GCAIoT)*, 2023: p. 148–152.
3. Bedia, C., Metabolomics in environmental toxicology: Applications and challenges. *Trends in Environmental Analytical Chemistry*, 2022. **34**: p. e00161.
4. Sweety, P., Impacts of soil pollution on environment. *NVEO-Natural Volatiles & Essential Oils*, 2021. **8**: p. 3314–3320.
5. Amorim, M.J.B., et al., Systems toxicology to advance human and environmental hazard assessment: A roadmap for advanced materials. *Nano Today*, 2023. **48**: p. 101735.
6. Nathanson, J.A., Air pollution control, in *Encyclopedia Britannica*. 16 May. 2024.
7. The Editors of Encyclopedia Britannica, Ocean fertilization, in *Encyclopedia Britannica*. 21 Apr. 2023.
8. The Editors of Encyclopedia Britannica, Emissions trading, in *Encyclopedia Britannica*. 11 Dec. 2023.
9. Nathanson, J.A., Environmental engineering, in *Encyclopedia Britannica*. 20 Feb. 2020.
10. Tortella, G., et al., Are white-rot fungi a real biotechnological option for the improvement of environmental health? *Critical Reviews in Biotechnology*, 2015. **35**(2): p. 165–172.
11. Lee, J., An overview of phytoremediation as a potentially promising technology for environmental pollution control. *Biotechnology and Bioprocess Engineering*, 2013. **18**: p. 431–439.

12. Fan, T., et al., Research on the purification of environmental pollution by biomaterials and its treatment methods in the development of rural cultural and creative industries. *Biotechnology and Bioprocess Engineering*, 2022. **2022**: p. 1594081.
13. Azubuike, C.C., C.B. Chikere, and G.C. Okpokwasili, Bioremediation techniques-classification based on site of application: Principles, advantages, limitations and prospects. *World Journal of Microbiology and Biotechnology*, 2016. **32**(11): p. 180.
14. Indu, S., Bioremediation Techniques for Polluted Environment: Concept, Advantages, Limitations, and Prospects, in *Trace Metals in the Environment*, M.-T. Mario Alfonso, S.- N. Hugo, and S. Agnieszka, Editors. 2020, IntechOpen: Rijeka. p. Ch. 12.
15. Silva-Castro, G., et al., Application of selected microbial consortia combined with inorganic and oleophilic fertilizers to recuperate oil-polluted soil using land farming technology. *Clean Technologies and Environmental Policy*, 2012. **14**: p. 719–726.
16. Coulon, F., et al., When is a soil remediated? Comparison of biopiled and windrowed soils contaminated with bunker-fuel in a full-scale trial. *Environmental Pollution*, 2010. **158**(10): p. 3032–3040.
17. Gomez, F. and M. Sartaj, Optimization of field scale biopiles for bioremediation of petroleum hydrocarbon contaminated soil at low temperature conditions by response surface methodology (RSM). *International Biodeterioration & Biodegradation*, 2014. **89**: p. 103–109.
18. Gidarakos, E. and M. Aivalioti, Large scale and long term application of bioslurping: The case of a Greek petroleum refinery site. *Journal of Hazardous Materials*, 2007. **149**(3): p. 574–581.
19. Singh, B., R. Kaur, and K. Singh. Degradation of organophosphate pesticide, malathion, by Bacillus spp. isolated from soil. *Proceedings of the 11th International Conference on Environmental Science and Technology*, Chania, Crete, Greece, 2009.
20. Haseena, M., Review of microorganisms role in bioremediation. Scholar: National School of Leadership, 2019. **8**(2.2).
21. Dell'Anno, F., et al., Bacteria, fungi and microalgae for the bioremediation of marine sediments contaminated by petroleum hydrocarbons in the omics era. *Microorganisms*, 2021. **9**(8): p. 1695.
22. Burgess, J.E., S.A. Parsons, and R.M. Stuetz, Developments in odour control and waste gas treatment biotechnology: A review. *Biotechnology Advances*, 2001. **19**(1): p. 35–63.
23. Philp, J., *Bioremediation: Applied Microbial Solutions for Real-World Environment Cleanup*. 2005, ASM Press.
24. Xu, B., et al., Characterization of acetamiprid biodegradation by the microbial consortium ACE-3 enriched from contaminated soil. *Frontiers in Microbiology*, 2020. **11**: p. 1429.
25. Bhatt, P., et al., New insights into the degradation of synthetic pollutants in contaminated environments. *Chemosphere*, 2021. **268**: p. 128827.
26. Kumar, A., et al., Recent advances in plasmonic photocatalysis based on TiO_2 and noble metal nanoparticles for energy conversion, environmental remediation, and organic synthesis. *Small*, 2022. **18**(1): p. 2101638.
27. Sharma, P., et al., Recent advancements in microbial-assisted remediation strategies for toxic contaminants. *Cleaner Chemical Engineering*, 2022. **2**: p. 100020.
28. Bailey, J.E., Toward a science of metabolic engineering. *Science*, 1991. **252**(5013): p. 1668–1675.
29. Markina, N., A. Kotlobay, and A. Tsarkova, Heterologous metabolic pathways: Strategies for optimal expression in eukaryotic hosts. *Acta Naturae (англоязычная версия)*, 2020. **12**(2 (45)): p. 28–39.
30. Flannagan, R.S., T. Linn, and M.A. Valvano, A system for the construction of targeted unmarked gene deletions in the genus Burkholderia. *Environmental Microbiology*, 2008. **10**(6): p. 1652–1660.

31. Zampolli, J., et al., Biodegradation of naphthenic acids: Identification of Rhodococcus opacus R7 genes as molecular markers for environmental monitoring and their application in slurry microcosms. *Applied Microbiology and Biotechnology*, 2020. **104**: p. 2675–2689.
32. Al-Maqdi, K.A., et al., Challenges and recent advances in enzyme-mediated wastewater remediation—A review. *Nanomaterials*, 2021. **11**(11): p. 3124.
33. Rao, M.A., et al., Role of enzymes in the remediation of polluted environments. *Journal of Soil Science and Plant Nutrition*, 2010. **10**(3): p. 333–353.
34. Bajaj, D., et al., Exploration of Microbial Enzymes in Bioremediation, in *Microbial Approaches for Sustainable Green Technologies*, J. Shankar, P. Verma, and M. P. Shah, Editors. 2024, CRC Press: Boca Raton, FL. p. 250–272.
35. Mandelbaum, R.T., D.L. Allan, and L.P. Wackett, Isolation and characterization of a Pseudomonas sp. that mineralizes the s-triazine herbicide atrazine. *Applied and Environmental Microbiology*, 1995. **61**(4): p. 1451–1457.
36. Webster, L.J., et al., A synthetic biology approach for the treatment of pollutants with microalgae. *Frontiers in Bioengineering and Biotechnology*, 2024. **12**: p. 1379301.
37. Zuo, Z., et al., Engineering Pseudomonas putida KT2440 for simultaneous degradation of organophosphates and pyrethroids and its application in bioremediation of soil. *Biodegradation*, 2015. **26**: p. 223–233.
38. Udaondo, Z., et al., Analysis of solvent tolerance in Pseudomonas putida DOT-T1E based on its genome sequence and a collection of mutants. *Febs Letters*, 2012. **586**(18): p. 2932–2938.
39. Zylstra, G.J. and D.T. Gibson, Toluene degradation by Pseudomonas putida F1: Nucleotide sequence of the todC1C2BADE genes and their expression in Escherichia coli. *Journal of Biological Chemistry*, 1989. **264**(25): p. 14940–14946.
40. Duque, E., et al., Providing octane degradation capability to Pseudomonas putida KT2440 through the horizontal acquisition of oct genes located on an integrative and conjugative element. *Environmental Microbiology Reports*, 2022. **14**(6): p. 934–946.
41. Liu, H., et al., An intelligent synthetic bacterium for chronological toxicant detection, biodegradation, and its subsequent suicide. *Advanced Science* (Weinh), 2023. **10**(31): p. e2304318.
42. Mishra, A.T. and R. Kaur, Techniques of bioremediation using bacteria for the treatment of polycyclic aromatic hydrocarbons: A Review. *Journal of Experimental Biology and Agricultural Sciences*, 2022. 10: p. 1318–1330.
43. Hussein, O.A. and J.A.-A. Ibrahim, Leachates recirculation impact on the stabilization of the solid wastes – A review. *Journal of Ecological Engineering*, 2023. **24**(4): p. 172–183.
44. Larki, I., et al., Mitigation approaches and techniques for combustion power plants flue gas emissions: A comprehensive review. *Science of the Total Environment*, 2023. **903**: p. 166108.
45. Cañete Vela, I., et al., Co-recycling of natural and synthetic carbon materials for a sustainable circular economy. *Journal of Cleaner Production*, 2022. **365**: p. 132674.
46. Sheoran, K., et al., Air pollutants removal using biofiltration technique: A challenge at the frontiers of sustainable environment. *ACS Engineering Au*, 2022. **2**(5): p. 378–396.
47. Septic Tank Polishing, *Decentralized Systems Technology Fact Sheet*. US Environmental Protection Agency.
48. Uekert, T., et al., Technical, economic, and environmental comparison of closed-loop recycling technologies for common plastics. *ACS Sustainable Chemistry & Engineering*, 2023. **11**(3): p. 965–978.
49. Kara, S., et al., Closed-loop systems to circular economy: A pathway to environmental sustainability? *CIRP Annals*, 2022. **71**(2): p. 505–528.

50. Yang, M., et al., Circular economy strategies for combating climate change and other environmental issues. *Environmental Chemistry Letters*, 2023. **21**(1): p. 55–80.
51. Zaman, A.U., Identification of key assessment indicators of the zero waste management systems. *Ecological Indicators*, 2014. **36**: p. 682–693.
52. Ibrahim, M.I.M. and N.A.E.M. Mohamed, Towards sustainable management of solid waste in Egypt. *Procedia Environmental Sciences*, 2016. **34**: p. 336–347.
53. Ghisellini, P., C. Cialani, and S. Ulgiati, A review on circular economy: The expected transition to a balanced interplay of environmental and economic systems. *Journal of Cleaner Production*, 2016. **114**: p. 11–32.
54. Jonidi Jafari, A., E. Charkhloo, and H. Pasalari, Urban air pollution control policies and strategies: A systematic review. *Journal of Environmental Health Science and Engineering*, 2021. **19**(2): p. 1911–1940.
55. Peng, W., et al., The critical role of policy enforcement in achieving health, air quality, and climate benefits from India's clean electricity transition. *Environmental Science & Technology*, 2020. **54**(19): p. 11720–11731.
56. Chavez, C.A., J.K. Stranlund, and W. Gomez, Controlling urban air pollution caused by households: Uncertainty, prices, and income. *Journal of Environmental Management*, 2011. **92**(10): p. 2746–2753.
57. Hao, J., et al., Air pollutants contribution and control strategies of energy-use related sources in Beijing. *Science in China, Series D: Earth Sciences*, 2005. **48**: p. 138.
58. Cui, L., et al., Analyses of air pollution control measures and co-benefits in the heavily air-polluted Jinan city of China, 2013–2017. *Scientific Reports*, 2020. **10**(1): p. 5423.
59. Hoban, T.J., *Public perceptions of the benefits and risks of biotechnology*. 1993, NABC.
60. Ezeonu, C.S., et al., Biotechnological tools for environmental sustainability: Prospects and challenges for environments in Nigeria-a standard review. *Biotechnology Research International*, 2012. **2012**: p. 450802.

5 The Potential of Biocatalysts for Pollutant Removal

Efficient Remediation and Toxicity Removal

Katarzyna Jankowska, Agnieszka Rybarczyk, Manuel Pinelo, Jakub Zdarta, and Teofil Jesionowski

5.1 INTRODUCTION

Biocatalysts, commonly known as enzymes, are biological molecules that accelerate specific chemical reactions (Wu et al., 2021). Unlike traditional chemical catalysts, which are often inorganic or synthetic, enzymes are predominantly derived from microorganisms, plants, and even animals. For instance, certain fungi, such as *Trametes versicolor*, produce enzymes like laccase, which is widely used in the degradation of phenolic compounds (Athanasiou et al., 2024). Similarly, *Papaya carica*, a plant species, produces the enzyme papain, which is employed in the removal of metal ions (Dutta et al., 2009). An enzyme that can be produced by animals is trypsin, which is widely used in the breakdown of proteins (Santos et al., 2023). Due to its unique catalytic properties, enzymes offer a sustainable and eco-friendly alternative to traditional environmental remediation methods. Unlike harsh chemical treatments, biocatalysts can operate under mild conditions, minimizing harmful by-products and energy consumption. However, the selection of a specific enzyme for the targeted pollutant is crucial for effective remediation. For example, oxidoreductases, such as laccase or horseradish peroxidase, are particularly effective in degrading phenolic compounds, such as dyes and pharmaceuticals (Varga et al., 2019), while hydrolases, such as esterases and lipases, are used in the breakdown of organic compounds like pesticides (Li et al., 2023) and oils (Cammarota and Freire, 2006). In addition to selecting the appropriate enzyme for a particular pollutant, optimizing the process conditions—such as temperature, pH, and enzyme concentration—can significantly enhance the efficiency of pollutant removal (Zdarta et al., 2019). Thus, the choice of enzyme, coupled with finely tuned environmental conditions, plays a vital role in maximizing the efficacy of biocatalytic remediation and toxicity removal of environmental pollutants. Figure 5.1 presents the specific compounds causing environmental pollution.

DOI: 10.1201/9781003536277-5

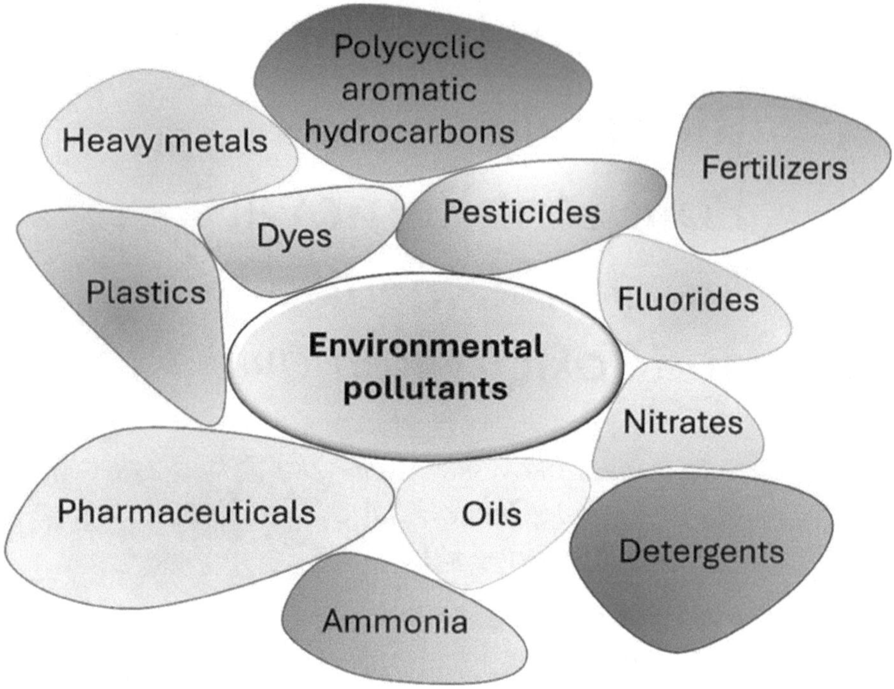

FIGURE 5.1 Main environmental pollutants.

5.2 MECHANISMS OF POLLUTANT REMOVAL BY ENZYMES

Pollutants can be classified as either biodegradable or nonbiodegradable, depending on how they interact with the environment. Biodegradable pollutants are substances that can naturally break down into simpler, nontoxic compounds over time, often with the help of microorganisms. Examples include organic materials like kitchen waste, paper, wood, agricultural residues, cattle waste, wool, and plant matter. In contrast, nonbiodegradable pollutants are resistant to natural degradation and persist in the environment as harmful substances. Plastics, pesticides, pharmaceuticals, detergents, dyes, metals, polycyclic aromatic hydrocarbons, etc., fall into this category, posing risks to ecosystems and human health (Abdel-Raouf et al., 2012). Many of these compounds, such as dyes, pharmaceuticals, pesticides, bisphenols, and detergents, are endocrine-disrupting chemicals, linked to various health issues in humans (Schug et al., 2011). Additionally, pollutants like plastics and heavy metals accumulate in ecosystems, interfering with natural cycles like animal reproduction. Enzymes offer a potential solution, as they can selectively convert certain pollutants into less toxic forms, thus reducing environmental and health hazards.

5.2.1 Oxidoreductases

Oxidoreductases, particularly laccase and peroxidases like horseradish peroxidase, manganese peroxidase, and lignin peroxidase, play a key role in pollutant remediation.

Belonging to the EC1 enzyme class, they catalyze oxidation and redox reactions across a diverse range of substrates, including phenolic compounds, amines, alcohols, and metal ions. These enzymes are found across various life forms, from bacteria and fungi to plants, animals, and even humans, highlighting their importance in cellular and complex system reactions (Xu, 2005). Although oxidoreductases primarily drive oxidation and redox processes, each subclass within EC1 is specialized for distinct reaction types. Their catalytic behavior is closely tied to their structural polypeptide makeup and active sites, where the molecular structure (often tertiary or quaternary) influences stability and specificity (Robinson, 2015).

5.2.1.1 Laccase

Laccase, known as an "oxygen oxidoreductase," is abundant in nature and present in numerous bacteria, fungi, and plants. This enzyme's main strength lies in its ability to catalyze the oxidation of compounds like phenols and aromatic amines. Known as "blue oxidoreductases," laccases have four copper atoms in their active site—one Type-1, one Type-2, and two Type-3. Each copper atom plays a specific role in the catalytic process. Type-1 contributes to the enzyme's blue color and electron transfer during substrate oxidation. The oxygen-dependent reaction cycle produces radical products, which are then transferred to the Type-2 and Type-3 atoms via a cysteine-histidine pathway, where oxygen is reduced to water (Arregui et al., 2019). The overall reaction scheme catalyzed by laccase is depicted in Figure 5.2.

5.2.1.2 Horseradish Peroxidase

Horseradish peroxidase (EC 1.11.1.7) is another enzyme with great potential for remediation. Averaging 40 kDa in size, it belongs to a subgroup of peroxidases made of protein and oligosaccharides and features two metal centers—a heme group (ferroporphyrin) and two calcium ions. The heme group, composed of four pyrrole rings linked by methane bridges and coordinating with iron ions, is critical to enzyme function, as Fe(IV) becomes an electron donor for the substrate after oxidation by hydrogen peroxide. Subsequently, hydrogen peroxide is broken down into water, and the oxidized horshradish peroxidase (HRP) returns to its original state, producing new compounds through aromatic ring condensation (Azevedo et al., 2003). Figure 5.3 presents the scheme of reaction catalyzed by HRP in the presence of H_2O_2.

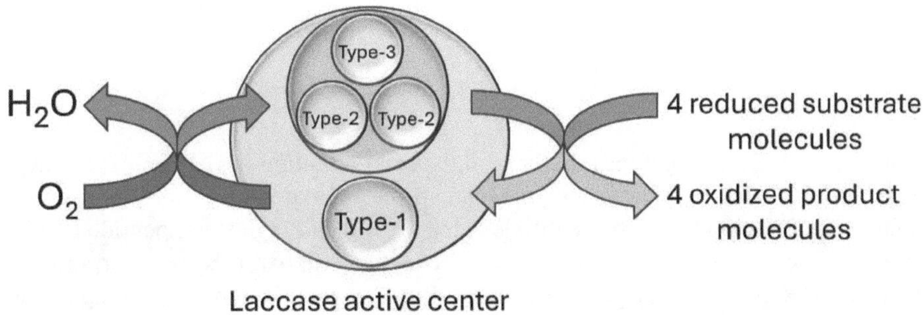

FIGURE 5.2 Redox reactions catalyzed by laccase indicating the presence of copper atoms.

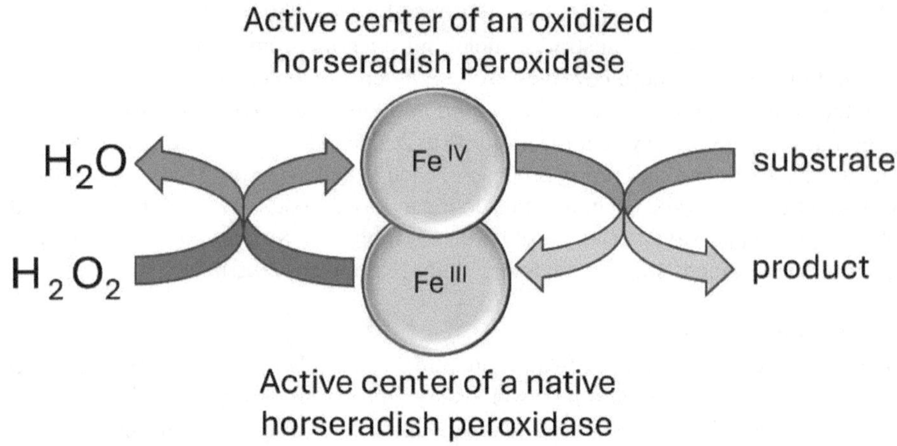

FIGURE 5.3 Reactions catalyzed by horseradish peroxidase.

5.2.1.3 Manganese Peroxidase

Another peroxidase that can be used in environmental remediation is manganese peroxidase (EC 1.11.1.13). This enzyme was first identified in 1984 by Kuwahara and colleagues in *Phanerochaete chrysosporium*, a species of white-rot fungus. This ligninolytic enzyme, secreted by various white-rot fungi, exists in several forms with molecular weights ranging from 38 to 62.5 kDa. It is an extracellular glycoprotein composed of approximately 350 amino acid residues, featuring a heme group as a prosthetic component. Additionally, manganese peroxidase contains a Mn(II) substrate ion, two structural calcium ions, two glycerol molecules, and 478 solvent molecules (Saikia et al., 2023).

5.2.1.4 Lignin Peroxidase

Lignin peroxidase is a heme-containing enzyme that facilitates the oxidative breakdown of lignin in the presence of hydrogen peroxide. Structurally, it is a monomeric hemoprotein. Initially identified in the white-rot fungus *Phanerochaete chrysosporium*, this enzyme is also found in other species, including *Trametes versicolor*, *Phlebia radiata*, and *Phanerochaete sordida*. Lignin peroxidase has a molecular weight ranging between 38 and 43 kDa and operates optimally at a notably low pH of about 3.0, setting it apart from other peroxidases (Falade et al., 2017).

5.2.2 Hydrolases

Hydrolases (EC3) are enzymes that warrant special attention in the context of environmental remediation. These enzymes aid in the breakdown of various essential biochemical compounds across multiple categories, including esters, peptide bonds, carbon-halide bonds, ureas, and thioesters. They accomplish this by incorporating an –H or –OH group from H_2O into each resulting product, thereby separating bonds from the substrate's surface (Athiappan et al., 2022). Among their notable advantages are their lack of cofactor requirements, stereoselectivity, and ability to function in water-miscible solvents. Extracellular hydrolytic enzymes, such as lipases and esterases, belong to the α/β hydrolase superfamily, which specializes in catalyzing the hydrolysis and synthesis of ester bonds. Esterases (EC 3.1.1.1) break down short-chain fatty esters (≤C8), while lipases (EC 3.1.1.3) act on long-chain acylglycerols (≥C8). They serve as highly efficient biocatalysts in reactions like esterification, transesterification, aminolysis, and alcoholysis (Sharma et al., 2019). Examples of esterases and lipases with properties suited to potential applications in bioremediating plastics, pesticides, and oily chemicals—the most prevalent environmental pollutants—are described below.

5.2.2.1 PETase

Polyethylene terephthalate hydrolases (PETases) are a recently identified and industrially significant group of enzymes that enable the enzymatic breakdown of polyethylene terephthalate (PET), one of the most widely used plastics globally. In 2016, Yoshida et al. discovered two enzymes from *Ideonella sakaiensis*, a bacterium capable of utilizing PET as its main carbon and energy source. These enzymes include PETase, which degrades PET into mono(2-hydroxyethyl) terephthalic acid (MHET), and mono(2-hydroxyethyl) terephthalate hydrolase (MHETase), which further breaks down MHET into terephthalic acid and ethylene glycol (Figure 5.4). Since this discovery, numerous PET hydrolases, classified as carboxylic ester hydrolases (EC 3.1.1), have been identified and studied (Di Rocco et al., 2023).

5.2.2.2 Organophosphate Pesticide Hydrolase

Organophosphate hydrolase, derived from *Flavobacterium* sp. and *Pseudomonas diminuta*, has demonstrated the ability to degrade organophosphorus pesticides. This enzyme exhibits a broad range of substrate specificity, enabling it to cleave various bonds, including P–O, P–S, P–F, and P–CN. It facilitates the breakdown of organophosphorus compounds, such as methyl and ethyl parathion, paraoxon, chlorpyrifos, coumaphos, cyanophos, and diazinon. The enzyme achieves catalytic hydrolysis of phosphotriesters/organophosphates through an SN2-type mechanism, where the phosphotriester bond is cleaved with the assistance of a water molecule bound to the α-metal ion. Interaction with the metal ion and a bridging hydroxide lowers the pKa of this water molecule, generating a hydroxide nucleophile optimally positioned to attack the electrophilic phosphorus atom.

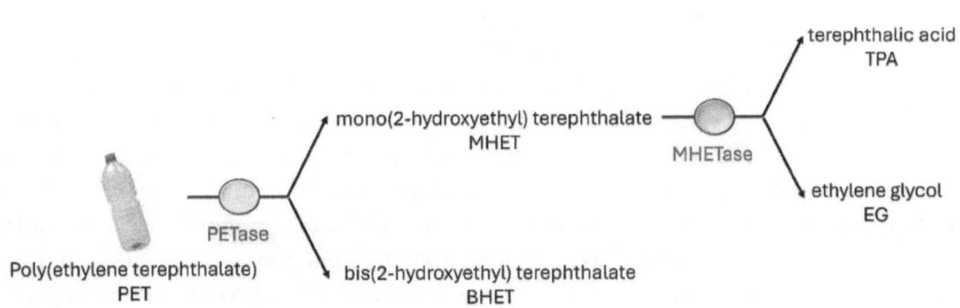

FIGURE 5.4 Conversion of PET using PETase and MHET using MHETase.

5.2.2.3 Lipases

Lipases, also known as triacylglycerol hydrolases (EC 3.1.1.3), are enzymes classified as serine hydrolases that facilitate both the synthesis and breakdown of ester bonds. These esters can form from reactions between glycerol and long-chain fatty acids, or they can be broken down into simpler molecules like di- and monoacylglycerols, free fatty acids, and glycerol. Lipases are commonly produced by plants, animals, and microorganisms, and these organisms use lipases to mineralize or degrade hydrocarbon-based contaminants, such as oils or lipids, taking up the resulting fatty acids as nutrients. The bacterial genus *Pseudomonas* is particularly prominent in bioremediation, with several strains of *Pseudomonas* species playing a key role in oil hydrolysis. Additionally, certain *Bacillus* species, *including Bacillus stearothermophilus* isolated from slaughterhouse waste and other *Bacillus* species from petroleum-contaminated soil, are also effective in bioremediation (Basha, 2021).

5.3 ADVANCES IN ENZYME ENGINEERING FOR POLLUTANT REMOVAL

Despite their advantages, such as the ability to catalyze reactions under mild conditions, enzymes also have certain drawbacks. These include the loss of catalytic activity, reduced stability under varying reaction conditions, and the inability to be reused. Therefore, to enhance the properties of enzymes used in environmental pollutant removal, genetic modification of biocatalysts and their immobilization are applied (Somu et al., 2022; Vandenberghe et al., 2022).

5.3.1 Genetic Engineering

Recent breakthroughs in molecular biology have introduced advanced techniques for the genetic modification of enzymes, aiming to enhance their structure, resilience, and catalytic efficiency. These molecular tools can be applied in various ways, such as gene recombination through DNA shuffling between species, genetic engineering to optimize heterologous protein production, and altering the enzyme's amino acid sequence to improve performance. Additionally, computational biology and omics technologies enable the analysis of large datasets to inform the design and refinement of novel enzymes (Vandenberghe et al., 2022).

Directed evolution, also known as DNA shuffling, is a highly effective mutagenesis technique that emulates natural gene evolution to redesign genes with greater efficiency. This approach is particularly powerful because it introduces multiple mutations into a gene, enabling the creation of new enzymatic activities, which can then be identified through specific selection methods. Predicting the exact amino acid changes required for these new activities remains challenging, if not impossible, to accomplish through rational design alone. Interestingly, these beneficial mutations can occur both close to and far from the enzyme's active site (Wood, 2008).

The application of DNA shuffling in enzyme improvement was demonstrated by Ouyang and Zhao (2019). The performance of bacterial CotA-laccases is hindered by low yields and catalytic efficiency. DNA shuffling produced a mutant (T232P/Q367R

5E29) with two amino acid changes, increasing catalytic efficiency by 1.21-fold. This mutant showed enhanced thermal stability and effectively decolorized dyes like Indigo Carmine and Congo Red at pH 9.0. Similarly, Yee et al. (2018) showed that recombinant manganese peroxidase from *Pichia pastoris* can degrade furfural, reducing its toxicity. In another study, Das et al. (2023) demonstrated that biogenic manganese peroxidases from genetically modified *Escherichia coli* could remove up to 96% of bisphenol A from an aqueous solution within 60 minutes.

Like oxidoreductases, hydrolase properties can also be modified through DNA shuffling and subsequently used in remediation. Luo et al. (2016) enhanced the degradation efficiency of malathion by applying iterative saturation mutagenesis and random mutagenesis to a phosphotriesterase enzyme. The beneficial mutations were then combined through DNA shuffling, resulting in an improved thermostable enzyme variant. After screening 9000 mutants, the optimal variant exhibited 25 times higher enzymatic activity than the wild-type enzyme, making it a highly effective and robust candidate for malathion detoxification.

Although the use of recombinant enzymes makes the removal of certain pollutants more efficient compared to nonrecombinant biocatalysts, the main disadvantage is the high production cost of such biocatalysts. This is primarily due to the use of expensive molecular biology tools in advanced laboratories authorized to handle genetically modified organisms. However, the application of recombinant biocatalysts in pollution removal may be the only effective method, and therefore, it should be considered a viable solution.

5.3.2 Enzyme Immobilization

Enzyme immobilization is one of the most commonly used methods aimed at increasing the stability of biocatalysts under harsh reaction conditions and enabling their repeated use (Maghraby et al., 2023). Immobilization involves attaching the biocatalyst to a solid support within the reaction medium. Several methods of enzyme immobilization include adsorption, covalent bonding, encapsulation, entrapment, and crosslinking (Zdarta et al., 2018) (Figure 5.5.).

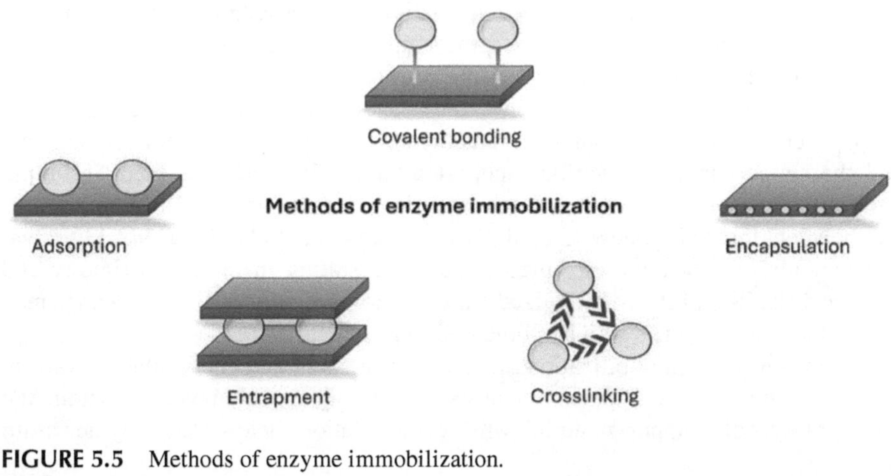

FIGURE 5.5 Methods of enzyme immobilization.

The nature of the bond between the biocatalyst and the support material plays a critical role in determining enzyme stability and activity. This interaction is influenced by the characteristics of the support, such as its porosity and other physical properties as well as chemical affinity, in particular the presence of functional groups. Additionally, enzymes can be attached to various parts of the support depending on the immobilization method. Techniques like adsorption and covalent bonding generally involve fixing the enzyme to the outer surface of the support. In contrast, encapsulation and entrapment place the enzyme within the support matrix. In some cases, enzymes can also be immobilized without a support material through direct crosslinking between the proteins themselves (Mulinari et al., 2020).

The most widely used technique for enzyme immobilization is adsorption. This method is straightforward and cost-effective because it does not require additional chemicals, apart from the enzyme and support material. The bond between the enzyme and the support is formed through weak interactions, such as hydrogen bonds, van der Waals forces, or electrostatic attractions (Jesionowski et al., 2014). Therefore, the support materials should have functional groups, like hydroxyl or amine groups, that can form these bonds with the enzyme. Additionally, characteristics like porosity and a large surface area are beneficial as they help protect the enzyme from harsh conditions. Although enzymes immobilized via adsorption can be easily detached from the support, minimal structural changes occur, allowing the enzyme to retain high activity (Datta et al., 2013). Enzymes immobilized via adsorption are among the most frequently used biocatalysts for environmental pollutant removal and reducing toxicity. Zhou et al. (2022) immobilized laccase on graphene oxide using the adsorption method, and then applied this biocatalyst to effectively remove Malachite Green dye from an aqueous solution. Similarly, Xiao et al. (2020) demonstrated that horseradish peroxidase adsorbed onto Fe_3O_4 particles was able to degrade over 90% of bisphenol A and 60% of 17α-ethinylestradiol, both of which are endocrine-disrupting chemicals.

Another enzyme immobilization method that requires special attention is covalent bonding. This approach involves selecting a suitable support material, focusing particularly on the presence of functional groups compatible with the biomolecule. Commonly used supports include biopolymers and polymers that feature functional groups like amine and carbonyl, which facilitate the formation of strong covalent bonds with the enzyme. In some cases, the support material may lack the functional groups needed for direct covalent bonding with the enzyme; however, it can still be effective for enzyme immobilization due to benefits like enhanced stability under reaction conditions. To promote interactions between the immobilized enzyme and the support material, various support modifiers are often applied, including aminosilane, glutaraldehyde, polyethyleneimine, (3-aminopropyl)triethoxysilane, and polydopamine (Jesionowski et al., 2014; Jankowska et al., 2021). Notably, covalent bonding can alter the enzyme's structure, resulting in increased rigidity and improved stability of the immobilized enzyme. However, this structural change may also reduce the enzyme's activity compared to its native form.

Among enzyme immobilization methods, entrapment and encapsulation warrant particular attention. Entrapment involves embedding the biomolecule within the pores or layers of a support material, while encapsulation encloses the enzyme within a semipermeable layer composed of organic or inorganic substances. A common limitation of both methods is the potential reduction in mass transfer between the substrate and enzyme active sites, often due to steric hindrances (Datta et al., 2013). For enzyme entrapment, a crucial characteristic of the support material is porosity. Materials with a high surface area, such as oxides, gels, polymers, and ceramics, are especially suitable for this type of protein immobilization (Imam et al., 2021). In the case of encapsulation, the porosity of the semipermeable layer is also essential. The pore size must be tailored to the dimensions of the biomolecule, substrates, and products to prevent enzyme leakage while allowing the efficient transfer of substrates and reaction products between the external environment and the encapsulated enzyme (Nguyen et al., 2017).

Crosslinking is an enzyme immobilization method that forms heterogeneous biocatalysts by bonding enzymatic biomolecules through intermolecular crosslinks. This process can be done either without support material or by using a support to enhance stability. When no support is used, enzymes are crosslinked directly, often with bifunctional organic compounds like glutaraldehyde (Jesionowski et al., 2014). In this approach, proteins may be pre-aggregated through precipitation before crosslinking, creating stable biocatalysts like crosslinked enzyme aggregates or crystals (CLECs). CLECs, however, are costly as they require crystallized enzyme forms. Alternatively, enzymes can be crosslinked with each other and with support, increasing stability under harsh conditions and reducing enzyme elution. This approach improves the durability of the enzyme over multiple uses but can also lead to structural changes that might limit activity due to reduced mass transport.

5.4 APPLICATION OF ENZYMES IN WASTEWATER TREATMENT

Enzymes are widely employed for the removal and detoxification of contaminants, particularly in wastewater treatment. They can be used in their native and immobilized forms, as well as within various bioreactors, allowing for efficient and selective action against a range of pollutants.

One group of enzymes utilized for contaminant removal is oxidoreductases, which include laccase, horseradish peroxidase, lignin peroxidase, and manganese peroxidase. Due to their oxidative-reductive properties with phenolic compounds, they can effectively convert these substances, including dyes and pharmaceuticals. Lassouane et al. (2019) demonstrated that free laccase can remove 100% of bisphenol A under optimal conditions (pH 5, 30°C, 20 mg/L BPA, and 1500 U/L laccase). However, while the use of free enzymes appears relatively straightforward in remediation processes, these biocatalysts cannot be reused multiple times and quickly lose their catalytic activity. This limitation has led to the application of enzyme immobilization (Rafeeq et al., 2022). Bilal and Asgher (2015) immobilized manganese peroxidase using calcium alginate beads. They assessed the reusability, thermal stability, and storage stability of the immobilized biocatalyst and compared it with that of the free enzyme. Both free and immobilized manganese peroxidase demonstrated potential in decolorizing various textile dyes (Sandal reactive dyes), with the immobilized enzyme capable of removing up to 100% of these chemicals. The cytotoxicity of the solution after biocatalytic treatment was measured through erythrocyte lysis and brine shrimp lethality tests, showing a reduction to around 5% following manganese peroxidase action. Pharmaceuticals, such as nonsteroidal anti-inflammatory drugs, antibiotics, and hormones, are particular contaminants that warrant special attention and can be remediated by oxidoreductases (Naghdi et al., 2018). Previous research has provided ample information on the application of both native and immobilized oxidoreductases. However, studies that focus on the toxicity of solutions after catalytic conversion are particularly noteworthy (Varga et al., 2019). In the work by Zdarta et al. (2023), laccase was immobilized on a $CaCO_3$ support for the removal of 17α-ethynylestradiol (EE2), a persistent synthetic estrogen. Optimal conditions—12-hour treatment with 100 mg of the biosystem at pH 5 and 25°C—achieved 100% removal. Additionally, a reduction of over 60% in solution toxicity and estrogenic activity was recorded. Other interesting results regarding solution toxicity after enzymatic conversion are presented in the study by Neiverth de Freitas et al. (2017), which used crude laccases from *Pleurotus ostreatus* and *Pleurotus pulmonarius* to degrade BPA in water. Both laccases removed 100 mg/L and 200 mg/L of BPA by 100% and 85%, respectively, in 1 hour at 8 U/L. Thirteen metabolites were identified, including *p*-isopropenylphenol and methylpent-3-oic acid. *Pleurotus ostreatus* laccase reduced BPA toxicity from 85% to below 5%, while *P. pulmonarius* did not significantly decrease toxicity.

Another group of enzymes that effectively contribute to pollutant remediation are hydrolases, which represent a large class of biocatalysts enabling the removal of contaminants such as pesticides, oils, and plastics. For instance, lipase from *Candida rugosa* efficiently hydrolyzes triolein in a two-phase oil-water system, adsorbing at the oil–water interface. The enzyme successively breaks down triolein's ester bonds, producing diolein, monoolein, and glycerol while releasing omega-9 fatty acids during each catalytic step (Sarmah et al., 2018). Parathion hydrolase is another hydrolase of special interest, used for pesticide detoxification and offering a practical alternative to conventional treatment methods. It is produced by various bacteria, such as *Pseudomonas* and *Flavobacterium*, and breaks down widely used organophosphate pesticides (Sarkar et al., 2020). However, in recent years, the most attention has focused on PETase enzymes due to their ability to degrade PET plastic. PET is a synthetic polymer produced in large quantities and accumulating in the environment at an alarming rate. Its remarkable resistance to natural degradation means PET could persist in nature for centuries. Wang et al. (2024) developed a genetically modified PETase for PET degradation, yielding monomeric products (255.7 μM), approximately 25.5% higher than the yield achieved after 50 hours of non-genetically modified organism PETase-catalyzed PET degradation.

5.5 CHALLENGES AND FUTURE PERSPECTIVES IN ENZYME-BASED POLLUTANT REMOVAL

Although biocatalysts play a significant role as an alternative method for removing environmental pollutants, there are many challenges that need special attention. One of the primary challenges in commercializing biocatalysts for pollutant removal is the cost of production compared to traditional methods. While enzymes often have the advantage of operating under milder conditions than traditional chemical catalysts, their production—particularly at scale—can be expensive. The fermentation processes used to produce enzymes or whole-cell biocatalysts are costly, and downstream processing for purification adds to the overall expense (Mousavi et al., 2021). Another challenge is that many enzymes lack the thermal and chemical stability required for large-scale pollutant removal. However, further research into enzyme engineering and biocatalyst immobilization is essential for improving their robustness and maintaining efficiency under various industrial conditions (Bernal et al., 2018). Moreover, transitioning from lab-scale to industrial-scale production is challenging. The scalability of biocatalyst production must be addressed, as small-scale systems often do not translate seamlessly into larger operations. Issues such as maintaining enzyme activity, yield, and consistent quality at scale can hinder commercial viability.

Despite these challenges, enzymes have significant potential as tools for environmental pollutant removal. Biocatalysts typically operate under milder conditions—lower temperatures and pressures—than traditional chemical catalysts, reducing energy consumption and making industrial processes more energy-efficient and environment-friendly, which aligns with sustainability goals (Zdarta et al., 2021). Additionally, the specificity of biocatalysts often results in fewer side reactions and less waste generation compared to chemical processes. Unlike many synthetic chemical catalysts, which can be toxic and persistent in the environment, biocatalysts are often biodegradable and less harmful (Alcalde et al., 2006). It is also worth noting that enzyme engineering is improving the performance of biocatalysts. Innovations in this area are enabling the design of more robust, efficient, and versatile biocatalysts, which can be applied in the removal of hard-to-degrade pollutants.

The future of biocatalysts is promising, particularly in the context of sustainability and environmental protection. Despite challenges related to cost, stability, scalability, and market adoption, ongoing advances in biotechnology and the increasing focus on sustainable industrial processes position biocatalysts as key players in replacing traditional methods of pollutant removal. The shift toward cleaner, more energy-efficient production methods aligns biocatalysts with the goals of a circular and low-carbon economy.

ACKNOWLEDGMENTS

This work was supported by a research grant (VIL69618) from VILLUM FONDEN. This work was also supported by the National Science Centre, Poland under the Research Grant number 2021/43/B/ST8/01854.

REFERENCES

Abdel-Raouf, Nazim, Al-Homaidan, Ali, and Ibraheem, Ibraheem Borie. 2012. "Microalgae and Wastewater Treatment." *Saudi Journal of Biological Sciences* 19, no. 3: 257–275. https://doi.org/10.1016/j.sjbs.2012.04.005.

Alcalde, Miguel, Manuel Ferrer, Francisco J. Plou, and Antonio Ballesteros. 2006. "Environmental Biocatalysis: From Remediation with Enzymes to Novel Green Processes." *Trends in Biotechnology* 24, no. 6: 281–287. https://doi.org/10.1016/j.tibtech.2006.04.002.

Arregui, Leticia, Marcela Ayala, Ximena Gómez-Gil, Guadalupe Gutiérrez-Soto, Carlos Eduardo Hernández-Luna, Mayra Herrera de los Santos, Laura Levin, Arturo Rojo-Domínguez, Daniel Romero-Martínez, Mario C. N. Saparrat, and Mauricio A. Trujillo-Roldán. 2019. "Laccases: Structure, Function, and Potential Application in Water Bioremediation." *Microbial Cell Factories* 18, no. 1: 200. https://doi.org/10.1186/s12934-019-1248-0.

Athanasiou, Panagiotis E., Christina I. Gkountela, Michaela Patila, Renia Fotiadou, Alexandra V. Chatzikonstantinou, Stamatina N. Vouyiouka, and Haralambos Stamatis. 2024. "Laccase-Mediated Oxidation of Phenolic Compounds from Wine Lees Extract towards the Synthesis of Polymers with Potential Applications in Food Packaging." *Biomolecules* 14, no. 3: 323. https://doi.org/10.3390/biom14030323.

Athiappan, Murugan, Saul Dinesh Kumar, Sankaran Umamaheswari, and Muhammad Rajaprabu. 2022. "Chapter 13: Rhizosphere Engineering through Pesticides-Degrading Beneficial Bacteria." In *Rhizosphere Engineering*, edited by Ramesh Chandra Dubey and Pankaj Kumar, 239–257. Academic Press. https://doi.org/10.1016/B978-0-323-89973-4.00014-4.

Azevedo, Ana M., Verónica C. Martins, Duarte M. F. Prazeres, Vojislav Vojinović, Joaquim M. S. Cabral, and Luís P. Fonseca. 2003. "Horseradish Peroxidase: A Valuable Tool in Biotechnology." *Biotechnology Annual Review* 9: 199–247. https://doi.org/10.1016/S1387-2656(03)09003-3.

Basha, Pinjari Aleem. 2021. "Oil Degrading Lipases and Their Role in Environmental Pollution." In *Recent Developments in Applied Microbiology and Biochemistry*, edited by Buddolla Viswanath, 269–277. Academic Press. https://doi.org/10.1016/B978-0-12-821406-0.00025-4.

Bernal, Claudia, Karen Rodríguez, and Ronny Martínez. 2018. "Integrating Enzyme Immobilization and Protein Engineering: An Alternative Path for the Development of Novel and Improved Industrial Biocatalysts." *Biotechnology Advances* 36, no. 5: 1470–1480. https://doi.org/10.1016/j.biotechadv.2018.06.002.

Bilal, Muhammad, and Muhammad Asgher. 2015. "Dye Decolorization and Detoxification Potential of Ca-Alginate Beads Immobilized Manganese Peroxidase." *BMC Biotechnology* 15, no. 1: 111. https://doi.org/10.1186/s12896-015-0227-8.

Cammarota, Magali, and Denise Maria Guimarães Freire. 2006. "A Review on Hydrolytic Enzymes in the Treatment of Wastewater with High Oil and Grease Content." *Bioresource Technology* 97, no. 17: 2195–2210. https://doi.org/10.1016/j.biortech.2006.02.030.

Das, Ranjit, Pengzhao Yao, Hongliang Yin, Zhishu Liang, Guiying Li, and Taicheng An. 2023. "BPA Degradation Using Biogenic Manganese Oxides Produced by an Engineered *Escherichia coli* with a Non-Blue Laccase from *Bacillus sp. GZB*." *Chemosphere* 326: 138407. https://doi.org/10.1016/j.chemosphere.2023.138407.

Datta, Sumitra, Louis Rene Christena, and Yamuna Rani Sriramulu Rajaram. 2013. "Enzyme Immobilization: An Overview on Techniques and Support Materials." 3 Biotech 3, no. 1: 1–9. https://doi.org/10.1007/s13205-012-0071-7.

Di Rocco, Giulia, Henry N. Taunt, Marcello Berto, Harry O. Jackson, Daniele Piccinini, Alan Carletti, Giulia Scurani, Niccolò Braidi, and Saul Purton. 2023. "A PETase Enzyme Synthesised in the Chloroplast of the Microalga *Chlamydomonas reinhardtii* Is Active against Post-Consumer Plastics." *Scientific Reports* 13, no. 1: 10028. https://doi.org/10.1038/s41598-023-37227-5.

Dutta, Susmita, Aparupa Bhattacharyya, Parameswar De, Parthasarathi Ray, and Srabanti Basu. 2009. "Removal of Mercury from Its Aqueous Solution Using Charcoal-Immobilized Papain (CIP)." *Journal of Hazardous Materials* 172, no. 2–3: 888–896. https://doi.org/10.1016/j.jhazmat.2009.07.085.

Falade, Ayodeji O., Uchechukwu U. Nwodo, Benson C. Iweriebor, Ezekiel Green, Leonard V. Mabinya, and Anthony I. Okoh. 2017. "Lignin Peroxidase Functionalities and Prospective Applications." *MicrobiologyOpen* 6, no. 1: e00394. https://doi.org/10.1002/mbo3.394.

Imam, Hasan T., Patricia C. Marr, and Andrew C. Marr. 2021. "Enzyme Entrapment, Biocatalyst Immobilization without Covalent Attachment." *Green Chemistry* 23, no. 14: 4980–5005. https://doi.org/10.1039/D1GC01852C.

Jankowska, Katarzyna, Ziran Su, Sigyn Björk Sigurdardóttir, Maciej Staszak, Manuel Pinelo, Jakub Zdarta, and Teofil Jesionowski. 2021. "Tailor-Made Novel Electrospun Polystyrene/Poly(d,l-Lactide-Co-Glycolide) for Oxidoreductases Immobilization: Improvement of Catalytic Properties under Extreme Reaction Conditions." *Bioorganic Chemistry* 114: 105036. https://doi.org/10.1016/j.bioorg.2021.105036.

Jesionowski, Teofil, Jakub Zdarta, and Barbara Krajewska. 2014. "Enzyme Immobilization by Adsorption: A Review." *Adsorption* 20, no. 5: 801–821. https://doi.org/10.1007/s10450-014-9623-y.

Lassouane, Fatiha, Hamid Aït-Amar, Saïd Amrani, and Susana Rodriguez-Couto. 2019. "A Promising Laccase Immobilization Approach for Bisphenol A Removal from Aqueous Solutions." *Bioresource Technology* 271: 360–367. https://doi.org/10.1016/j.biortech.2018.09.129.

Li, Mengya, Lei Li, Yue Sun, Hengyan Ma, Hui Zhang, and Feng Li. 2023. "Facile Synthesis of Dual-Hydrolase Encapsulated Magnetic ZIF-8 Composite for Efficient Removal of Multi-Pesticides Induced Pollution in Water." *Chemosphere* 314: 137673. https://doi.org/10.1016/j.chemosphere.2022.137673.

Luo, Xiao-Jing, Jian Zhao, Chun-Xiu Li, Yun-Peng Bai, Manfred T. Reetz, Hui-Lei Yu, and Jian-He Xu. 2016. "Combinatorial Evolution of Phosphotriesterase toward a Robust Malathion Degrader by Hierarchical Iteration Mutagenesis." *Biotechnology and Bioengineering* 113, no. 11: 2350–2357. https://doi.org/10.1002/bit.26012.

Maghraby, Yasmin R., Rehan M. El-Shabasy, Ahmed H. Ibrahim, and Hassan Mohamed El-Said Azzazy. 2023. "Enzyme Immobilization Technologies and Industrial Applications." *ACS Omega* 8, no. 6: 5184–5196. https://doi.org/10.1021/acsomega.2c07560.

Mousavi, Seyyed Mojtaba, Seyyed Alireza Hashemi, Seyed Mohammad Iman Moezzi, Navid Ravan, Ahmad Gholami, Chin Wei Lai, Wei-Hung Chiang, Navid Omidifar, Khadije Yousefi, and Gity Behbudi. 2021. "Recent Advances in Enzymes for the Bioremediation of Pollutants." *Biochemistry Research International* 2021: 5599204. https://doi.org/10.1155/2021/5599204.

Mulinari, Jéssica, J. Vladimir Oliveira, and Dachamir Hotza. 2020. "Lipase Immobilization on Ceramic Supports: An Overview on Techniques and Materials." *Biotechnology Advances* 42: 107581. https://doi.org/10.1016/j.biotechadv.2020.107581.

Naghdi, Mitra, Mehrdad Taheran, Satinder Kaur Brar, Azadeh Kermanshahi-pour, Mausam Verma, and Rao Y. Surampalli. 2018. "Removal of Pharmaceutical Compounds in Water and Wastewater Using Fungal Oxidoreductase Enzymes." *Environmental Pollution* 234: 190–213. https://doi.org/10.1016/j.envpol.2017.11.060.

Neiverth de Freitas, Emanuelle, Gisele Adriana Bubna, Tatiane Brugnari, Camila Gabriel Kato, Mariene Nolli, Thalita G. Rauen, Regina de Fátima Peralta Muniz Moreira, Rosely Aparecida Peralta, Adelar Bracht, Cristina G.M. de Souza, and Rosane Marina Peralta. 2017. "Removal of Bisphenol A by Laccases from *Pleurotus ostreatus* and *Pleurotus pulmonarius* and Evaluation of Ecotoxicity of Degradation Products." *Chemical Engineering Journal* 330: 1361–1369. https://doi.org/10.1016/j.cej.2017.08.051.

Nguyen, Hoang Hiep, and Moonil Kim. 2017. "An Overview of Techniques in Enzyme Immobilization." *Applied Science and Convergence Technology* 26, no. 6: 157–163. https://www.e-asct.org/journal/view.html?doi=10.5757/ASCT.2017.26.6.157.

Ouyang, Fengju, and Min Zhao. 2019. "Enhanced Catalytic Efficiency of CotA-Laccase by DNA Shuffling." *Bioengineered* 10, no. 1: 182–89. https://doi.org/10.1080/21655979.2019.1621134.

Rafeeq, Hamza, Sarmad Ahmad Qamar, Syed Zakir Hussain Shah, Syed Salman Ashraf, Muhammad Bilal, Tuan Anh Nguyen, and Hafiz M.N. Iqbal. 2022. "Biodegradation of Environmental Pollutants Using Horseradish Peroxidase." In *Biodegradation and Biodeterioration at the Nanoscale*, edited by Hafiz M. N. Iqbal, Muhammad Bilal, Tuan Anh Nguyen, and Ghulam Yasin, 603–633. Elsevier. https://doi.org/10.1016/B978-0-12-823970-4.00022-1.

Robinson, Peter. 2015. "Enzymes: Principles and Biotechnological Applications." *Essays in Biochemistry* 59: 1–41. https://doi.org/10.1042/bse0590001.

Saikia, Shilpa, Meera Yadav, Rohida Amin Hoque, and Hardeo Singh Yadav. 2023. "Bioremediation Mediated by Manganese Peroxidase – An Overview." *Biocatalysis and Biotransformation* 41, no. 3: 161–173. https://doi.org/10.1080/10242422.2022.2113517.

Santos, Mateus P. F., Matheus A. Ferreira, Evaldo C. S. Junior, Renata C. F. Bonomo, and Cristiane M. Veloso. 2023. "Functionalized Activated Carbon as Support for Trypsin Immobilization and Its Application in Casein Hydrolysis." *Bioprocess and Biosystems Engineering* 46, no. 11: 1651–1664. https://doi.org/10.1007/s00449-023-02927-9.

Sarkar, Jayita, Arindam Dutta, Piyali Pal Chowdhury, Joydeep Chakraborty, and Tapan K. Dutta. 2020. "Characterization of a Novel Family VIII Esterase EstM2 from Soil Metagenome Capable of Hydrolyzing Estrogenic Phthalates." *Microbial Cell Factories* 19, no. 1: 77. https://doi.org/10.1186/s12934-020-01336-x.

Sarmah, Nipon, D. Revathi, G. Sheelu, K. Yamuna Rani, S. Sridhar, V. Mehtab, and C. Sumana. 2018. "Recent Advances on Sources and Industrial Applications of Lipases." *Biotechnology Progress* 34, no. 1: 5–28. https://doi.org/10.1002/btpr.2581.

Schug, Thaddeus T., Amanda Janesick, Bruce Blumberg, and Jerrold J. Heindel. 2011. "Endocrine Disrupting Chemicals and Disease Susceptibility." *The Journal of Steroid Biochemistry and Molecular Biology* 127, no. 3–5: 204–215. https://doi.org/10.1016/j.jsbmb.2011.08.007.

Sharma, Abhishek, Taruna Sharma, Tanvi Sharma, Shweta Sharma, and Shamsher Singh Kanwar. 2019. "Role of Microbial Hydrolases in Bioremediation." In *Microbes and Enzymes in Soil Health and Bioremediation*, edited by Ashok Kumar and Swati Sharma, 149–164. Singapore: Springer Singapore. https://doi.org/10.1007/978-981-13-9117-0_7.

Somu, Prathap, Saranya Narayanasamy, Levin Anbu Gomez, Saravanan Rajendran, Yong Rok Lee, and Deepanraj Balakrishnan. 2022. "Immobilization of Enzymes for Bioremediation: A Future Remedial and Mitigating Strategy." *Environmental Research* 212 (Part D): 113411. https://doi.org/10.1016/j.envres.2022.113411.

Vandenberghe, Luciana Porto de Souza, Nelson Libardi Junior, Kim Kley Valladares-Diestra, Susan Grace Karp, Joyce Gueiros Wanderley Siqueira, Cristine Rodrigues, and Carlos Ricardo Soccol. 2022. "Enzymatic Bioremediation: Current Status, Challenges, Future Prospects, and Applications." In *Development in Wastewater Treatment Research and Processes*, edited by Susana Rodriguez-Couto and Maulin P. Shah, 355–381. Elsevier. https://doi.org/10.1016/B978-0-323-85839-7.00002-5.

Varga, Béla, Viola Somogyi, Mónika Meiczinger, Nóra Kováts, and Endre Domokos. 2019. "Enzymatic Treatment and Subsequent Toxicity of Organic Micropollutants Using Oxidoreductases - A Review." *Journal of Cleaner Production* 221: 306–322. https://doi.org/10.1016/j.jclepro.2019.02.135.

Wang, Tao, Wen-tao Yang, Yu-ming Gong, Ying-kang Zhang, Xin-xin Fan, Guo-cheng Wang, Zhen-hua Lu, Fei Liu, Xiao-huan Liu, and You-shuang Zhu. 2024. "Molecular Engineering of PETase for Efficient PET Biodegradation." *Ecotoxicology and Environmental Safety* 280: 116540. https://doi.org/10.1016/j.ecoenv.2024.116540.

Wood, Thomas K. 2008. "Molecular Approaches in Bioremediation." *Current Opinion in Biotechnology* 19, no. 6: 572–578. https://doi.org/10.1016/j.copbio.2008.10.003.

Wu, Shuke, Radka Snajdrova, Jeffrey C. Moore, Kai Baldenius, and Uwe T. Bornscheuer. 2021. "Biocatalysis: Enzymatic Synthesis for Industrial Applications." *Angewandte Chemie International Edition* 60, no. 1: 88–119. https://doi.org/10.1002/anie.202006648.

Xiao, Feng, Ping Xiao, Wei Jiang, and Dongsheng Wang. 2020. "Immobilization of Horseradish Peroxidase on Fe_3O_4 Nanoparticles for Enzymatic Removal of Endocrine Disrupting Chemicals." *Environmental Science and Pollution Research* 27, no. 19: 24357–24368. https://doi.org/10.1007/s11356-020-08824-y.

Xu, Feng. 2005. "Applications of Oxidoreductases: Recent Progress." *Industrial Biotechnology* 1, no. 1: 38–50. https://doi.org/10.1089/ind.2005.1.38.

Yee, Kelsey L., Lauren E. Jansen, Curtis A. Lajoie, Michael H. Penner, Lettie Morse, and Christine J. Kelly. 2018. "Furfural and 5-Hydroxymethyl-Furfural Degradation Using Recombinant Manganese Peroxidase." *Enzyme and Microbial Technology* 108: 59–65. https://doi.org/10.1016/j.enzmictec.2017.08.009.

Zdarta, Jakub, Anne S. Meyer, Teofil Jesionowski, and Manuel Pinelo. 2018. "A General Overview of Support Materials for Enzyme Immobilization: Characteristics, Properties, Practical Utility." *Catalysts* 8, no. 2: 92. https://doi.org/10.3390/catal8020092.

Zdarta, Jakub, Anne S. Meyer, Teofil Jesionowski, and Manuel Pinelo. 2019. "Multi-Faceted Strategy Based on Enzyme Immobilization with Reactant Adsorption and Membrane Technology for Biocatalytic Removal of Pollutants: A Critical Review." *Biotechnology Advances* 37, no. 7: 107401. https://doi.org/10.1016/j.biotechadv.2019.05.007.

Zdarta, Jakub, Katarzyna Jankowska, Karolina Bachosz, Oliwia Degórska, Karolina Kaźmierczak, Luong N. Nguyen, Long D. Nghiem, and Teofil Jesionowski. 2021. "Enhanced Wastewater Treatment by Immobilized Enzymes." *Current Pollution Reports* 7, no. 2: 167–179. https://doi.org/10.1007/s40726-021-00183-7.

Zdarta, Jakub, Filip Ciesielczyk, Muhammad Bilal, Katarzyna Jankowska, Karolina Bachosz, Oliwia Degórska, Agnieszka Rybarczyk, Teofil Jesionowski, and Anne S. Meyer. 2023. "Inorganic Oxide Systems as Platforms for Synergistic Adsorption and Enzymatic Conversion of Estrogens from Aqueous Solutions: Mechanism, Stability and Toxicity Studies." *Journal of Environmental Chemical Engineering* 11, no. 2: 109443. https://doi.org/10.1016/j.jece.2023.109443.

Zhou, Wenting, Wenxiang Zhang, and Yanpeng Cai. 2022. "Enzyme-Enhanced Adsorption of Laccase Immobilized Graphene Oxide for Micro-Pollutant Removal." *Separation and Purification Technology* 294: 121178. https://doi.org/10.1016/j.seppur.2022.121178.

6 Microbiomes in Fight with Pollutant Remediation

From Single Strain to Omics Approach to Bioremediation

Fatima Azeem, Muhammad Osama, Aleena Sumrin, Tazeen Malik, and Nazim Hussain

6.1 INTRODUCTION

The startling increase in the population levels globally in recent years has us pay heed to the urgent issue of environmental preservation. The results of the ever-growing population and industrialization have been a threat to our planet and it includes all kinds of pollution such as soil, water, plastic, and air. All of these pose a major danger to the environment as well as human health.

Increased population means increased demand for natural resources, which is one of the obnoxious effects of population growth. Land, food, water, energy, and other resources are limited and are being used at unfeasible rates (Hunter 2000). It can be exemplified by the huge increase in agricultural production to support the growing population, which ultimately leads to damage the habitat, deforestation, and loss of biodiversity. The forests are essential to keep climate change in check because they are the carbon sinks. Additionally, its destruction also reduces the ability of our planet to absorb greenhouse gases thereby leading to intensified global warming (Hunter 2000).

The regions facing more rapid population expansion tend to have higher waste and population levels and it often finds it difficult to handle the trash, which results in contamination of water, air, and land. In addition to this, freshwater supplies are being depleted due to the continuous contamination and they need to be refilled or decontaminated as people need water for sanitation, drinking, and agriculture (Panayotou 2000). The contaminated water and soil resulting from this contain petroleum hydrocarbon products, which are widely used as the source of energy and this has been now poisoned making all life forms endangered including those of humans (Coo, Oviasogie, and Ikhajiagbe 2022).

 DOI: 10.1201/9781003536277-6

Nowadays, the majority of cleaning processes use chemical and physical technology and the most famous ones are not simple or safe for the environment (Soriano 2014). The chemical dispersants and sorbents are labeled as hazardous compounds for the environment (Soriano 2014). Hence, we can rely on bioremediation technology. The utilization of biological organisms to break down or eradicate contaminants from the areas that are polluted is termed bioremediation. It is known as an effective management method for the recovery of contaminated soil and water and overall decreasing environmental pollution (Biswas 2015).

The late 1800s is known to be the golden time of bioremediation. The research in this field involves the identification followed by isolation of the microbes that have the potential to degrade the pollutants, for example, *Candidatus accumulibacter* that accumulates an extra amount of phosphorus as polyphosphates in its cells from the sewage water when the water runs through the sewage water treatment plants (Saha et al. 2021). One of the main tenets of bioremediation is using the innate capacity of plants and microbes to degrade and metabolize toxic substances into less or non-toxic chemicals. This can be accomplished through various methods such as microbial bioremediation, phytoremediation, and bioaugmentation (Chibuike and Obiora 2014). The technique of employing microorganisms, in bioremediation to eliminate pollutants and toxins from the environment, and microbiomes is important. The principal method that bacteria support bioremediation involves the pathways that have enzymes that lead to the breakdown, immobilization, detoxification, and transformation of the contaminants. This occurs because the microbes use them as their source of nutrition and energy (Abatenh et al. 2017).

One of the key approaches is single-strain bioremediation, which involves the usage of a single specific strain of either bacteria or fungi for the degradation of the targeted pollutant. It has certain limitations and to improve the efficacy and efficiency of this process, single-strain bioremediation has transformed into multiomics bioremediation, which uses many strains of different microbes in sophisticated analytical methods (Saha et al. 2021). By overcoming the shortcomings of single-strain bioremediation, this shift to a multiomics approach creates various new possibilities for sustained cleaning of the environment.

In this chapter, we see the bioremediation approach for the cleaning of the environment, its usage, and pathways involved in microbial bioremediation, and finally the multiomic approach. By the end of the chapter, you will have the practical knowledge of pollutants that can be cleaned by bioremediation, mechanisms of microbial bioremediation, structural and functional insights from multiomics, practical applications, case studies, and future challenges.

6.2 ENVIRONMENTAL POLLUTION AND ITS IMPACT

The significant by-product of activities of human is waste that leads to environmental pollution. Economic growth and ever-increasing living standards have drastically increased the amount and complexity of waste production, diversification of industry, and expansion of healthcare amenities. It has added to the significant amounts of hazardous waste that does not only have adverse effects on the environment but also on human health. There are various bases on which we can categorize the type of waste and environmental pollutants that need bioremediation. Following are a few of the important ones.

6.2.1 Plastics

Plastic as an environmental pollutant has been a serious problem with an estimated 9.2 billion tons of plastic generated between 1950 and 2017. It is a polymeric substance that has a high molecular weight and is frequently composed of long chains. Although there are many naturally occurring plastics as well, examples being rubber or silk, they do not negatively affect the environment as they do not contaminate it because they are biodegradable (Moore 2024). However, people today consume a wide range of plastic materials, which are mostly produced from petroleum after being spun, molded, cast, and then created to surpass the process of natural deterioration.

As these are nonbiodegradable, they pose many hazards and it is projected that the amount of plastic waste products will keep on increasing in the future as shown in Table 6.1. The macroplastics and microplastics in the environment are accumulated and pose a serious adverse effect on human health and animals as well. The toxic emissions are also released into the air during manufacturing and insufficient disposal steps lead to air pollution (Moore 2024). Humans suffer from exposure to chemicals such as Bisphenol A and phthalates found in plastics which can seep into the food or drinks leading to difficulties with development, reproduction, and an increased risk of malignancy. Moreover, the ingestion of microplastics has been reported to accumulate in the body and can be a cause of many cancers (2024).

TABLE 6.1
Current Statistics and Future Projections of Different Pollutants

Pollutant	Current Levels (2023)	Projection Levels (2030)	Projection Levels (2050)	Sources
Carbon dioxide (CO_2)	417 ppm (parts per million)	430–450 ppm	470–500 ppm	IPCC, NOAA
Sulfur dioxide (SO_2)	80 million metric tons/year	70–75 million metric tons/year	60–65 million metric tons/year	EPA, WHO
Plastic pollution	8 million metric tons/year entering oceans	10–12 million metric tons/year	15–20 million metric tons/year	UNEP, National Geographic
Particulate matter (PM2.5)	35 μg/m³ (micrograms per cubic meter) in urban area	30–32 μg/m³	25–28 μg/m³	WHO, EPA
Nitrogen oxides (NOx)	110 million metric tons/year	100–105 million metric tons/year	90–95 million metric tons/year	EPA, WHO
Methane (CH_4)	1890 ppb (parts per billion)	2100–2300 ppb	2400–2700 ppb	IPCC, EPA

IPCC, Intergovernmental Panel on Climate Change; UNEP, United Nations Environment Programme; EPA, Environmental Protection Agency; NOAA, National Oceanic and Atmospheric Administration.

6.2.2 Fossils and Oils

Coal, oil, and natural gas, which are used extensively for transportation, energy production, and other activities, are known as fossil fuels. During the extraction, refinement, and use of fossil fuel and oils, many particulate matter, carbon monoxide, and other pollutants are released. This is contributing to the environmental deterioration and climate change. The particulate matter is produced after the burning of fossil fuels that can penetrate deep into the lungs resulting in asthma, several lung diseases, and cardiovascular diseases. When fossil fuels burn, tiny particles (PM2.5 and PM10) are released into the atmosphere, which might continue to increase if there is no bioremediation of different products (Table 6.1). Moreover, sulfur dioxide (SO_2), nitrogen oxides, and carbon monoxide are also released, which aggravates pre-existing diseases like asthma and can lead to respiratory issues. Additionally, carbon monoxide also causes headaches, lightheadedness, and poisoning known as carbon monoxide poisoning (CDC 2024; WHO 2024).

6.2.3 Synthetic Pollutants and Human Wastes

Artificial waste can lead to environmental pollution. Air, soil, and water pollution is a result of synthetic chemicals. Plasticizers, industrial solvents, and insecticides are examples of substances that can be dangerous for human health and linger in the environment for very long periods.

Moreover, the human detritus and sewage that is improperly handled can pollute water sources and cause ailments that are transmitted through the water such as polio and hepatitis. In many developing countries that lack the necessary amenities for sanitary disposal, untreated and mishandled human waste is a serious problem. The current statistics are shown in Table 6.1.

6.3 MICROBIOMES AND MECHANISM OF MICROBIAL BIOREMEDIATION

The term microbiome includes microbes and their genetic material. It is a collection of all microorganisms including bacteria, archaea, fungi, protists, and viruses that inhabit a specific environment. A microbiome can be defined as a characteristic microbial community occupying a reasonably well-defined habitat that has distinct physio-chemical properties. It includes the microorganisms involved and their theater of activity.

Microbiomes are classified based on the characteristics of the microorganisms they contain or the particular environment they inhabit. Based on the oxygen requirement of microorganisms, they are classified into two types:

a. Aerobic Microbiomes

   Microorganisms that require oxygen for their growth and metabolism are classified as aerobic microbiomes. These microorganisms produce energy by using oxygen as the final electron acceptor in cellular respiration. Examples include soil biomes, human skin microbes, and some aquatic ones. Some aerobic bacteria such as *Pseudomonas* and *Bacillus* are common in well-aerated soils, where they contribute to the decomposition of organic matter and play an important role in nutrient cycling (Fierer and Jackson 2006). Some aerobic bacteria such as *Staphylococcus epidermidis* and *Corynebacterium* help in the protection against pathogenic microbes on human skin (Grice and Segre 2011). Aerobic bacteria generally habitat in oxygen-rich water bodies, such as lakes and rivers. Aerobic bacteria such as *Nitrosomonas* species, a nitrifying bacterium, play an important role in nitrogen cycling by converting ammonia to nitrite.

b. Anaerobic Microbiomes

Microorganisms that do not require oxygen for growth and metabolism or are even killed or inhibited in the presence of oxygen are classified under anaerobic microbiomes. Other electron acceptors such as nitrate, sulfate, or carbon dioxide help these microorganisms in respiration. These organisms may ferment organic substrates for the production of energy.

Anaerobes are further divided into several types depending on their specific characteristics. *Obligate Anaerobes*: The microorganisms that cannot survive in the presence of oxygen are known as obligate anaerobes. *Clostridium* species, found in deep soil layers and human intestines, are examples of these anaerobes. *Facultative Anaerobes*: The microorganisms that can grow in both the presence and absence of oxygen are known as facultative anaerobes. *Escherichia coli*, a bacterium, commonly found in the human gut, is capable of switching between aerobic and anaerobic metabolism. *Aerotolerant Anaerobes*: The microorganisms that do not require oxygen for growth but can tolerate oxygen are known as aerotolerant anaerobes. *Lactobacillus* species that are used in the production of yogurt are aerotolerant anaerobes.

Human gut, sediments, and landfills are a few examples of microbiomes having anaerobes. The human gastrointestinal tract hosts a diverse number of anaerobic microbiomes, including *Bacteroides* and *Firmicutes.* These microbiomes play essential roles in digestion and immune functioning (Grice and Segre 2011). *Desulfovibrio* species, an anaerobic bacteria, usually habitat aquatic sediments where they are involved in sulfur cycling by converting sulfate to sulfide (Muyzer and Stams 2008). The decomposition of organic matter in the absence of oxygen by methanogenic archaea in landfill environments helps in the production of methane.

6.3.1 Importance of Microbial Diversity in Bioremediation

Microbial diversity plays a vital role in bioremediation, the utilization of microorganisms in the detoxification of hazardous substances from the environment. Individual microorganisms are known to possess different metabolic activities, which enable them to metabolize various pollutants. A diverse microbial community enhances the prospects of identifying microorganisms with the ability to degrade certain contaminants of a specific locality (35). Furthermore, in a complex ecosystem with various microorganisms, the same functions can be provided by different species as shown in Figure 6.1. This functional overlap guarantees that, if one species is impacted by a pressure derivative to the environment, other species

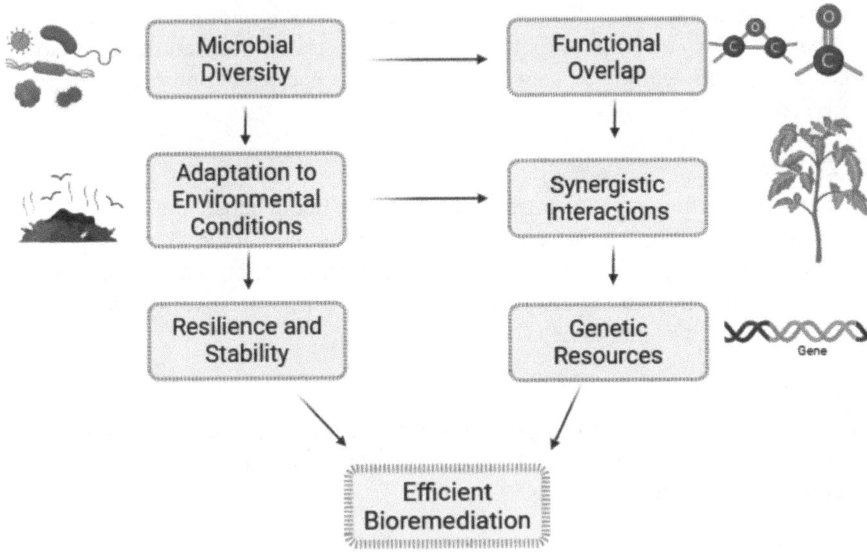

FIGURE 6.1 Importance of microbial diversity. Diversified microbes can contribute to resilience and stability in addition to synergistic interaction and functional overlap leading to effective bioremediation. (Created with www.biorender.com.)

can assume obligations of degrading chemicals, thus supporting the efficiencies of bioremediation (Paul et al. 2005).

Additionally, adaptation to environmental conditions microbial diversity allows for the presence of species adapted to various environmental conditions, such as pH, temperature, and nutrient availability as shown in Figure 6.1. This acclimatization assures that the process of bioremediation can be done efficiently at different locations having different conditions (Khanna et al. 2023). There are synergistic interactions among the various communities of microbes where they can work as a single unit against the contaminants, for example, one species that breaks down the pollutant into the intermediate can be further metabolized by another microbe leading to complete degradation in a better way (Khanna et al. 2023). The resilience to the disturbances of the environment is also increased when more microbes are present in a community. It also helps in maintaining the stability of the process of bioremediation over a long time (Paul et al. 2005). In addition to this, it also represents a vast genetic resource that can be used for exploring and exploiting the genetic potential for more effective strategies of bioremediation (Paul et al. 2005). Preserving and harnessing microbial diversity is crucial for developing sustainable and efficient bioremediation solutions.

6.3.2 Functional Roles of Different Microbes in Bioremediation

6.3.2.1 Bacteria

Bacteria can have a functional role in breaking down complex organic contaminants via various enzymes such as degrading hydrocarbons. Afterward, they use these products as a source of carbon and energy. It is also essential to reduce sulfate into sulfide and thus immobilize them so they cannot further contaminate the environment. They also turn atmospheric nitrogen into other forms that can be taken up by the trees and plants leading to vegetation growth, hence providing further help in bioremediation.

6.3.2.2 Fungi

Fungi, especially white-rot fungi, are well known for being capable of decomposing lignin and other complex organic compounds, allowing them to break pollutants like pesticides and dyes. This process is known as mycoremediation. Additionally, some fungi can absorb heavy metals from contaminated environments, effectively reducing their bioavailability and toxicity.

6.3.2.3 Archaea

These microorganisms can decompose organic substances in anaerobic conditions, releasing methane as a by-product. They are particularly effective in treating wastewater and organic waste in landfills.

6.3.2.4 Algae

Microalgae can absorb and accumulate heavy metals and other toxins in water, helping in the detoxification of aquatic environments.

6.3.3 Mechanisms of Microbial Bioremediation

The main mechanisms of bioremediation include breakdown (biodegradation), immobilization, and detoxification. Each mechanism uses diverse microbial processes and has unique applications in environmental remediation.

6.3.3.1 Breakdown

It is the most prevalent mechanism of bioremediation. It entails the enzymatic breakdown of organic pollutants into simpler, less toxic substances. Microorganisms such as bacteria, fungi, and actinomycetes play an essential role in this process. In aerobic biodegradation, microorganisms use oxygen to degrade organic pollutants. This mechanism is effective in the breakdown of petroleum hydrocarbons, pesticides, and other organic compounds, e.g., *Pseudomonas putida* degrades benzene, toluene, ethylbenzene, and xylene (BTEX) in the presence of oxygen (Margesin and Schinner 2001). Anaerobic biodegradation takes place in the absence of oxygen. This process involves microorganisms such as methanogens, sulfate-reducing bacteria, and nitrate-reducing bacteria. Anaerobic biodegradation is critical for the treatment of chlorinated solvents, explosives, and other persistent organic pollutants, e.g., the reductive dichlorination of tetrachloroethylene to ethene by *Dehalococcoides* species (Holliger et al. 1997).

6.3.3.2 Immobilization

The transformation of the pollutants to reduced mobility and less form for bioavailability is called Immobilization. In this process, the spread of pollutants is reduced due to the absorption of contaminants into the biofilms of the microbes or the microbial cells (see Figure 6.2). Biosorption is a passive process that uses the biomass of

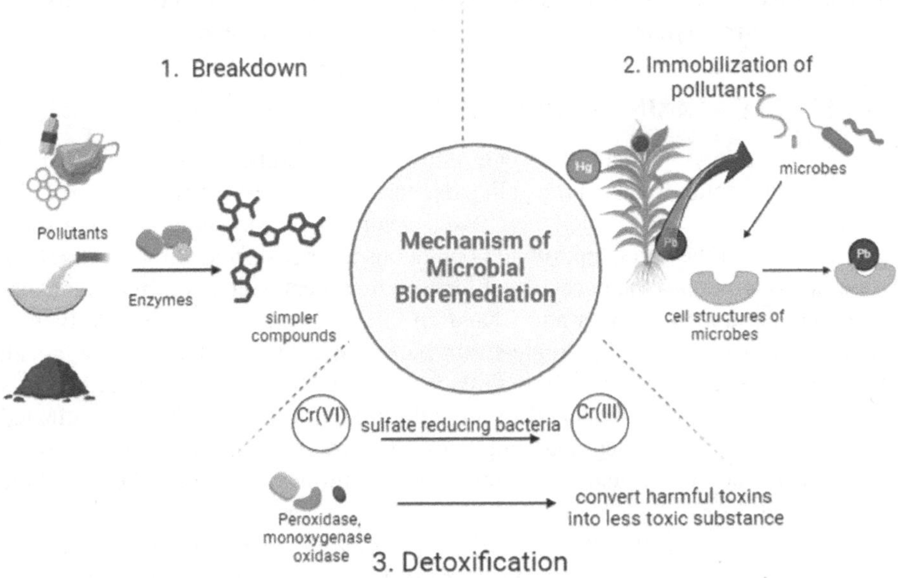

FIGURE 6.2 Mechanism of microbial bioremediation. First, the pollutants are broken down with the help of enzymes and then they are immobilized. Finally, they are detoxified. (Created with www.biorender.com.)

the microorganism for the absorption of contaminants. The whole process is useful in the treatment of contaminated water bodies including groundwater. *Aspergillus niger* has been used to remove metals like lead and cadmium from the contaminated bodies (Gadd 2016).

6.3.3.3 Detoxification

As the name shows, this process is turning the toxic contaminants into the less toxic ones. The microbial activities via enzymes such as including oxidation, reduction, and methylation are useful in this process. Microbial enzymes such as peroxidases, laccases, and monooxygenases are exploited to oxidize harmful organic compounds to make them less harmful. It is also used in the treatment of effluents from industries and soils contaminated by various organic compounds. White-rot fungi and their enzymes known as the laccase enzymes are used to oxidize and detoxify polycyclic aromatic hydrocarbons (Couto and Herrera 2006). In reductive transformation, contaminants are reduced to less toxic forms through enzymatic reduction. The halogenated compounds and heavy metals are detoxified by this process effectively. For example, sulfate-reducing bacteria assist in the reduction of hexavalent chromium (Cr(VI)) to the less toxic trivalent chromium (Cr(III)) as shown in Figure 6.2. The addition of methyl group to a contaminant reduces its toxicity or changes its environmental behavior, in a process called methylation. Microbial methylation is generally seen in the detoxification of heavy metals and metalloids. One such example of microbial methylation is the methylation of arsenic by certain microorganisms, transforming inorganic arsenic into less toxic methylated arsenic species (Bentley and Chasteen 2002).

6.4 SINGLE-STRAIN BIOREMEDIATION

In single-strain bioremediation, a single strain of the microbe is used for the detoxification of the contaminants in the polluted areas. This process utilizes the metabolic potential of the microbes, which they have developed over time to break down some pollutants. When the native microbial populations are not sufficient for this or when a specific pollutant has the necessity of focused treatment, single-strain bioremediation is the key approach (Das and Chandran 2011). By definition, a single type of microorganism is utilized by single-strain bioremediation and that microorganism has been identified or genetically modified to metabolize a specific pollutant. This microorganism is then introduced in an environment that has been contaminated and it can proliferate the degradation process. The compatibility of the strain in the environmental conditions and the nature of the contaminant show how successful the technique will be.

Some examples include that of *Pseudomonas putida* and *Deinococcus radiodurans*. One of the most well-known cases of a microorganism utilized in single-strain bioremediation is Pseudomonas putida. It is highly adaptable and can break down various organic pollutants, which include pesticides, hydrocarbons, and aromatic compounds. It effectively cleans up oil spills by degrading the highly complex and long chains on hydrocarbons into less toxic substances. Its effective capability to utilize carbon sources makes it of significant use in bioremediation. In addition to this, another example is *Deinococcus radiodurans* which is known for its significant resistance to different radiations and capability to stay alive in hard environments. It has been modified to break the harmful heavy metals and organic solvents. One study suggested that if this microorganism is engineered to express some specific enzymes it can degrade the mercury and hence have strong potential to be used in those environments that have been polluted with either or both chemical pollutants or radioactivity (Brim et al. 2000). A marine bacterium, *Alcanivorax borkumensis*, specializes in degrading compounds mostly hydrocarbons and is useful in detoxifying marine oil spills. Due to this, it has become a significant option to be used in oil-contaminated marine environments. It also degrades the wide range of alkanes found in crude oil (Yakimov et al. 1998).

6.4.1 Principles of Single-Strain Bioremediation

The main strategy of having single-strain bioremediation is to use a particular strain of microbes that have an innate ability to change hazardous and toxic substances to less toxic forms. For successful implementation of the bioremediation techniques, the following is important.

6.4.1.1 Isolation and Characterization

The first approach in this bioremediation process is finding the right microbial stain that can be used to degrade the particular contaminant. In this step, the samples from polluted areas are collected where microorganisms have undergone the evolutionary process to have a defense against different pollutants. The strains that require characteristics are first identified and then isolated using techniques including selective enrichment and screening. Selective enrichment entails establishing habitats that have the required nutrients to support the development of microorganisms that can lead to the degradation of the intended contaminant (Johnsen, Wick, and Harms 2005). For example, when the soil or water samples from the targeted land are cultured in a medium, the pollutant is the only available source of energy and carbon. Therefore, only the microorganisms that can utilize the pollutant will grow over time. Various screening techniques can be employed after enrichment, which differentiates which strains are the most effective and productive one. Other metabolic processes may be the part of this whole process after the evaluation of the growth rate and rate of contaminant degradation (Johnsen, Wick, and Harms 2005).

After the isolation, the strain is carefully scrutinized to learn its metabolic and breakdown pathways. By the examination of enzymatic activity, resilience to the stress related to the environment, and ideal circumstances for the growth of the strain, it is characterized. It is important to characterize the enzymes that will be part of the degradation process. Techniques such as enzymatic tests and genomic analysis are used for the clarification of the pathways of the enzymes that will break down the pollutant (Purohit, Dayananda, and Prashant 2019). By knowing the ideal parameters of the pH, growth, temperature, and availability of the nutrients, it can be assured that the strain is good to grow and use successfully. The prediction of strain performance in practical words involves the evaluation of several environmental stresses such as heavy metals and fluctuations of temperature (Purohit, Dayananda, and Prashant 2019).

6.4.1.2 Analysis of Metabolic Pathways

Comprehending the metabolic pathways of the strain facilitates the process of bioremediation and assures the efficient operation of the strain in the intended polluted environments. It can also help in our understanding of how the pollutant was broken down. Enzymes, such as oxygenases, reductases, and hydrolases, play an important role in the whole process. The characterization and identification of these enzymes can provide valuable information about any intermediate that might arise in the degradation route (Vidali 2001). Moreover, the scrutinization of the environmental tolerance and growth conditions ensures that it can remain active and continuously thrive at the contaminated site. Field trials and laboratory studies help us determine the condition for efficient activity of the strain.

6.4.2 Applications

The procedure of bioremediation starts by applying the isolated and characterized microbial strain at the site of pollution.

6.4.2.1 Oil Spill Remediation

It has been demonstrated that the strains of *Alcanivorax borkumensis* can break down hydrocarbons found in oil spills. Due to the efficient working of this strain to break down the complex molecules into simpler ones, the impact on the environment is diminished (Dhar et al. 2022).

6.4.2.2 Pesticide Degradation

Based on their ability to degrade and break down insecticides and herbicides, specific bacterial strains have been isolated. It can be exemplified by various *Bacillus* and *Pseudomonas* that have been shown to degrade pesticides like organophosphate pesticides, alleviating their toxic effects on the environment.

6.4.2.3 Removal of Heavy Metals

Certain strains of microorganisms can bioaccumulate and change the heavy metals in polluted streams and soils, thereby reducing their toxicity. The capability of *Bacillus subtilis* to combat and eradicate heavy metals like cadmium and lead from pollution has been scrutinized.

6.4.2.4 Wastewater Treatment

Single strains can also be used in wastewater bioremediation to target some pollutants, for example, strains of *Nitrobacter* and *Nitrosomonas* are used to change ammonia into less harmful substances in the nitrification process.

6.4.3 Mechanisms of Single-Strain Bioremediation

The mechanisms by which single-strain bioremediation occurs fall into different types. The main ones are as follows:

6.4.3.1 Biodegradation

In this process, microorganisms metabolize the pollutants into less toxic and simpler molecules. Mostly, this mechanism includes the enzymatic process and the enzymes target and degrade contaminants like pesticides, medications, and hydrocarbons. There are various enzymes involved such as esterases and lipases that break down ester bonds in pollutants. Moreover, the dioxygenases and monooxygenases are important in degrading hydrocarbons by commencing the breakdown of long hydrocarbon chains and peroxidases metabolize those pollutants that have phenolic groups. Furthermore, dehydrogenases facilitate the breakdown of pollutants by transferring electrons from one molecule to another and nitroreductases are useful in turning nitroaromatic compounds into less harmful compounds (Juhasz and Naidu 2000).

6.4.3.2 Sorption

The ability of some microbial strains to efficiently bind the impure substance to their structure at the cellular level is significant in limiting the mobility of heavy metals in the environment as shown in Figure 6.3. This procedure is known as sorption and can speed up the natural mechanism of breakdown to aid in preventing the spread of pollutants.

6.4.3.3 Enzymatic Detoxification

The production of enzymes is also induced by specific pollutants and this response is called enzymatic induction, which allows them to acclimatize to their environment to

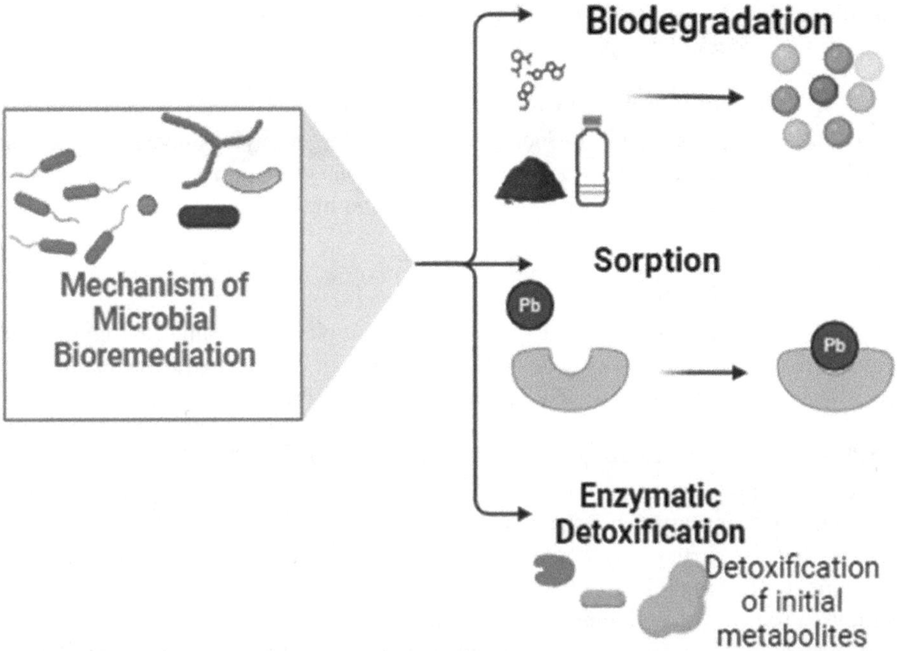

FIGURE 6.3 The main mechanism of microbial bioremediation. It includes biodegradation as its first step followed by sorption with the help of lead (Pb) and enzymatic detoxification of initial metabolites. (Created with www.biorender.com.)

effectively detoxify the pollutants. These enzymes are highly specific, which means that they attack specific functional groups or bonds within the contaminant. This ensures efficient reactions by these enzymes. The initial metabolites are still toxic so they are accessible to further reactions by additional enzymes to degrade them until they are completely nontoxic. The final products are usually water, carbon dioxide, and inorganic salts, which are benign to the environment (Wackett and Hershberger 2001).

6.4.4 Limitations

Although it has great potential in bioremediation, it also faces many limitations that include the following.

6.4.4.1 Contaminant Bioavailability

One of many limitations is that microbial attack has low availability for the contaminants. The pollutants may exist in a form that is not easily accessible to the microorganisms or else may be bound tightly to the soil particles, which limits its ability to break down substances in a more effective way (Jabbar et al. 2022).

6.4.4.2 Strain Selection

How successful the bioremediation with a single strain depends greatly on the selection of the microbial strain and not all strains are efficient enough to degrade the specific pollutants. Isolation of effective strains can be a complicated and time-consuming process (Thompson et al. 2005).

6.4.4.3 Environmental Conditions

Single strains cannot perform efficiently under different environmental conditions, which lead to lower degradation rates as not all strains work optimally under various environmental factors such as pH, temperature, and nutrient availability (Bala et al. 2022).

6.4.4.4 Lack of Synergistic Effects

Limitations to the efficiency of degradation of a single strain cannot be ignored as they cannot work synergistically like mixed consortia. Multiple strains can work on a broader range and enhance the degradation process (Bala et al. 2022).

6.4.4.5 Resistance to Toxicity and Less Functional Diversity

Some pollutants can be toxic to specific microbial strains, inhibiting their growth and metabolic activity. This toxicity can reduce the effectiveness of single-strain bioremediation, particularly in environments with high contaminant concentrations (Sharma 2020). Single strains may lack the functional diversity required to degrade complex mixtures of pollutants. This limitation can hinder the bioremediation process, especially in environments contaminated with multiple types of pollutants (Sharma 2020).

6.5 INTRODUCTION TO MULTIOMICS APPROACH

Multiomics is an integrated approach that provides a comprehensive understanding of biological systems by combining data from multiple "omics" fields, such as genomics, transcriptomics, proteomics, and metabolomics. Multiomics aims to discover the complex relationships and interactions that contribute to normal development, cellular response, and disease, by studying the entire complement of biomolecules within an organism. Genomics, transcriptomics, proteomics, and metabolomics are part of multiomics. Genetic variations and their potential roles in disease susceptibility are focused on by genomics, which is a complete set of DNA in an organism. What environmental factors, such as pollutants, interact with genetic predispositions to influence health outcomes can be uncovered by genomic data. Transcriptomics looks into the complete set of RNA transcripts produced by the genome, providing an understanding of gene expression levels due to the influence of environmental stimuli. Changes in transcriptomic profiles can show how organisms adapt to environmental stressors. The entire set of proteins expressed in a cell or organism can be explored by proteomics, which reveals protein interaction and function within biological pathways affected by environmental factors. Alterations in protein expression in response to environmental exposures can be identified by proteomic analyses. The focus of metabolomics is on the study of metabolites, the small molecules involved in metabolic processes. Metabolomic profiling can help to get insights into the biochemical changes that occur as a result of environmental factors, helping in identifying the potential biomarkers of exposure or disease.

TABLE 6.2
Description of Different Fields of Omics That Include Proteomics, Genomics, Transcriptomics, and Metabolomics Along with Its Description, Usage, Main Points, and Various References

Field	Description	Main points	References
Proteomics	All of the proteins that are expressed in a cell of an organism are studied and show how these proteins interact and undergo their functions in biological pathways	Identification of the alterations in the expression of proteins that are influenced by environmental factors	Madeira and Costa (2021)
Genomics	Study of the complete set of DNA with a special focus on genetic variation	How predispositions of genetic materials interact with environmental factors, such as pollutants	Olden and Wilson (2000)
Transcriptomics	Examination of the complete set of RNA transcripts that are produced by the genome. Provides valuable information into gene expression levels	Gives indications on how organisms acclimatize to environmental stresses of the environment through changes in transcriptomic profiles	Ghoussaini et al. (2021)
Metabolomics	The study that focuses on the metabolites formed during metabolic processes	Give information about biochemical changes in response to environmental factors	Kim, Hollinger, and Radke (2022)

6.5.1 Multiomics in Understanding Microbial Functions

By integrating different types of data, multiomics allows for a comprehensive characterization of microbial communities. Genomics has its main focus on the genetic composition of microbes while, on the other hand, transcriptomics focuses on the expression of genes that can be influenced by various environmental changes and factors. Additional information about the function of proteins including their metabolic activities is provided by proteomics and metabolomics (O'Donnell, Ross, and Stanton 2020).

Using the multiomics approach, we can study and explore the influence of different environmental factors on the functions of the microbes; for instance, we can reveal the effect of pollutants and the availability of nutrients on microbial metabolism, and community structure by using metabolomics data with both the genomic and transcriptomic analyses as shown in Figure 6.4. In an agricultural setting, it is significant where the sustainability and resilience of the crop are achieved by interactions between plants and microbes present in the soil. Researchers can make strategies to exploit this interaction by understating the beneficial functions of the microbes to mitigate the harmful effects (Jain et al. 2024). Researchers can unveil the mechanisms through which environmental stressors influence these complex relationships and their implications for host health by integrating omics data from both the host and microbial community (Jain et al. 2024).

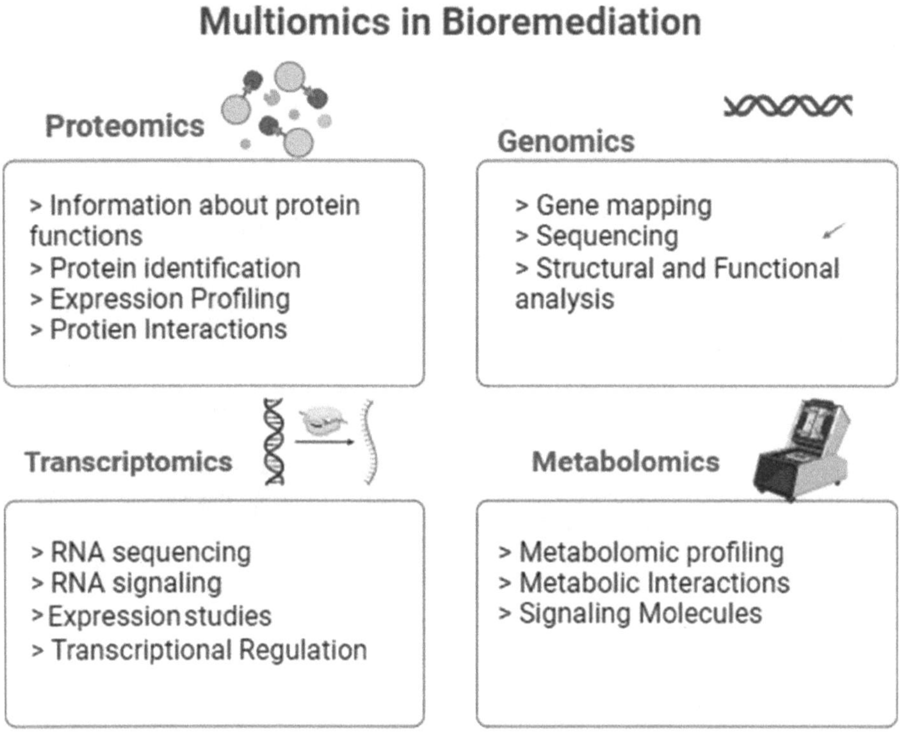

FIGURE 6.4 Multiomics in bioremediation. Proteomics gives us information about protein functions, genomics informs us about the structural and functional analysis of genes, transcriptomics shows the gene expression analysis, and metabolomics shows metabolic interactions. (Created with www.biorender.com.)

To manage and remediate issues of the environment, strategies can be made with the help of multiomics studies, for example, scrutinizing the nutrient cycle can be used to make targeted strategies to enhance the beneficial activities of microbes in ecosystems (Koh and Hwang 2019).

6.5.2 Comparison of Multiomics with Single-Omics Approaches

Multiomics approaches have proved itself important in biological research. This methodology differs sharply from single-omic approaches, which explore one type of biological data at a time. Below is a comparison of multiomics and single-omic approaches, highlighting their respective benefits and limitations.

6.5.3 Multiomics in Bioremediation

To integrate and analyze data from different biological layers, various techniques and tools are employed by multiomics studies. High-throughput sequencing technologies,

TABLE 6.3
Comparison of Single-Omics and Multiomics Approach

Single-Omics Approach	Multiomic Approach
It mainly focuses on the data from one type of biological strain at a time. Moreover, it also hinders the capability to focus on the complexity of the responses of microbes to pollutants.	It integrates the data of genomics, transcriptomics, proteomics, and metabolomics to provide us with a complete understanding of microbial functions and their bioremediation potential.
Can identify biomarkers within a specific layer but may not capture the full spectrum of microbial responses to environmental pollutants.	Facilitates the discovery of novel microbial biomarkers associated with pollutant degradation or bioremediation efficiency by analyzing multiple layers simultaneously.
May identify specific genes or proteins associated with pollutant degradation but cannot elucidate dynamic interactions between microbial communities and their environment.	Allows exploration of interactions among different microbial species and their collective bioremediation capabilities. Provides insights into how microbes adapt and respond to environmental pollutants.
It is valuable to identify genes or proteins that are involved in the degradation of the pollutants, but it limits the understanding of the multifaceted nature of microbial bioremediation processes.	Significant applications in understanding the mechanisms of microbial bioremediation, identifying key microbial players, and optimizing bioremediation strategies by integrating data across various biological layers.
It does not give us detailed information on the microbial bioremediation process but it is very economical, and in addition to this it also ignores the important response of the microbes toward the pollutants.	Faces challenges in data integration and the need for advanced computational tools to analyze large datasets. Inherent variability and noise in multiomics data can complicate analyses and interpretations

The single-omic approach focuses on the data from only one biological strain while the integration of different sources of genomics, transcriptomics, proteomics, and metabolomics is the multiomics approach.

such as next-generation sequencing, are significant in multiomics studies. Rapid sequencing of DNA and RNA can be done by using these techniques, allowing researchers to understand entire genomes and transcriptomes at a large scale. Whole-genome sequencing, RNA sequencing for gene expression analysis, and recognizing genetic variations associated with diseases or environmental responses are some applications. Moreover, mass spectrometry (MS) is a powerful analytical technique and is widely used in proteomics and metabolomics. It allows the recognition and quantification of proteins and metabolites in complex biological samples. With the aid of MS in proteomics, the identification and quantification of proteins can be performed and the field metabolomics can be used to make metabolite profiles as shown in Figure 6.5. It can make us better understand the changes in the metabolic products influenced by environmental factors. Microarray technology is one of the most important technologies that has been used to measure gene expression levels. To understand how gene expression patterns change, microarrays are useful to check

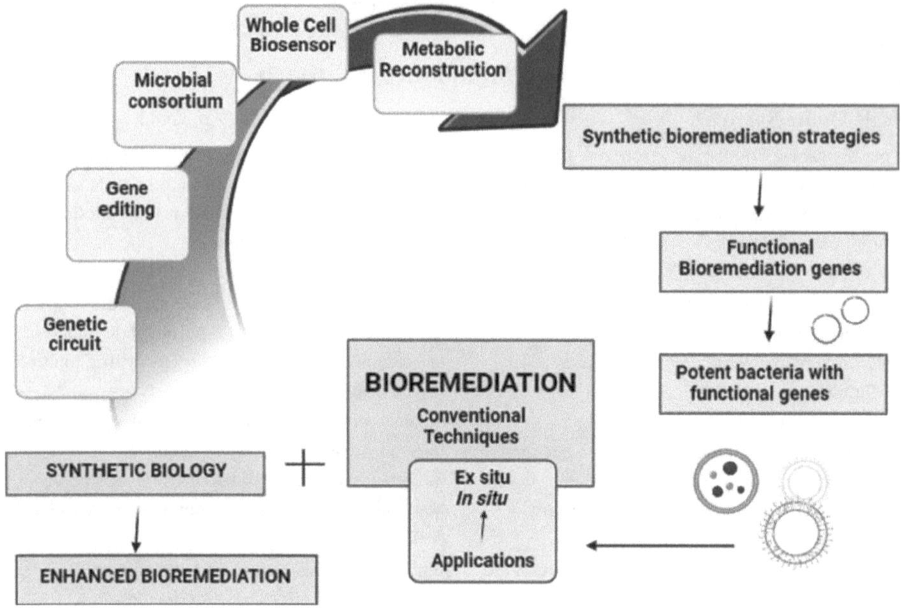

FIGURE 6.5 Multiomics in bioremediation. The multiomic approach is used in synthetic bioremediation strategies to create bacteria with functional genes, which are then used in synthetic biology to enhance bioremediation. (Created with www.biorender.com.)

various stimuli that can trigger these changes, and a snapshot of transcriptomes can also be given by this technique in different conditions.

The importance of bioinformatics in multiomics studies cannot be ignored because it helps in the integration, analysis, and visualization of the given data with the help of various software. Bioinformatics tools like OmicsAnalyst, OmicsNet 2.0, and Hiplot have been used to allow us to understand correlation, visualization of the network, and statistical analysis of multiomics sets of data. Interpretation of complicated data can be done with the help of these tools. Spatial transcriptomics is an emerging technique that allows researchers to explore the expression of genes within samples of tissues by combining transcriptomics with this spatial information providing context to the molecular data in understanding the relation of microbes to mitigate the environmental pollutants.

In addition to this, advanced computational approaches that are rapidly used to analyze and integrate multiomics data can be used for the identification of patterns and relationships across different omics layers. Lastly, multiomics studies often involve cohort-based studies; hence, it can capture environmental exposure data in addition to omics data. It is important to understand the relationships between the pollutants, microbial communities, different strains, and molecular profiles leading us to draw meaningful conclusions about the influence of the environment on the biological systems. By understanding all of this, we can make better bioremediation methods using these multiomics techniques.

6.5.4 Key Benefits in Bioremediation and Dynamics of Microbial Communities

The key benefit of multiomics is to identify key microbial strains involved in bioremediation and their interactions that can act synergistically; therefore, the process of bioremediation is enhanced (Jaiswal, Singh, and Shukla 2019). Due to multiomics, the researchers can isolate the specific markers that have good microbial activity for pollutant degradation. This is crucial as it makes us monitor the bioremediation strategies and metabolism of the contaminants (Yunusa and Umar 2021).

Traditional single-omic approaches often focus on one aspect of microbial function, such as gene expression or protein activity, which may not provide a complete picture of the bioremediation process. Multiomics overcomes these limitations by integrating various data types, enabling researchers to capture the complexity of microbial interactions and responses to environmental stressors (Koh and Hwang 2019).

Microbial communities in polluted environments are dynamic and can change over time in response to fluctuations in pollutant levels, nutrient availability, and other environmental factors. These temporal dynamics can affect community stability and resilience. For instance, studies have shown that microbial communities can exhibit rapid shifts in composition following pollution events, emphasizing the need for continuous monitoring to capture these changes (Widder et al. 2016). Various environmental factors, such as pH, temperature, and moisture, can influence microbial community dynamics. In polluted environments, these factors may interact with contaminants to shape community structure and function. For example, the presence of heavy metals can be selected for metal-resistant microbial taxa, altering the overall community composition and functional potential. The ability of microbial communities to adapt to polluted environments is crucial for their survival and functionality. Resilience refers to the community's capacity to recover from disturbances, while adaptation involves genetic and phenotypic changes that allow microbes to thrive in adverse conditions. Understanding these processes can inform strategies for enhancing bioremediation efforts and restoring ecosystem health (Konopka, Lindemann, and Fredrickson 2015).

6.6 PRACTICAL APPLICATIONS OF MULTIOMICS AND FUTURE DIRECTIONS

Bioremediation based on the multiomics approach exploits different technologies including genomics, transcriptomics, proteomics, and metabolomics. Due to this, the effectiveness of the degradation process is enhanced. Some of the practical applications of this approach are as follows:

6.6.1 Enhanced Oil Spill Remediation

By analyzing the communities of the microbes involved in the degradation of hydrocarbons, the whole process can be improved. Important microbial taxa can be identified by the researchers by using genomics and metabolomic data that can be useful in environments where there are incidents of oil spills and metabolic pathways can also be understood. The whole information can be of great importance for increasing the efficiency of the process of cleaning the oil spills at the water bodies (Jaiswal, Singh, and Shukla 2019).

6.6.2 Heavy Metal Detoxification

Multiomics can help in metal detoxification by using metagenomic analysis to reveal the genetic potential of the communities of microorganisms. It can be done by biosorption or biotransformation at the sites contaminated with heavy metals. Using the proteomics data, researchers can also evaluate specific proteins for their expressions for both metal resistance and detoxification (Yunusa and Umar 2021).

6.6.3 Wastewater Treatment Optimization

The analysis of the microbial structure and function through transcriptomics and metagenomics can help researchers identify the metabolic pathways that utilize the microbes in the process of treatment. The operational conditions and removal of pollutants in wastewater treatment facilities can be optimized using this information (Atlas and Philp 2005).

6.6.4 Pharmaceutical and Industrial Waste Remediation

Pharmaceutical and industrial waste usually containing complex-chained organic compounds can use multiomics techniques to combat the challenges of treating the waste. The use of data from both genomics and proteomics helps in the identification of specific strains having specific enzymes capable of degrading the pollutants. This approach can be utilized to target specific strains and develop bioremediation strategies efficiently that address the unique challenges of different types of waste (Jaiswal, Singh, and Shukla 2019).

6.6.5 Soil Remediation in Agricultural Settings

To remediate the soils that are contaminated with pesticides and fertilizer multiomics can be of great help. In agricultural communities, the dynamics of the microbial community can be understood leading to the degradation of harmful chemicals and restoration of soil health. This is important in promoting good practices in agriculture and reducing the impact of pollutants in the environment in farming activities (Atlas and Philp 2005).

6.6.6 Microbial Fuel Cells for Bioremediation

This approach can also be used simultaneously to treat wastewater and generate electricity. Researchers can understand microbial communities to optimize the conditions for the production of bioenergy and the degradation of pollutants. Hence, the dual-purpose application of multiomics can provide solutions for the remediation of the environment (Atlas and Philp 2005).

6.6.7 Challenges in Multiomics

Although multiomics is of great importance in the field of bioremediation, it still faces many limitations and challenges. One of the main issues stems from the data quality being used. Various sources are used for the datasets that may lead to inconsistency in the quality and scaling. The key factor that requires us to have quality control and assurance for laboratories is the variability in omics data (Oldoni et al. 2022). Moreover, the integration of multiomics data that has heterogeneity requires transformation for each dataset. This must be handled properly as the omics data is comprised of heterogeneous data from various modalities that vary data distributions. The datasets for Omics usually have some missing value that hampers downstream analyses. One such example where the variable largely outnumbers the sample is the high-dimension low sample size problem that leads to overfitting the datasets by the machine learning algorithms.

In addition to this, another significant challenge is that algorithms that can preserve the global and local features are not viable as the number of integrated datasets grows. To provide annotations of a biological system, various applications of artificial intelligence (AI) are necessary because in this way the researchers can choose the correct picture that is not biased by the user himself (Odenkirk, Reif, and Baker 2021).

6.6.8 Future Directions

Significant challenges need to be overcome before multiomics techniques can be successfully implemented for bioremediation in developing countries. These include a lack of infrastructure, high costs, and limited expertise. Strategies are needed to make these technologies more accessible and affordable in resource-limited settings. Moreover, combining data from metagenomics, transcriptomics, proteomics, and metabolomics with the help of AI will provide a more comprehensive understanding of microbial bioremediation processes as shown in Figure 7. However, integrating these diverse datasets remains challenging. Improved computational tools and machine learning approaches are needed to extract meaningful insights from integrated multiomics data (Yunusa and Umar 2021).

Multiomics can elucidate how microbial community composition and interactions change in response to pollutants and bioremediation conditions. This knowledge is crucial for maintaining effective consortia. Future research should focus on using multiomics to model and predict microbial community responses for optimizing bioremediation. By identifying key genes, proteins, and metabolic pathways involved in pollutant degradation, multiomics guides the engineering of microbes with enhanced bioremediation capabilities. Synthetic biology approaches can be used to construct designer microbes with optimized bioremediation functions. Multiomics data can identify specific molecular biomarkers that indicate bioremediation efficiency in real time. Developing rapid, cost-effective biomonitoring tools based on these biomarkers will enable dynamic optimization of bioremediation processes (Yunusa and Umar 2021).

6.7 CONCLUSION

The integration of microbiomes in pollutant remediation has evolved significantly, transitioning from the use of single microbial strains to sophisticated multiomics approaches. This evolution reflects a deeper understanding of microbial ecology and the complexities involved in bioremediation processes. Microbial bioremediation leverages the natural metabolic capabilities of microorganisms to degrade or transform hazardous pollutants into less harmful substances. Initially, the focus was primarily on isolating specific strains known for their degradation abilities. However, this approach often fell short in addressing the multifaceted nature of environmental pollutants and the dynamic interactions within microbial communities. The limitations of single-strain applications highlighted the need for a more comprehensive understanding of microbial ecosystems.

With advancements in omics technologies—such as genomics, transcriptomics, proteomics, and metabolomics—researchers can now analyze microbial communities at unprecedented levels of detail. It facilitates researchers to identify the important metabolic pathways and how they interact among the diversified populations of microbes. It allows the optimization of bioremediation strategies. Genomics and transcriptomics provide us with insights into the genetic potential of the microorganism and the changes in gene expression in response to the pollutant, respectively. Additionally, metabolomics and proteomics provide information about the metabolic products and functional abilities of the microbes.

The future of bioremediations aims to engineer and mold microbial strains that have greater efficiency and capability to degrade pollutants. With the understanding of the genetics and metabolic frameworks, the designing of microbes to tailor specific contaminants will become very useful. This strategy will broaden the pollutant range that needs to be addressed by bioremediation.

Moreover, multiomics integration can help to monitor real-time and manage the bioremediation process. The identification of biomarkers to decompose pollutants allows us to effectively and promptly assess bioremediation.

In conclusion, shifting to multiomics approaches from single-strain ones represents a shift in the management of environmental pollutants, as it increases our understanding of interactions and functions of microbes with innovative solutions. With the continuation of the research, the potential bioremediation using multiomics to address complication challenges of the environment will expand that will offer pathways to restore the ecosystem.

REFERENCES

Abatenh, Endeshaw, Birhanu Gizaw, Zerihun Tsegaye, and Misganaw Wassie. 2017. "The role of microorganisms in bioremediation: A review." *Open Journal of Environmental Biology* 2 (1):038–046.

Atlas, Ronald M, and Jim Philp. 2005. *Bioremediation. Applied Microbial Solutions for Real-World Environmental Cleanup*. John Wiley & Sons.

Bala, Saroj, Diksha Garg, Banjagere Veerabhadrappa Thirumalesh, Minaxi Sharma, Kandi Sridhar, Baskaran Stephen Inbaraj, and Manikant Tripathi. 2022. "Recent strategies for bioremediation of emerging pollutants: A review for a green and sustainable environment." *Toxics* 10 (8):484.

Bentley, Ronald, and Thomas G Chasteen. 2002. "Microbial methylation of metalloids: Arsenic, antimony, and bismuth." *Microbiology and Molecular Biology Reviews* 66 (2):250–271.

Biswas, Karabi. 2015. "Biological agents of bioremediation: A concise review." *Frontiers in Environmental Microbiology* 1 (3):39–43.

Brim, Hassan, Sara C McFarlan, James K Fredrickson, Kenneth W Minton, Min Zhai, Lawrence P Wackett, and Michael J Daly. 2000. "Engineering Deinococcus radiodurans for metal remediation in radioactive mixed waste environments." *Nature Biotechnology* 18 (1):85–90.

Centres of Disease Control and Prevention (CDC). 2024. "Carbon monoxide poisoning basics." Accessed April 17, 2024. https://www.cdc.gov/carbon-monoxide/about/?CDC_AAref_Val=https://www.cdc.gov/co/faqs.htm.

Chibuike, Grace U, and Smart C Obiora. 2014. "Heavy metal polluted soils: effect on plants and bioremediation methods." *Applied and Environmental Soil Science* 2014 (1):752708.

Coo, Ayotunde E, Faith E Oviasogie, and Beckley Ikhajiagbe. 2022. "Microbial effects of manure from poultry droppings and pig dung in diesel-contaminated soil." *BioTechnologia* 103 (4):385.

Couto, Susana Rodríguez, and José Luis Toca Herrera. 2006. "Industrial and biotechnological applications of laccases: A review." *Biotechnology Advances* 24 (5):500–513.

Das, Nilanjana, and Preethy Chandran. 2011. "Microbial degradation of petroleum hydrocarbon contaminants: An overview." *Biotechnology Research International* 2011 (1):941810.

Dhar, Kartik, Logeshwaran Panneerselvan, Kadiyala Venkateswarlu, and Mallavarapu Megharaj. 2022. "Efficient bioremediation of PAHs-contaminated soils by a methylotrophic enrichment culture." *Biodegradation* 33 (6):575–591.

Fierer, Noah, and Robert B Jackson. 2006. "The diversity and biogeography of soil bacterial communities." *Proceedings of the National Academy of Sciences* 103 (3):626–631.

Gadd, GM. 2016. "5: Fungi and industrial pollutants." In: Druzhinina, I, and Kubicek, C (eds) *Environmental and Microbial Relationships*:99–125. Springer.

Ghoussaini, Maya, Edward Mountjoy, Miguel Carmona, Gareth Peat, Ellen M Schmidt, Andrew Hercules, Luca Fumis, Alfredo Miranda, Denise Carvalho-Silva, and Annalisa Buniello. 2021. "Open Targets Genetics: Systematic identification of trait-associated genes using large-scale genetics and functional genomics." *Nucleic Acids Research* 49 (D1):D1311–D1320.

Grice, Elizabeth A, and Julia A Segre. 2011. "The skin microbiome." *Nature Reviews Microbiology* 9 (4):244–253.

Holliger, Christof, Sarra Gaspard, Guy Glod, Cornelis Heijman, Wolfram Schumacher, René P Schwarzenbach, and Francisco Vazquez. 1997. "Contaminated environments in the subsurface and bioremediation: Organic contaminants." *FEMS Microbiology Reviews* 20 (3–4):517–523.

Hunter, Lori M. 2000. "Population and environment: A complex relationship." RAND Corporation. https://www.rand.org/pubs/research_briefs/RB5045.html.

Jabbar, Noor M, Saja M Alardhi, Alaa K Mohammed, Issam K Salih, and Talib M Albayati. 2022. "Challenges in the implementation of bioremediation processes in petroleum-contaminated soils: A review." *Environmental Nanotechnology, Monitoring & Management* 18:100694.

Jain, Archana, Surendra Sarsaiya, Ranjan Singh, Qihai Gong, Qin Wu, and Jingshan Shi. 2024. "Omics approaches in understanding the benefits of plant-microbe interactions." *Frontiers in Microbiology* 15:1391059.

Jaiswal, Shweta, Dileep Kumar Singh, and Pratyoosh Shukla. 2019. "Gene editing and systems biology tools for pesticide bioremediation: A review." *Frontiers in Microbiology* 10:87.

Johnsen, Anders R, Lukas Y Wick, and Hauke Harms. 2005. "Principles of microbial PAH-degradation in soil." *Environmental Pollution* 133 (1):71–84.

Juhasz, Albert L, and Ravendra Naidu. 2000. "Bioremediation of high molecular weight polycyclic aromatic hydrocarbons: A review of the microbial degradation of benzo [a] pyrene." *International Biodeterioration & Biodegradation* 45 (1–2):57–86.

Khanna, Namrata, Tanushri Chatterji, Suruchi Singh, and Poonam Chaturvedi. 2023. "Microbial metabolism in bioremediation: A review." *AIP Conference Proceedings* 2800:020218.

Kim, Stephanie, Hillary Hollinger, and Elizabeth G Radke. 2022. "'Omics in environmental epidemiological studies of chemical exposures: A systematic evidence map." *Environment International* 164:107243.

Koh, Eun Jung, and Seung Yong Hwang. 2019. "Multi-omics approaches for understanding environmental exposure and human health." *Molecular & Cellular Toxicology* 15:1–7.

Konopka, Allan, Stephen Lindemann, and Jim Fredrickson. 2015. "Dynamics in microbial communities: Unraveling mechanisms to identify principles." *The ISME Journal* 9 (7):1488–1495.

Madeira, Carolina, and Pedro M Costa. 2021. "Proteomics in systems toxicology." *Advances in Protein Chemistry and Structural Biology* 127:55–91.

Margesin, R, and F Schinner. 2001. "Biodegradation and bioremediation of hydrocarbons in extreme environments." *Applied Microbiology and Biotechnology* 56:650–663.

Moore, Charles. 2024. "Platic pollution." Accessed July 18, 2024. https://www.britannica.com/science/plastic-pollution.

Muyzer, Gerard, and Alfons JM Stams. 2008. "The ecology and biotechnology of sulphate-reducing bacteria." *Nature Reviews Microbiology* 6 (6):441–454.

O'Donnell, Shane Thomas, R Paul Ross, and Catherine Stanton. 2020. "The progress of multi-omics technologies: Determining function in lactic acid bacteria using a systems level approach." *Frontiers in Microbiology* 10:3084.

Odenkirk, Melanie T, David M Reif, and Erin S Baker. 2021. "Multiomic big data analysis challenges: increasing confidence in the interpretation of artificial intelligence assessments." *Analytical Chemistry* 93 (22):7763–7773.

Olden, Kenneth, and Samuel Wilson. 2000. "Environmental health and genomics: Visions and implications." *Nature Reviews Genetics* 1 (2):149–153.

Oldoni, Emanuela, Gary Saunders, Florence Bietrix, Maria Laura Garcia Bermejo, Anna Niehues, Peter AC 't Hoen, Jessica Nordlund, Marian Hajduch, Andreas Scherer, and Katja Kivinen. 2022. "Tackling the translational challenges of multi-omics research in the realm of European personalised medicine: A workshop report." *Frontiers in Molecular Biosciences* 9:974799.

Panayotou, Theodore. 2000. "Population and environment." *CID Working Paper Series*.

Paul, Debarati, Gunjan Pandey, Janmejay Pandey, and Rakesh K Jain. 2005. "Accessing microbial diversity for bioremediation and environmental restoration." *Trends in Biotechnology* 23 (3):135–142.

Purohit, Hemant J, S Dayananda, and Phale Prashant. 2019. "Microbe-assisted biodegradation, bioremediation and metabolic engineering." *Proceedings of the Indian National Science Academy* 85:963–983.

Saha, Lala, Jaya Tiwari, Kuldeep Bauddh, and Ying Ma. 2021. "Recent developments in microbe–plant-based bioremediation for tackling heavy metal-polluted soils." *Frontiers in Microbiology* 12:731723.

Sharma, Indu. 2020. "Bioremediation techniques for polluted environment: Concept, advantages, limitations, and prospects." In: Mario Alfonso Murillo-Tovar, Hugo Saldarriaga-Noreña and Agnieszka Saeid (eds) *Trace Metals in the Environment: New Approaches and Recent Advances*. IntechOpen.

Soriano, Maria C Hernandez. 2014. *Environmental Risk Assessment of Soil Contamination*: BoD–Books on Demand.

Thompson, Ian P, Christopher J Van Der Gast, Lena Ciric, and Andrew C Singer. 2005. "Bioaugmentation for bioremediation: The challenge of strain selection." *Environmental Microbiology* 7 (7):909–915.

Vidali, Marisa. 2001. "Bioremediation. An overview." *Pure and Applied Chemistry* 73 (7):1163–1172.

Wackett, Lawrence Philip, and C Douglas Hershberger. 2001. *Biocatalysis and Biodegradation: Microbial Transformation of Organic Compounds*. ASM Press.

WHO. 25 June 2024 "Health consequences of air pollution on populations." https://www.who.int/news/item/25-06-2024-what-are-health-consequences-of-air-pollution-on-populations.

Widder, Stefanie, Rosalind J Allen, Thomas Pfeiffer, Thomas P Curtis, Carsten Wiuf, William T Sloan, Otto X Cordero, Sam P Brown, Babak Momeni, and Wenying Shou. 2016. "Challenges in microbial ecology: building predictive understanding of community function and dynamics." *The ISME Journal* 10 (11):2557–2568.

Yakimov, Michail M, Peter N Golyshin, Siegmund Lang, Edward RB Moore, Wolf-Rainer Abraham, Heinrich Lünsdorf, and Kenneth N Timmis. 1998. "Alcanivorax borkumensis gen. nov., sp. nov., a new, hydrocarbon-degrading and surfactant-producing marine bacterium." *International Journal of Systematic and Evolutionary Microbiology* 48 (2):339–348.

Yunusa, YR, and ZD Umar. 2021. "Effective microbial bioremediation via the multi-omics approach: An overview of trends, problems and prospects." *UMYU Journal of Microbiology Research (UJMR)* 6 (1):127–145.

7 Biomass Production

A Sustainable Approach for Waste Recovery and Its Conversion into New Products

Huma Tahreem, Maham, Tallha Sultan, and Nazim Hussain

7.1 INTRODUCTION

In today's world, environmental challenges are increasingly pressing, with waste management standing out as a significant concern. Conventional methods of waste disposal include landfilling and incineration, which, apart from polluting the environment, deplete natural resources, hence worsening the state of the environment globally [1]. Such challenges call for green waste management strategies that go beyond providing solutions for waste disposal and embrace the recycling and reuse of waste into useful commodities.

Thus, biomass production stands out as a possible solution in this case. This is a process of converting wastes such as agricultural and animal waste, municipal waste, and industrial waste into useful products such as fuel, plastic, and chemicals. This not only helps in dealing with waste disposal problems but also adds to cutting emissions of greenhouse gases and the use of fossil fuels [2]. Based on biomass production, we can close the loop by reusing waste as a valuable form of energy.

7.1.1 Background and Context

Sources of waste disposal have undergone tremendous change in the past few decades. First, waste was just dumped in open dumps, which became a severe threat to people's health and the environment. Over time, sanitary landfills and incinerators have become more widespread; they have certain advantages, but they keep having negative impacts on the environment. Some methods, like incineration, for example, let dangerous impurities escape into the air if they have not been treated correctly [1]. Such cultural practices that, in one way or another, were relevant at one time have been found to lack efficient and sustainable solutions to waste management.

Technical advancements in biomass production systems have been defined as a transition from conventional practices. The first forms of biomass use can be categorized

DOI: 10.1201/9781003536277-7

purely under energy use, where biomass is burned to produce energy. However, such methods were not very efficient and were some of the causes of air pollution. Newer methods like anaerobic digestion, pyrolysis, and gasification have been enhanced to be more efficient and eco-friendly [3]. These technologies allow the transformation of agricultural residues, food waste, and forestry residues into useful energy and products.

By observing the differences between conventional and contemporary techniques of waste disposal, it is possible to note the advantages of biomass production. Disposal techniques like landfilling and incineration particularly dispose of valuable resources and increase pollution of the environment. On the other hand, contemporary biomass production technologies enable easy collection and recycling of organic wastes, reducing harm to the surroundings as well as improving resource utilization and sustainability [2]. Additionally, the biorefinery idea represents the interaction of multiple conversion processes of biomass to obtain the maximum yield with minimal waste, which is known as sustainable development goals [3]. Figure 7.1 shows the detailed process of waste management.

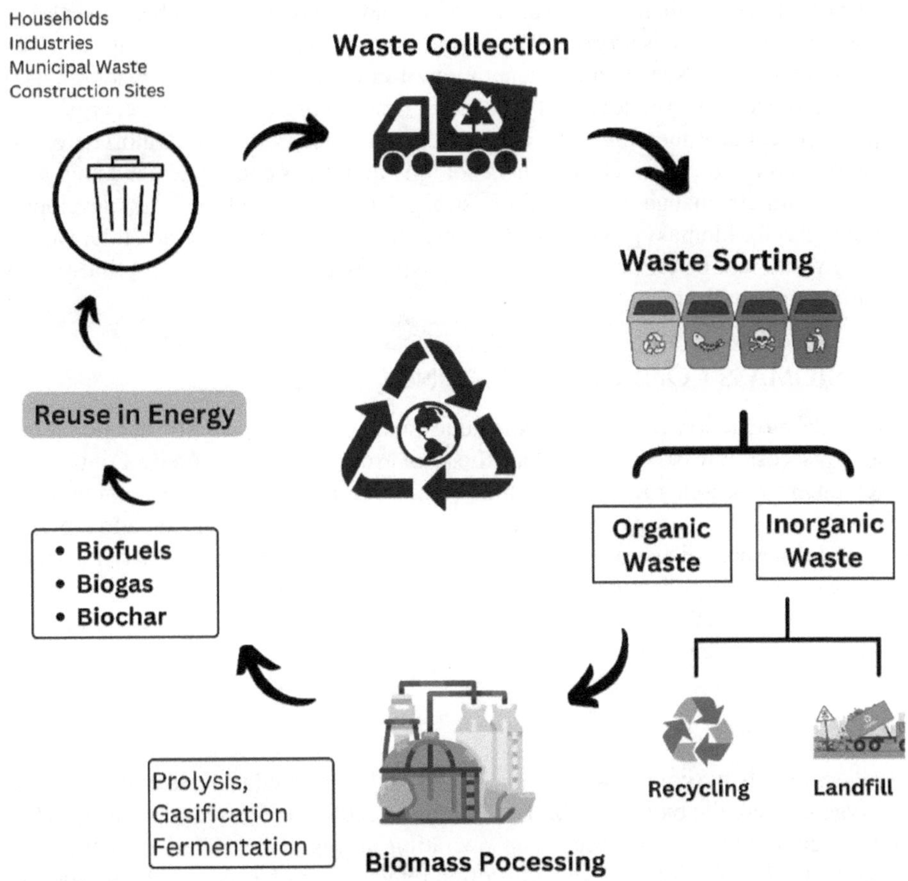

FIGURE 7.1 Illustration of the waste management process from waste generation to biomass conversion.

7.1.2 Applications in Water and Soil Remediation

Biomass conversion technologies are most appropriate in water and soil treatment since they harness waste material to restore the environment with measures of sustainability. An excellent example is phytoremediation, which uses plants, which are later used to produce biofuels by processes like pyrolysis and gasification after being harvested for their accumulated heavy metals within the soil. This is a two-fold approach of purging pollutants while at the same time also promoting a circular economy by transforming waste into energy vectors [4,5]. More so, engineered biomass materials add to the efficiency of soil remediation in the following ways; engineered biomass has good chemical resistance and sorption, which enable it to immobilize the contaminants and increase soil fertility [6]. Also, the process of pyrolysis of agricultural waste and manure biochar is crucial for the immobilization of heavy metals, minimizing soil toxicities, and enhancing crop yields [7]. Nonetheless, some issues are harder to solve while integrating and improving these technologies with a view of reducing impacts on the environment as well as increasing efficiency. Research should continue to be carried out to enhance their sustainability and eventual use on a bigger scale.

Biomass production is an effective method for waste recovery in response to modern environmental issues and the generation of new opportunities for resource use. Utilizing waste biomass contributes to carbon neutrality and cleaner energy production, potentially meeting 25% of global energy needs [8]. The valorization of waste biomass into chemicals and fuels not only addresses energy demands but also mitigates climate change impacts. From straight waste disposal to highly developed technologies for biomass production, we can therefore form part of a healthier world, turning waste into better products in society with lesser and lesser susceptibilities to polluting our environment.

7.2 BIOMASS CONVERSION TECHNOLOGIES

Bioenergy production from biomass residues and waste holds significant potential, though its costs are still higher than those of traditional fossil fuels. Continuous development research focuses on refining conversion processes more comprehensively to increase efficiency and reduce the costs of processes like gasification, pyrolysis, and anaerobic digestion. All these advancements seek to ensure that bioenergy as a source of energy becomes more lucrative by ensuring that as much energy as possible is recovered from different types of waste without compromising on issues of waste management or greenhouse gases [9].

7.2.1 Pyrolysis

Pyrolysis entails the conversion of biomass through a thermal process in the absence of oxygen, where the biomass is broken into liquid, solid, and gaseous products. This process commonly takes place when operating at slightly elevated temperatures, roughly 500°C. The liquid produced is often referred to as bio-oil, while the solid is char, and the gas includes various chemical compounds [10]. A simplified process of pyrolysis is illustrated in Figure 7.2.

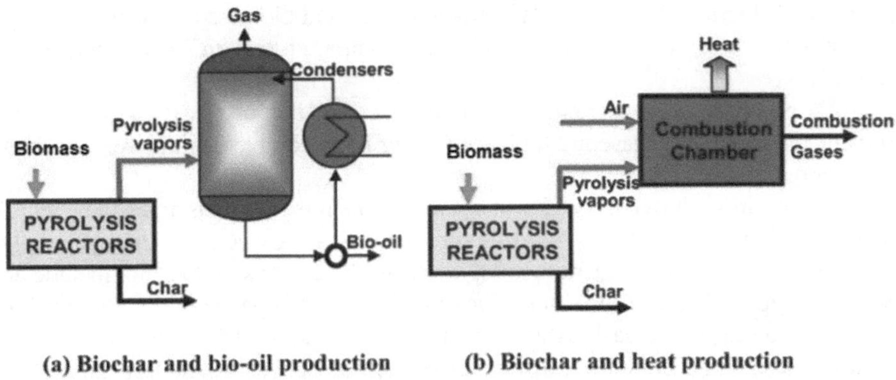

FIGURE 7.2 Simplified flow diagram for a typical pyrolysis unit. (a) Biochar and bio-oil production. (b) Biochar and heat production [11].

Types of Pyrolysis

- **Slow pyrolysis:** Slow pyrolysis aims mainly at making a solid product, namely char, with lesser proportions of liquids and gases. This process is carried out at lower temperatures, and the residence time is much longer, in hours to days
- **Fast pyrolysis:** This is because fast pyrolysis is carried out at moderate temperatures with a very minimal residence time that does not exceed 2 seconds, and it results in bio-oil. This type of pyrolysis is of huge commercial value since the yield of liquid hydrocarbon, which can be utilized in several processes, is high.
- **Flash pyrolysis:** Both fast and flash pyrolysis involve moderate heating rates and short residence times, but the rates of heating in flash pyrolysis are much higher than in fast pyrolysis. This method also seeks to optimize liquid production, as in the case of fast pyrolysis [12].

Applications and benefits: Several uses of the products resulting from pyrolysis have been indicated, as follows: Bio-oil can be used in the production of biofuels, energy carriers, and the chemists' industries. The slow pyrolysis produces charcoal, which can be utilized in agriculture as a soil amendment. Gases that are produced can be utilized for heating and energy purposes [10].

7.2.2 Gasification

Gasification thermochemically converts organic materials into a gaseous form, mainly comprising CO, H_2, CO_2, and CH_4. This process is carried out at high temperatures above 1200°C, which means limited oxidation. Biomass gasification can also convert CO_2 into useful gases, achieving up to 93% CO_2 conversion efficiency [13].

Gasification occurs at elevated temperatures (around 700°C–1000°C) and includes an oxidation step, converting biomass into syngas, which can be used for energy or chemical production [14,15]. Direct gasification of solid biomass has been shown to achieve higher energy and exergy efficiencies compared to indirect methods using bio-oil [14].

While pyrolysis is advantageous for liquid fuel production, gasification is more efficient for generating gaseous fuels, reflecting their complementary roles in biomass energy conversion.

Comparison with pyrolysis: Pyrolysis mainly results in liquids and solids, while gasification is more inclined toward the production of a gaseous product generally called syngas. Pyrolysis is carried out at lower temperatures and does not include an oxidation step, as is the case with gasification. Pyrolysis is particularly employed in producing liquids, while gasification is more focused on producing gaseous fuel [12]

Applications and benefits: Gasification offers a flexible energy source that can be used for heat, power, and the production of synthetic fuels. The generated syngas can be combusted to produce electricity or transformed into liquid biofuel. Gasification also permits the use of numerous supplies, such as waste material, making it very effective in the production of energy because of its flexibility [2].

7.2.3 Fermentation

Fermentation is a process of breakdown of complex organic chemicals with the help of bacteria, yeast, or fungi, and it mainly takes place in carbohydrates, where they are converted into simpler products like alcohol or acids. During this process, carbon dioxide is also produced. This process can be done aerobically and nonaerobically, though, for biomass conversion, the latter is commonly used, namely through fermentation [16].

Types of Fermentation

Anaerobic fermentation: During anaerobic fermentation processes, the sugars and the starches are converted into ethanol and carbonation, in the absence of oxygen. This process is widely used for the fermentation of feeds that contain sugarcane, corn, and virtually any feed material containing carbohydrates. The most preferred microorganism used in this process is the yeast cell of *Saccharomyces cerevisiae* because the rate of transformation is high when the sugars are being fermented to ethanol [17].

Aerobic fermentation: Aerobic fermentation is a bit less used in biofuel contexts and requires the presence of oxygen, though it is used, for instance, in the manufacture of other types of biochemicals rather than biofuels. For instance, in the aerobic mode, certain bacteria and fungi creating the organic acids, vitamins, and antibiotics are used [17].

Applications and benefits: The first product of anaerobic fermentation is ethanol, which is renewable and can be mixed with gasoline or consumed by flexible fuel vehicles. The ethanol produced through the process of anaerobic fermentation is one of the sustainable sources of energy in replacement with fossil fuels, which collectively results in reduced emissions of carbon [18]. It thus minimizes the pollution of the environment by the use

TABLE 7.1
Overview of Biomass Conversion Technologies: Key Processes, Outputs, and Applications

Process Name	Temperature Range	Main Products	Advantages	Applications
Pyrolysis	350°C–600°C	Bio-oil, Char, Syngas	Efficient conversion of biomass to liquid fuels; produces valuable by-products like biochar; reduces waste volume	Biofuels, soil amendment (biochar), energy generation, and production of chemicals
Gasification	800°C–1200°C	Syngas (CO, H_2, CH_4)	High energy efficiency; flexible feedstock use; syngas can be used for electricity generation or as a chemical feedstock	Power generation, production of synthetic fuels, hydrogen, and various chemicals
Fermentation	Ambient (30°C–40°C)	Bioethanol, biogas	Converts a wide range of biomass types; low energy input; reduces greenhouse gas emissions	Production of bioethanol as fuel, biogas for energy, and various biochemical products such as organic acids

of agricultural co-products and food waste [19]. This process also enables the fight against greenhouse emissions, which are caused by the limitation of fossil-based energy sources and also has a great contribution to the agricultural value chain to balance the surplus production. Furthermore, the fermentation products, for example, distillers' grains, can be utilized for feeding animals, hence expanding the total value chain [17] (Table 7.1)

7.3 BIOFUELS

Biofuels are the products obtained from biomass conversion processes, which make them better than fossil fuels. Based on the structure, the major types of biofuels are biodiesel fuel, bioethanol fuel, and biogas, and each of them has its own method of preparation and usage.

- **Biodiesel:** Biodiesel is prepared with the process of transesterification of fats and oils, including vegetable oil and animal fat, which reacts with alcohol, resulting in fatty acid methyl esters and glycerol as the by-product [20]. These can be applied to diesel engines without or with little modification, thereby providing renewable energy that has low greenhouse gas emissions and dependence on petroleum-based diesel.
- **Bioethanol:** Bioethanol is received mainly as the product of the fermentation of sugars derived from crops such as corn and sugarcane [21]. New methods allow the conversion of cellulosic ethanol from biomass, which is

not used for the production of food, agricultural residues, or wood [17,22]. Bioethanol is usually added to fuel in its ability to reduce emissions, particularly carbon monoxide that is produced by gasoline engines.

- **Biogas:** Biogas is produced via the breakdown of organic materials through anaerobic digestion of wastes from agricultural and municipal activities [20]. This process generates a mixture of methane and carbon dioxide, which is useful for the generation of electricity and heat and as vehicle fuel. Biogas production has the added advantages of generating renewable energy and decomposing waste in a way that decreases greenhouse gas emissions to the environment [3].

Environmental and economic benefits: Biofuels are mostly produced and consumed to achieve reductions in the emission of greenhouse gases, thus controlling climate change. Furthermore, applications of biofuels create energy security because of the decreased reliance on imported hydrocarbons and also by supporting bio-energy industries as well as biomass feedstock production, which leads to employment opportunities [2].

7.4 BIOFERTILIZERS

Biofertilizers are organic fertilizers, which use microbial nutrients to improve the nutrient value of the soil. Some of these inoculants include bacteria, fungi, and algae that facilitate the breakdown of nutrients such as nitrogen, phosphorus, and potassium to be available to plants. Table 7.2 describes various types of biofertilizers and their impact on soil fertility and crop yields.

- **Types and composition:** Some popular biofertilizer organisms include nitrogen-fixing bacteria (e.g., *Rhizobium*), phosphate-solubilizing bacteria (e.g., *Bacillus* species), and mycorrhizal fungi that form symbiotics with plant roots enhancing the uptake of nutrients as well as water [10].

TABLE 7.2
Comparison of Biofuels, Energy Output, Production, and Applications

Type of Biofuel	Energy Output (MJ/L)	Production Method	Typical Applications
Bioethanol	21.1–26.8	Fermentation of sugars from crops like corn and sugarcane, or lignocellulosic biomass	Added to gasoline as a blend for vehicles, industrial solvents
Biodiesel	35.8	Transesterification of vegetable oils or animal fats with alcohol	Used in diesel engines for transportation, power generation
Biogas	20–25 (methane-rich)	Anaerobic digestion of organic waste, agricultural residues, or municipal waste	Used for electricity and heat production, as vehicle fuel

TABLE 7.3
Types of Biofertilizers and Their Impact on Soil Fertility and Crop Yields

Type of Biofertilizer	Microorganism	Function	Impact on Soil Fertility	Impact on Crop Yields
Nitrogen-fixing bacteria	Rhizobium, Azotobacter, Anabaena	Fix atmospheric nitrogen into ammonia	Enriches soil nitrogen, reduces need for chemical fertilizers	Increased plant growth and higher yields
Phosphate-solubilizing bacteria	*Bacillus, Pseudomonas, Aspergillus*	Convert insoluble phosphorus to soluble forms	Improves phosphorus availability, enhances root development	Better nutrient uptake and stronger roots
Mycorrhizal fungi	*Glomus* spp., *Rhizophagus* spp.	Enhance water and nutrient absorption via symbiosis	Improves soil structure and moisture retention	Boosted yields, especially under stress

- **Production and application methods:** Biofertilizers involve the use of the culture of those beneficial microorganisms, which are grown in a nutrient media and then incorporated into the carrier-based products that can be used directly on seeds, soil, or plants. The given products are biodegradable, and there is no requirement for chemical fertilizers; moreover, the possibilities of pollution of the soil and water resources are minimized [17].
- **Impact on soil health and crop yield:** Biofertilizers perform such functions as modification of the structure of soil enhancement of its organic component and support of sustainability in agriculture. This is because biofertilizers improve nutrient availability that enhances crop yields accompanied by the replenishment of soil fertility in the long run, a factor of importance in feeding the ever-growing population and land-use efficiency [1] (Table 7.3).

7.5 VALUE-ADDED PRODUCTS

Biomass, therefore, goes beyond the energy and fertilizers to embrace products, which can be manufactured to make other valuable products in the industrial and commercial sectors that are circulatory. Figure 7.3 shows the whole process of biomass conversion to value-added products.

1. **Bioplastics:** Bioplastics are made from renewable sources of raw materials including corn starch, sugarcane, or cellulose. These plastics are biodegradable and hence can be used as an environmentally friendly substitute for the common petroleum-based plastics, thus lowering plastic pollution and environmental degradation from plastics and other related materials [23].
2. **Biochar:** Biochar is the charcoal that is obtained through the burning of biomass in the absence of oxygen. This is applied as a fertilizer to improve the soil's fertility and maintenance of moisture in the soil; it also acts as a carbon sink in the fight against climate change. Biochar is also used in

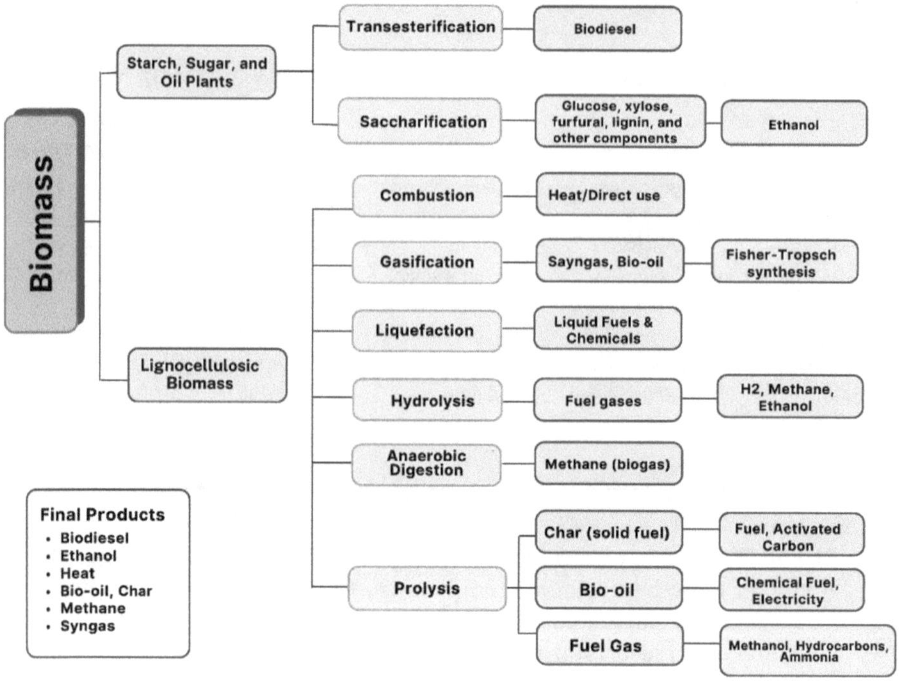

FIGURE 7.3 Biomass transformation to value-added products.

water treatment since it can selectively remove specific substances, which can enhance the water quality [17].

3. **Industrial chemicals and other by-products:** The processes of converting biomass yields crucial chemical compounds, some of which include organic acids, solvents, polymers, and others used in the production line of various industries including branded pharmaceuticals and biodegradable packing industries among others [24]. Suntec catalytic processes additionally improve the efficient conversion of biomass into valuable products such as ethanol and diglycerol that are used in industries as fuel substitutes [8]. However, continuous improvement of its results is possible only by carrying out further work, which raises certain economic and technological problems within this area.

7.6 ENVIRONMENTAL AND SOCIOECONOMIC IMPACTS

7.6.1 Environmental Benefits

The benefits of bioenergy derived from biomass conversion technologies are well known and are due for the most part to the environmental impacts of concerns such as pollution, resource conservation, and restoration of ecosystems.

1. **Reduction in pollution and greenhouse gas emissions:** Biomass conversion technologies, more specifically biofuel generation, are known to emit

fewer pollutants and greenhouse gases than traditional fossil fuels. Some of the biochar management practices include the following; for instance, the use of biochar for carbon sequestration in the soil lowers the concentration of carbon dioxide in the atmosphere [23].

2. **Enhancement of soil quality:** Biochar and biofertilizers supplement the soils and have a beneficial impact on the soil structure, nutrient status, and microbial health, hence better soils are produced. It also fosters eco-friendly farming methods and little reliance on chemicals that harm the soils [10].
3. **Conservation of natural resources:** Making use of environmentally clean technologies, waste materials and renewable biomass can be used to extend depleted useful resources such as fossil fuels and minerals. Compared to the traditional linear economy model, which relates to the take-make-dispose model, the principles of circular economy improve resource use efficiency and management. This strategy is strongly important because technology can reduce the negative effects on the environment while proposal options for generating energy and entailing socioeconomic advantages can be achieved [25].

7.6.2 Socioeconomic Advantages

Besides environmental advantages, biomass conversion technologies provide society with key socioeconomic values to support growth, investment, and access to energy.

1. **Job creation and new income sources:** The biomass industry plays a crucial role in employment and income generation; several companies are engaged in the biomass industry, especially in the regions that have abundant biomass resources. This creates employment opportunities in the supply chain starting from the producers of biomass, manufacturers, and distributors contributing to the country's economy in the agricultural, forestry, and waste management sectors [26]. These opportunities are good for rural communities as they turn on some economic issues. Sustainable biomass adds to employment opportunities in the creation process without significant exploitation of resources in environment conservation and energy policy [27]. But to do this, legal barriers and misconceptions have to be resolved.
2. **Energy security and independence:** Thus, the production of energy from renewable biomass resources locally can help in the decentralization of energy resources and decrease the use of fossil fuels, hence improving energy security and the ability to cope with energy market shocks [3].
3. **Contribution to the circular economy:** Biomass conversion ensures that wastes become valuable products and not a nuisance; thus, it supports the circular economy by avoiding using raw materials from natural resources. It complies with concepts of sustainable development goals and supports sustainable economic development in the long term [23].

7.7 CHALLENGES AND LIMITATIONS IN BIOMASS CONVERSION TECHNOLOGIES

Even though there are huge benefits in biomass conversion for waste disposal and sustainable energy supply, there are various technical, economic, and societal issues that limit its application. It makes sense that these are, in essence, the barriers to reaching a goal and understanding them is important to construct better solutions that are both more affordable and more effective.

7.7.1 Technical Barriers in Biomass Conversion

Several technical challenges confront biomass conversion processes including pyrolysis, gasification, and fermentation. One potential challenge is the *heterogeneous nature of biomass*, which can consist of various components and compounds based on the type of feedstock used [28]. The feedstocks such as agricultural residues, wood, and municipal solid waste have different carbon content, moisture, and ash characteristics; therefore, the prevailing conversion systems cannot be standardized to accommodate the diverse feedstocks [29]. For instance, the presence of a considerable amount of moisture in biomass can reduce the combustion rates and add energy expenditure for removing moisture. In the same manner, pyrolysis and gasification of municipal waste contain impurities that contribute to the corrosion of reactors such as chlorine and alkali metals [30].

Another challenge is *process optimization*. Pyrolysis and gasification involve setting the temperature and time of stay for the biomass feed to achieve high product yields such as bio-oil and syngas. Nonetheless, attaining this control on a commercial scale is quite challenging and this results in suboptimal yields [17]. Further, the resistance that arises from the structure of the cellulose, hemicellulose, and lignin in the biomass of lignocellulosic does not permit quick enzymatic or microbial hydrolysis during fermentation processes. Such recalcitrance enhances the use of pretreatment technologies that include chemical or steam explosion methods, which increase the overall cost and complexity of production [3].

7.7.2 Economic Challenges and Investment Requirements

Biomass conversion technologies tend to be capital-intensive as they are likely to require a huge initial investment in the establishment of plants and machinery. For example, biomass conversion biorefineries that convert biomass into fuels and chemicals involve high capital costs on such equipment as reactors, separators, and storage systems [31]. The large-scale facility requirements for achieving economies of scale once again push up the costs; it is almost impossible to imagine that these technologies could be implemented in small enterprises or developing nations. In addition, the cost incurred to collect, transport, and store biomass is high occasioned by the fact that most feedstocks particularly the low-density ones such as agri-residues require a complex logistics chain [30].

7.7.3 Market Competition

Technological rivalry is another major economic constraint for the uptake of bio-based products since fossil fuels are cheaper and easily accessible in the market and thus tend to capture the market more than biofuels. In the absence of adequate financial support from the government, for instance, through subsidies or the introduction of differential carbon pricing policies, biomass-derived fuels, and chemicals cannot compete with those derived from petroleum [32]. Also, the fluctuation of biomass, caused by seasonal variations and crops, brings instability to the manufacturing and pricing strategy and prevents investors from making long-term commitments to biomass technologies [33].

7.7.4 Regulatory and Policy Constraints

Another issue associated with biomass conversion's application concerns regulatory factors in implementing the technologies. Most countries have policies that regulate the utilization of biomass that are overlapping and/or inconsistent, hence complicated policies that pose challenges to investors and entrepreneurs [34]. For instance, some regions do not have clearly stated guidelines for biofuels and bio-related chemicals and this hampers their market development and penetration into the existing energy markets. In addition, the government set nationwide environmental standards concerning lighting emissions and waste disposal during biomass processing, which forces them to raise their operating costs and hence reduces the possibility of these technologies turning into popular large-scale commodities [35].

Further, the cross-subsidization of fossil fuels and the lack of effectiveness of carbon prices through taxes in some areas make bio-based products economically unsustainable. Since adequate policy environments that reward the utilization of renewable energy and penalize carbon use are lacking, biomass conversion technologies cannot be considered to compete with entrenched fossil fuel industries [23].

7.7.5 Social Acceptance and Awareness Issues

The work undertaken also shows that public acceptance and awareness are important in the success of biomass conversion technologies. While on the positive impact on the environment, members of the community still lack information or knowledge concerning the gains of bio-based products and renewable energy from biomass [34]. Social resistance may also be reported due to mere misunderstandings regarding changes in land use or competition with food production as observed in the first-generation biofuels. Concerns by the public regarding the effects on food security, land use change, and the environment will help to slow down the use of biomass technologies [1].

Also, there are few awareness campaigns and educational programs on the part of biomass in the quest to attain sustainability objectives, thereby putting little demand on bio-based products. These challenges can only be met by engendering cooperation between the governments, industries, and the education sector in disseminating the values of biomass conversion to the public and also by formulating a policy that would affect change [2].

7.8 CASE STUDIES AND REAL-WORLD EXAMPLES

To illustrate the potential and challenges of biomass conversion, several successful and failed projects from around the world can be examined. These case studies highlight the lessons learned and the opportunities for improvement in the field.

7.8.1 Success Stories of Biomass Conversion Projects Worldwide

A successful example of biomass-based power generation at a regional level is the well-implemented project Bioenergy Village Jühnde in Germany. The next idea is to apply the technology of anaerobic digestion of agricultural waste in the given village to produce biogas for electricity and heat generation and provide job opportunities for farmers in the region. The local farmers' and mayors' engagement also increases social acceptance, democratic control, and social innovation, which positively affect system sustainability [36,37]. Jühnde's CHP systems that are dependent on biomass bring down operating expenses as well as emissions since the company uses cheap, renewable regional biomass materials. There is a reduction in energy expenses and enhanced capacity to recover, which proves that localized biomass production is environment-friendly. However, improving the scale of the system remains difficult mainly in terms of resource supply and fluctuation in biomass supply throughout the year [38].

Another example is the *Novozymes biorefinery in Denmark*, which shows a successful development of innovative enzymatic solutions to make efficient conversion of lignocellulosic biomass to bioethanol. Through the use of cellulases and laccases for the degradation of the difficult-to-convert complex lignocellulosic structures to fermentable sugars, Novozymes has a way to diversify and amplify the saccharification process, hence increasing the efficiency and sustainability of biofuel production [39]. Immobilization of enzymes enhances the stability of the enzyme and also the lifespan making the operation costs cheap [40]. This biorefinery does not only refine bioethanol but also has other by-products that make the process profitable, thus making it more economical. Moreover, by using waste lignocellulosic biomass, Novozymes aligns itself with the circular economy model in tackling both energy requirements and waste issues [41]. Such improvements have placed Denmark as the second-generation biofuel producer that provided a non-food-based solution that could help to address global carbon emissions too.

7.9 CHALLENGES IN BIOMASS CONVERSION: INSIGHTS FROM FAILED PROJECTS

Nonetheless, there are various success stories; however, some projects have not achieved their goals because of economic and technical issues. For instance, the Kior project in the United States that intended to use fast pyrolysis to produce bio-oil from biomass is a typical example of the problems associated with bioenergy mega projects. Major technical challenges faced by the project were challenges in the scalability of the pyrolysis process, which put a drag on the bio-oil yield and catalyst deactivation which in turn down the economic efficiency of the project [42]. These technical challenges together with the problem of a discontinuous source of funding precipitated the shutdown of the project [29]. This failure escalates the need for adequate financial analyses, technical economic viability, and stakeholder involvement to enhance the success rates of future biomass conversion undertakings.

7.9.1 Analysis of Failed Projects and Opportunities for Improvement

Inadequate technical know-how, poor infrastructure facilities, and dismal commitment toward stakeholders are some of the key reasons for the failures in biomass-specific ventures. For example, intimately the Kior project did not involve the local communities and this encouraged them to resist hence lacking the social license to operate [42]. Future projects may only aim at the new vision of the community and they have to make sure that all the key stakeholders of that particular project are in parallel to that vision.

Furthermore, a large number of conversion technologies for biomass remain economically unfeasible without subsidies. Subsidies for these technologies that often raise funding from the public have been useful in the short term but there is much that can be done in the long term through investing in research that would shave off costs and increase the efficiency of these technologies to make them competitive. Other studies of failed projects also reflect the necessity of pilot-scale implementation before going to the full commercial scale [34].

7.10 FUTURE PROSPECTS AND TRENDS IN BIOMASS CONVERSION TECHNOLOGIES

The growth and development of biomass conversion technologies are therefore expected to be steady as innovations and trends set to prevail as key indicators to improved systems. Biomass conversion can meet the increasing global demand for renewable energy and bio-based products to ensure that some of the impending environmental and socioeconomic issues are solved.

7.10.1 Emerging Technologies in Biomass Conversion

Some of the technical and economic limitations have been highlighted regarding the traditional biomass conversion processes, and through the recent developments in technology, advanced solutions are being obtained. The conversion technologies that are receiving attention due to operational efficiency and optimized resource and produce utilization are the hybrid conversion technologies, which involve both thermochemical and biochemical conversion. For instance, integrated gasification and fermentation systems facilitate the transition of syngas into biofuels or chemicals by adopting the efficiency of gasification's energy density and the versatility of fermentation in choosing the kind of final product [17]. This hybrid approach optimizes the biomass conversion process by the use of more modes, thus minimizing wastage. Another promising trend that is currently yielding great potential is the improved catalytic conversion technologies where complex catalysts can be employed to boost the efficacy of certain reactions that occur in biomass conversion. Catalysts of this type can reduce the amount of energy needed for the processes as well as the reaction time of pyrolysis, gasification, and hydrothermal liquefaction [31]. For instance, catalytic pyrolysis can yield bio-oil of a better quality and with lesser impurities, serving as a better replacement for petroleum-based fuels. In the same way, catalysts are applied in hydrothermal carbonization for the transformation of wet biomass to biochar and biochemical through low-energy input [28].

7.10.2 Innovations in Bioenergy and Bio-Based Products

Recent advances in techniques of converting biomass have a broad impact on industries other than the energy sector including bioplastic where plastic waste and fossil fuel dependence are minimized. Novel advanced biopolymers with C–O and C–N bonds have better performance than petroleum-based polymers, which leads to increased usage of such polymers [43]. Furthermore, the utilization of biowaste as a resource in a circular economy improves resource efficiency in the production of sustainable bioproducts via microbial processes; even though high production costs and entrenched market trends persist as key drivers to biowaste adaption [44].

Another important field of innovation is the production of second-generation biofuels. Second-generation biofuels include biofuels from second-generation feedstocks such as algae, municipal solid waste, and lignocellulosic residues; third-generation biofuels which include algae biofuels are gradually becoming cheaper due to the application of advanced enzymes, integration of processes, and improvements in feedstock supply chains [34]. Cellulosic ethanol together with algal biofuels has been seen to have a strong possibility of reducing greenhouse emissions and acting as a substitute for fossil fuels [29].

Biomass utilization for the synthesis of high-added-value chemicals is yet another trend that is emerging at a gradual pace. These chemicals recognized as platform chemicals are used as basic components for many industrial applications such as plastics, pharmaceuticals, and cosmetics, among others [23]. For instance, succinic acid, levulinic acid, and furfural which are platform chemicals today are being produced from biomass at a commercial level and are renewable sources, thereby replacing the petrochemical-based chemicals [17]. Recent advancements in biomass conversion technologies are summarized in Table 7.4, highlighting key innovations and their impacts.

7.10.3 Potential for Large-Scale Adoption and Integration into Global Markets

The successful application of biomass conversion technologies is determined by some factors such as technological developments, policies in energy and economic deeds, and market demands on bioproducts. Another factor that can be attributed to

TABLE 7.4
Summary of Recent Advances in Biomass Conversion Technologies and Their Impacts

Innovation Area	Details	Impacts/Benefits
Bioplastics	Advanced biopolymers with C–O and C–N bonds	Reduced plastic waste; improved performance
Circular economy	Use of biowaste in microbial processes	Better resource efficiency; challenges include high costs
Second-generation biofuels	Biofuels from algae, waste, lignocellulosic residues	Lower greenhouse emissions; cheaper biofuels
Platform chemicals	Chemicals like succinic acid, levulinic acid, furfural from biomass	Replacement for petrochemicals; renewable sources

extensive integration is a decrease in the cost of renewable energy technologies making bio-based products cheaper compared with fossil-based counterparts [31]. Thus, benefits from solar, wind, and biomass energy are cheaper today and as these conversion technologies develop further, biomass conversion industries and communities globally will open up.

It is also important to note that there are also policy structures and governmental stimulus for the utilization of biomass conversion technologies. Globalization has seen many nations adopt renewable energy standards, carbon taxes, and bio-based subsidies that will promote an efficient low-carbon economy [34]. For example, the renewable energy directive of the European Union has provided target 2030 bioenergy obligations to its member states and this has promoted investments in biomass conversion technologies [29]. Similarly, countries like Brazil and the United States of America have put in place blender pumps that encourage the usage of biofuels in the transportation sector [30].

The fact that biomass conversion technologies are in a position to be incorporated into existing industrial value chains is a strong factor that will result in large applications. Comprehensive conversion strategies that yield biofuels and chemicals as well as materials from a single biomass source are economically feasible and sustainable as compared to other integrated biorefineries [10]. These co-production systems enable the production of both low-quantity and high-quantity, but high-value bioresources alongside the low-value bulk volumes, enhancing the efficiency of biomass conversion [3].

In addition, there will be a need to foster cooperation with other international entities and to share knowledge when it comes to expanding the applications of biomass conversion technologies across the globe. Synchronization of Innovation activities by successive governments, research centers, and industries helps to speed up research and development of new technologies, solve such barriers, and reduce costs. Presently, there is the Global Bioenergy Partnership and International Renewable Energy Agency that has created the foundation that is encouraging collaborative partnerships and capacity building in the bioenergy industry toward the commercial utilization of biomass conversion technologies [1].

7.11 CONCLUSION

Using biomass development and conversion provides a viable potential for waste recovery and generating high-value-added products. While the problems of pollution and climate change remain persistent issues, biomass conversion technologies remain among the most effective solutions to the problems of waste disposal, natural resource preservation, and the reduction of greenhouse emissions. If those technologies are to be successfully integrated into global markets, issues such as technical, economic, and social barriers, innovation, and collaboration must be addressed.

Biomass technologies such as pyrolysis, gasification, and fermentation can transform waste biomass into biofuels, biofertilizers, bioplastics, and industrial chemicals. Further, emerging conditions like catalytic conversion technology, hybrid systems, and biorefineries mark progress in biomass conversion technology for commercial processes. They do not only enhance the sustainability of waste management but also new economic prospects and energy security are here.

However, the implementation still faces some barriers, specifically, the availability of relatively cheaper technologies, favorable policies, and enhanced public awareness. This is why through research, investment, and policy changes to tackle these challenges biomass conversion technology remains the single most important factor in the global shift to a circular economy and sustainability targets.

Call to Action: To make biomass conversion technologies a reality, there must be a collaboration of related professionals, policymakers, industry and institutions, and researchers. Thus, with the further development of innovative technologies, the improvement of market conditions, and increased public awareness, we can develop biomass to the full to become a crucial energy source for the future.

REFERENCES

1. Usmani Z, Sharma M, Awasthi AK, Sivakumar N, Lukk T, Pecoraro L, et al. Bioprocessing of waste biomass for sustainable product development and minimizing environmental impact. *Bioresource Technology*. 2021;322:124548.
2. Clauser NM, González G, Mendieta CM, Kruyeniski J, Area MC, Vallejos ME. Biomass waste is a sustainable raw material for energy and fuels. *Sustainability*. 2021;13(2):794.
3. Dessie W, Luo X, Wang M, Feng L, Liao Y, Wang Z, et al. Current advances in waste biomass transformation into value-added products. *Applied Microbiology and Biotechnology*. 2020;104(11):4757–70.
4. Kick C, Kidikas Ž, Kasiulienė A, Maletić S, Zeremski T, Rubežius M, et al. Feasibility of using phytoremediation biomass for sustainable biofuel production via thermochemical conversion. *Biofuels, Bioproducts and Biorefining*. 2024;18:1010–1026.
5. Koutsonikolas D, Karagiannakis G, Plakas K, Chatzis V, Skevis G, Giudicianni P, et al. Membrane and electrochemical based technologies for the decontamination of exploitable streams produced by thermochemical processing of contaminated biomass. *Energies*. 2022;15(7):2683.
6. Das S, Newar R, Saikia A, Baruah A. Biomass-Based Engineered Materials for Soil Remediation. In: Sarma H, Joshi S (eds.), *Land Remediation and Management:* Bioengineering Strategies: Springer; 2023. p. 253–93.
7. Tan S, Zhou G, Yang Q, Ge S, Liu J, Cheng YW, et al. Utilization of current pyrolysis technology to convert biomass and manure waste into biochar for soil remediation: A review. *Science of the Total Environment*. 2023;864:160990.

8. Pal DB, Yadav S, Jana SK, Tiwari AK. Waste Biomass Conversion in Useful Various Chemicals. In: Pal DB, Yadav S, Jana SK, Tiwari AK (eds.), *Sustainable Clean Energy Production Using Waste Biomass: Sustainable Energy Production and Utilization*: Springer; 2024. p. 75–100.
9. Lee SY, Sankaran R, Chew KW, Tan CH, Krishnamoorthy R, Chu D-T, et al. Waste to bioenergy: A review on the recent conversion technologies. *BMC Energy*. 2019;1(1):1–22.
10. Adams P, Bridgwater T, Lea-Langton A, Ross A, Watson I. Biomass conversion technologies. In: Thornley P, Adams P (eds.), *Greenhouse Gas Balances of Bioenergy Systems*: Elsevier; 2018. p. 107–39.
11. Zaman CZ, Pal K, Yehye WA, Sagadevan S, Shah ST, Adebisi GA, et al. Pyrolysis: A sustainable way to generate energy from waste. *Pyrolysis*. 2017;1:3–36.
12. Basu P, Kaushal P. *Biomass Gasification, Pyrolysis, and Torrefaction: Practical Design, Theory, and Climate Change Mitigation*: Elsevier; 2023.
13. Müller FJ, Fuchs J, Müller S, Winter F. CO_2 conversion to CO by fluidized bed biomass gasification: Measuring CO_2 utilization via stable carbon isotope ratios. *Journal of CO_2 Utilization*. 2024;83:102792.
14. Buelvas A, Quintero-Coronel DA, Vanegas O, Ortegon K, Bula A, Mesa J, et al. Gasification of solid biomass or fast pyrolysis bio-oil: Comparative energy and exergy analyses using AspenPlus®. *Engineering Reports*. 2024;6:e12825.
15. Shah HH, Amin M, Iqbal A, Nadeem I, Kalin M, Soomar AM, et al. A review on gasification and pyrolysis of waste plastics. *Frontiers in Chemistry*. 2023;10:960894.
16. Danladi MM, Istifanus MF, Makeri MS, Egbere JO, Danahap LS, Okechalu BO, et al. *Fermentation: A Broader Perspective*: InTech Open. 2024.
17. Liakakou E, Vreugdenhil B, Cerone N, Zimbardi F, Pinto F, André R, et al. Gasification of lignin-rich residues for the production of biofuels via syngas fermentation: Comparison of gasification technologies. *Fuel*. 2019;251:580–92.
18. Zhang Y, Wang X, Zhu W, Zhao Y, Wang N, Gao M, et al. Anaerobic fermentation of organic solid waste: Recent updates in substrates, products, and the process with multiple products co-production. *Environmental Research*. 2023;233:116444.
19. Bernet N, Escudié R, Steyer J-P. Anaerobic Digestion: Application to the Energy Recovery of Agricultural Co-products and Food Industry Waste. In: Baumberger S (ed.), *Green Chemistry and Agro-food Industry: Towards a Sustainable Bioeconomy*: Springer; 2024. p. 85–103.
20. Jamil F, Inayat A, Hussain M, Akhter P, Abideen Z, Ghenai C, et al. Valorization of waste biomass to biofuels for power production and transportation in optimized way: A comprehensive review. *Advanced Energy and Sustainability Research*. 2024;5:2400104.
21. Malik K, Capareda SC, Kamboj BR, Malik S, Singh K, Arya S, et al. Biofuels production: A review on sustainable alternatives to traditional fuels and energy sources. *Fuels*. 2024;5(2):157–75.
22. Sehgal M, Borah S. *Current Trends in Agriculture & Allied Sciences* (Volume 2). SP Publishing.
23. Barragán-Ocaña A, Merritt H, Sánchez-Estrada OE, Méndez-Becerril JL, del Pilar Longar-Blanco M. Biorefinery and sustainability for the production of biofuels and value-added products: A trends analysis based on network and patent analysis. *PLoS One*. 2023;18(1):e0279659.
24. Ventura M, Marinas A, Domine ME. Catalytic processes for biomass-derived platform molecules valorisation. *Topics in Catalysis*. 2020;63(9):846–65.
25. Strovsky V, Logvinenko O, Moor I, Pustokhina N, editors. Resource conservation as a condition for preserving non-renewable natural capital. *E3S Web of Conferences*; 2020: EDP Sciences.
26. Lam HL, Varbanov P, Klemeš J. Minimising carbon footprint of regional biomass supply chains. *Resources, Conservation and Recycling*. 2010;54(5):303–9.

27. Stevanović S, Stevanović S. Biomass as a renewable energy source. *Економика пољопривреде*. 2022;69(1):195–209.
28. Dhar P, Nickhil C, Pandiselvam R, Deka SC. Pineapple waste-based biorefinery for sustainable generation of value-added products. *Biomass Conversion and Biorefinery*. 2023;14:1–22.
29. Kataya G, Cornu D, Bechelany M, Hijazi A, Issa M. Biomass waste conversion technologies and its application for sustainable environmental development—A review. *Agronomy*. 2023;13(11):2833.
30. Muh E, Tabet F, Amara S. Biomass conversion to fuels and value-added chemicals: A comprehensive review of the thermochemical processes. *Current Alternative Energy*. 2021;4(1):3–25.
31. Solarte-Toro JC, Ortiz-Sanchez M, Inocencio-García P-J, Cardona Alzate CA. Sustainable biorefineries based on catalytic biomass conversion: A review. *Catalysts*. 2023;13(5):902.
32. Chen M, Smith PM. The US cellulosic biofuels industry: Expert views on commercialization drivers and barriers. *Biomass and Bioenergy*. 2017;102:52–61.
33. Rijal P, Carvalho H, de Sistemas Inteligentes LA, Matias J, Center GR, Garrido S, et al. Drivers and Barriers for Sustainable and Efficient Biomass Valorization Processes. *Proceedings of the 6th European Conference on Industrial Engineering and Operations Management*, Lisbon, Portugal, July 18–20, 2023.
34. Pandey S, Erbaugh JT. Driving sustainable uptake: A systematic review of global literature on policies governing woody biomass for energy. *Discover Sustainability*. 2024;5(1):28.
35. Zhang YS, Zhu HL, Yao D, Williams PT, Wu C, Xu D, et al. Thermo-chemical conversion of carbonaceous wastes for CNT and hydrogen production: A review. *Sustainable Energy & Fuels*. 2021;5(17):4173–208.
36. Ceglia F, Marrasso E, Roselli C, Sasso M. Energy and environmental assessment of a biomass-based renewable energy community including photovoltaic and hydroelectric systems. *Energy*. 2023;282:128348.
37. Guhl F, Zeigermann U. Local heat transitions—A comparative case study of five bioenergy villages in Northern and Southern Germany. *Zeitschrift für Vergleichende Politikwissenschaft*. 2024;18:157–179.
38. Sadr M, Esmaeili Aliabadi D, Avşar B, Thrän D. Assessing the impact of seasonality on bioenergy production from energy crops in Germany, considering just-in-time philosophy. *Biofuels, Bioproducts and Biorefining*. 2024;18:883–898.
39. Choudhury SR, Das A, Chakraborty R. Review on progress in cellulase catalyzed saccharification of agricultural lignocellulosic biomass towards fermentable sugar and bioethanol: Kinetics & reactor configurations. *Biocatalysis and Agricultural Biotechnology*. 2024;58:103142.
40. de Araújo B, Costa IO, de Brito HG, Rios NS, dos Santos ES. Enzyme technology in bioethanol production from lignocellulosic biomass: Recent trends with a focus on immobilized enzymes. *Bioresources*. 2023;18(4):8653.
41. Cai T, Liu C, Jiang J, Meng X, Ragauskas AJ, Wang K. Innovative biphasic solvent systems for lignocellulosic biorefinery. *Trends in Chemistry*. 2024;6:219–233.
42. Liu R, Sarker M, Rahman MM, Li C, Chai M, Cotillon R, et al. Multi-scale complexities of solid acid catalysts in the catalytic fast pyrolysis of biomass for bio-oil production – A review. *Progress in Energy and Combustion Science*. 2020;80:100852.
43. Cywar RM, Rorrer NA, Hoyt CB, Beckham GT, Chen EY-X. Bio-based polymers with performance-advantaged properties. *Nature Reviews Materials*. 2022;7(2):83–103.
44. Jain A, Sarsaiya S, Awasthi MK, Singh R, Rajput R, Mishra UC, et al. Bioenergy and bio-products from bio-waste and its associated modern circular economy: Current research trends, challenges, and future outlooks. *Fuel*. 2022;307:121859.

8 Sustainable Biomass Production from Microalgae Using Biorefinery Approach

New Prospects and Challenges

Swati Rani and Archana Tiwari

8.1 INTRODUCTION

Sustainable development is the pathway for addressing today's need without damaging nonrenewable resources so that future generation can also meet their requirements. It needs equilibrium of three dimensions: protection of the environment, social benefits, and economic benefits. Due to urbanization and an increase in population, the consumption of natural resources such as water, wood, and coal escalates, which is alarming for the need of sustainable development. Researchers are passionately working on preserving natural resources by implementing many green technologies such as rainwater preservation, electric vehicles, the use of solar and wind energy, bio-based fuels, wastewater treatment, and microalgae biorefineries (Orejuela-Escobar et al., 2021). Microalgae biorefinery is the phenomenon of using diverse biochemical composition microalgae to produce primary and secondary metabolites to reduce carbon foot printing and many other environmental issues. It helps to reduce the load of global energy demands and protect the environment by recycling waste, greenhouse gases reduction, higher growth rates, low land requirement, and generation of high nutritional value products (Alam et al., 2020). There are some key components that affect the efficacy of microalgae biorefinery such as cultivation and harvesting techniques, extraction procedures, biochemical processing of extracted biomass, and selected integrated approaches. The integrated approach of waste (nutrients) recycling and production of a large spectrum of functional molecules holds major potential in the microalgae biorefinery process (Vuppaladadiyam et al., 2018). Pollution, lifestyle, and animal-based food diet have massively increased the diabetic, high cholesterol, and cancer patients, which in turn make people more aware about nutrition, health, functional foods, and dietary supplements. Microalgae are one of the main producers of protein, lipids, and carbohydrates,

DOI: 10.1201/9781003536277-8

it is more efficient than terrestrial plants. There are many microalgae species such as *Arthrospira* sp., *Chlorella* sp., *Scenedesmus* sp., *Nannochloropsis* sp., *Dunaliella* sp., and *Synechococcus* sp., which can produce up to 50%–70% of lipids, proteins, and carbohydrates, which is way higher than the production through plants (Xin et al., 2024; Costa et al., 2021; Amorim et al., 2021. Algae-based products are gaining attention due to their flexibility, higher energy, and bioactive properties such as antioxidant, anti-inflammatory, antitumor, and antimicrobial (Wang et al., 2022;).That is why the global market for microalgae is increasing by 5.4% annually and presently it is estimated to be 1 billion USD (Eltanahy et al., 2021). They are a valuable source of bulk bioproducts as well as functional biomolecules as illustrated in Figure 8.1.

Despite having so many benefits of microalgae, several issues such as technological development, large-scale implementation, market acceptance, and economic

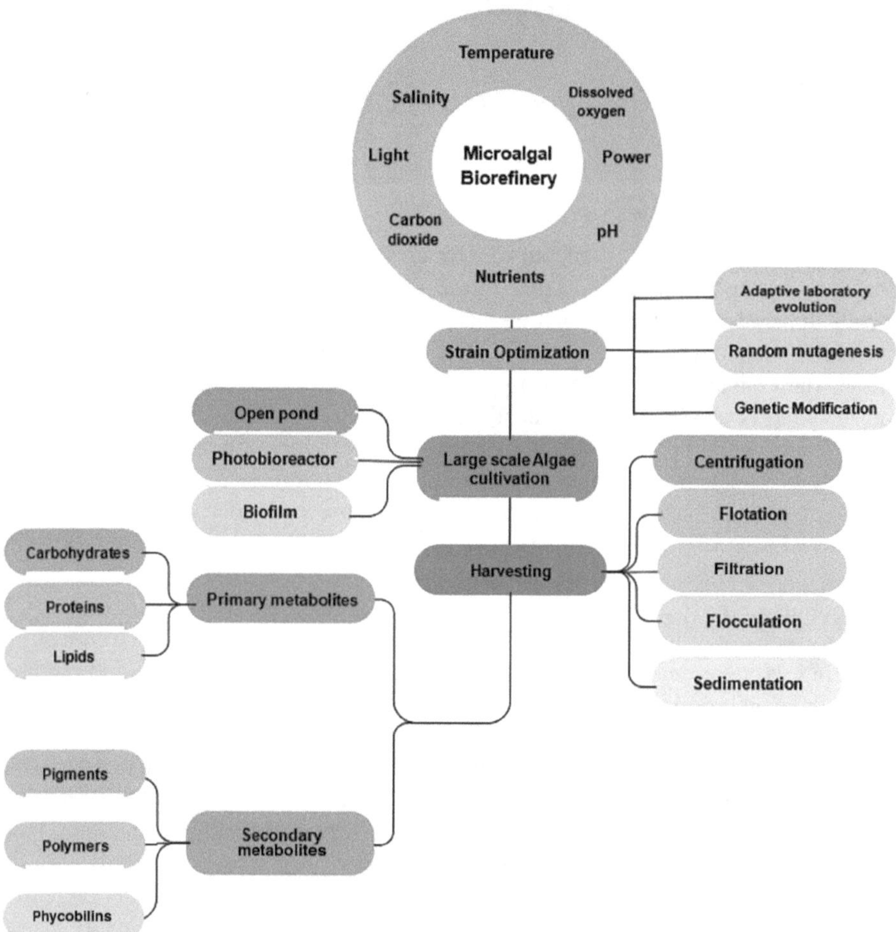

FIGURE 8.1 Sustainable development through microalgal biorefinery.

viability hinder the potential of microalgal biorefineries. That is why, addressing these challenges is also one of the major goals to achieve the sustainable production using the biorefinery approach. This chapter attempts to give a thorough review of the above-mentioned elements, including current research scenarios, technical advancements, challenges faced, and potential solutions that could lead to more effective and sustainable systems for producing microalgal biomass.

It also adds to the larger-scale production of renewable energy and bioproduct development by exploring key parameters such as advanced technology, socioeconomics, and sustainability.

8.2 STRATEGIES FOR DEVELOPING SUSTAINABLE MICROALGAE BIOREFINERIES

The commercialization of microalgae-based materials, such as the biomass or high-value compounds, is still limited to specific markets, despite the remarkable potential of microalgae for various applications in biotechnology. Microalgal cultivation faces many challenges such as low biomass conversion rates, low target productivities, inadequate light penetration in medium, contaminations, and media selection. Compared to closed systems, open systems emerged as more affordable cultivation methods. However, abiotic and biotic stresses have a negative impact on overall cultivation, unless a stress-tolerant strain is employed. A key step in developing large-scale microalgal production is strain selection and improvement. This field has not received sufficient attention, and more research is still needed. Only a small number of naturally occurring microalgal strains presently possess the qualities needed for commercially feasible industrial production. Therefore, it is vital to isolate and develop new strains that will overcome all these challenges (Trovão et al., 2022).

8.2.1 Adaptive Laboratory Evolution

Another economical method that has no need for a prior understanding of genetics of the algal strain is adaptive laboratory evolution. Additionally, it prevents the insertion of foreign genetic material into the desired cells, and the enhanced strains do not present any regulatory concerns because of the extensive history of biosafety of these techniques. Cell cultures are exposed to constant selective pressure for an extended length of time in these experiments. As a result, selection can influence all characteristics that lead to a suitable improvement in the relevant environmental systems. As a result, the genome is altered by spontaneous, adaptive, as well as nonspecific mutations that are transferred from mother to daughter cells. However, in order to achieve the desired phenotype, this technique is a very time-consuming and laborious method that requires many generations (Trovão et al., 2022).

8.2.1.1 Enhancing Chemical Composition and Yield through Engineering

The majority of genetically modified (GM) applications aim to address the rate-limiting steps, over-expression, and reduce catabolic reactions by regulating the synthesis of bioproducts from microalgae at the enzyme as well as at the gene level.

To attain the intended bioproduct yield, one main strategy is to regulate flux in competing pathways. Genes and pertinent proteins linked to central carbon metabolism can be functionally understood via genome engineering techniques. More advanced techniques such as artificial intelligence (AI) have also been used for screening and optimization of algal strain. Integrated approach of using AI with genome data can help in modifications of interested gene expression (Novoveská et al., 2023).

8.2.1.2 Random Mutagenesis

Mutagenesis is the phenomenon of creating mutants of particular organisms by changing desirable characteristics by using a mutagenic agent. To create mutants, random mutagenesis is a reliable, well-established, simple, and economical method. Random mutagenesis is the selection of mutants with the desirable characteristics after exposing cells to a strong physical or chemical mutagenic agent. This accelerates naturally occurring processes. The generation of these mutants at industrial facilities is permitted without the limitations placed on genetically modified organisms because these mutations are nonspecific, and no foreign genetic material is introduced. Random mutagenesis can produce a wide range of mutants, but these mutations are frequently fatal or mildly harmful to the organisms that are affected. Furthermore, mutant traits may also be reversible and unstable, demonstrating the significance of creating efficient selection strategies and offering phenotypic stability (Trovão et al., 2022).

8.2.2 Cultivation Systems: Advantages and Disadvantages

In the last few decades, researchers have been working on scaling up the microalgae biorefineries. Due to their high cost of cultivation, harvesting, and low biomass productivity, the treatment process is very challenging (Tan et al., 2020). Optimization of growth conditions, harvesting techniques, and extraction methods will help to reduce the cost and increase efficiency. Microalgae cultivation methods have their own pros and cons, and treatment efficiency is dependent upon the type of strain and the end product requirement. That is why, combinations of cultivation methods can be used to enhance biomass productivity (Novoveská et al., 2023). Each system has benefits as well as drawbacks, which may be reduced by optimization as illustrated in Figure 8.2.

8.2.2.1 Suspended Systems

Microalgae can be grown in open as well as in closed systems; for open systems, natural ponds and man-made open ponds have been used, whereas in closed systems, photobioreactors (PBRs) are used which allows precise control of growth conditions and can be used to culture pure strains of microalgae. Open pond is used for large-scale cultivation of microalgae; a simple system that is easy to construct and maintain. This system also has numerous advantages such as cost-effectiveness, low cost of operation, and low energy demand (Qin et al,.2023). Despite having many benefits, it also faces several challenges such as maintenance of cultural conditions and contamination as it is fully exposed to the environment (Tan et al 2020). Moreover, these systems also have longer light paths that lead to self-shading, limit

System	Advantages	Disadvantages	Optimization
Open ponds	• Economic viability. • Low energy consumption • Low operating costs	• High risk of contamination • Low biomass productivity • Culture exposed to weather conditions	• To adjust depth of pond • To adjusts cell conc. to ensure optimum growth
Photobioreactor	• High cell density • Low evaporation rate • Low risk of contaminants • Provides controlled conditions for growth	• Expensive to operate • Difficult to sterilize	• To optimize factors affecting growth • Use AI and theoretical models
Biofilm	• High cell density • Easy harvesting via scraping	• Biofilm sloughing • Microbial contamination	• To determine ideal growth conditions while maintaining biomass productivity

FIGURE 8.2 Types of cultivation systems: advantages and disadvantages.

cell density, and ultimately increase the cost of harvesting; require a large amount of land, and are not CO_2 efficient (Novoveská et al. 2023; Qin et al,.2023). Based on their design and characteristics, open ponds are categorized into high-rate algal ponds (HRAPs), raceway ponds, circulation cascades, and natural oxidation ponds. HRAPs and circulation cascades have many benefits over oxidation ponds as these systems prevent self-shading by generating turbulence, which mainly results in high biomass production but have the high cost of construction which makes it difficult to scale-up (Qin et al,.2023). In closed systems, various types of PBRs are used for the cultivation of microalgae such as tubular photobioreactors and vertical column bioreactors. This system consists of long tubes, which are made of transparent glass or plastic and may be arranged in a slanted, vertical, horizontal, or helix manner to increase the amount of light that is absorbed by culture. Culture is circulated by utilization of air lift system or mechanical pumps. However, due to the long tubes utilized in the system, there is the challenge of poor mass transfer due to a difference in the concentration of substrate along the tube whereas the vertical column PBRs system has a transparent vertical tube and a sparger. The sparger pumps in small air bubbles, which allow the exchange of gasses. This system has an efficient mass transfer due to the presence of a sparger, is easy to operate, and has lower energy demand. However, it has a high cost of construction, is difficult to sterilize, and does not provide an adequate amount of light.

8.2.2.2 Attached System

Microalgae biofilm consists of cells that are entrapped in self-secreted extracellular polysaccharides (EPS). A biofilm can be developed on any surface, which is supplied with nutrients and exposed to a light source. First, cells attach reversibly to the surface by adsorption, which is followed by irreversible adherence as EPS is secreted. Further replication occurs as the concentration of organic molecules increases.

The film slowly forms a multilayer 3D structure. However, after reaching a certain thickness, the biofilm will lose its integrity and it also collapses after dehydration, as it majorly consists of water. A biofilm comprises 10% microalgal cells and 90% EPS (Schnurr et al., 2015). They are composed of nucleic acids, proteins, polysaccharides, and phospholipids, along with some functional groups. They may be single-species or multispecies biofilm. Single-species biofilms exist very rarely in nature but have significant potential in the biotechnology industry. Multispecies biofilm has complex interactions between microalgae and bacteria that may range from mutualism to parasitism (Cheah et al., 2021). Reactors such as fluidized bed reactor and rotating algal film bioreactors with varying support materials such as spent coffee grounds, polyurethane foam, and beads bed for attached growth were used (Mohd-Sahib et al., 2017; Rosmahadi et al., 2022). Attached growth cultivation has several benefits over suspended growth systems like higher biomass production, low cost of harvesting, more efficient, and good light penetration. But it has some disadvantages too like it is costly to install the attached system initially, it requires regular surface cleaning, and its scalability is also challenging. Therefore, designing the best system for microalgae cultivation requires prior knowledge about the strain, its growth cycle, and system preference according to the need.

8.2.3 Optimization of System

Optimization of growth in the cultivation system requires maintaining mainly temperature and biomass concentration in media. These are species dependent and also require knowledge of weather conditions. During hot days, the depth of the pond should be kept low, which allows the temperature to reach optimal value faster. In the afternoon, more media may be added to maintain cell concentration at the optimum level and also control temperature. However, during winter if the temperature of the pond is lower than the media from the inlet then more media has to be added without causing washout. This allows the temperature of the pond to reach close to the optimum level. These conditions require knowledge about seasonal variations and also require a constant change to ensure culture conditions are close to optimum (Mantzorou et al 2019). Some essential nutrients such as nitrogen, phosphorus, and carbon along with other environmental factors such as pH, temperature, and light have a significant effect on growth as well as EPS production (Boonchai et al. 2014.). However, some studies suggest that nutrient depletion does not have a positive effect on biofilm (Schnurr et al., 2015). Temperature variations affect the microalgae cells rate of metabolism. Although the ideal temperature range for algal culture is typically between 20°C and 25°C, each species of microalgae has varied temperature needs (Mantzorou et al., 2019). Different microalgal species have different ideal pH values for biofilm formation. Microalgae cells typically generate more EPS to protect themselves from drastic pH fluctuations if the pH is greater than 9, which enhances attachment (Cheah et al., 2021). Similar to *Chlorococcum* sp. onto glass fiber-reinforced plastic, the ideal pH for the formation of its biofilm was 8 (Shen et al., 2014). In general, a pH of 6–9 that is close to neutral promotes the adhesion of microalgal biofilms. The hydrophobicity of cells will rise as the pH rises, and the concentration of dissolved CO_2 was at pH equilibrium without creating an extremely acidic environment (Schnurr et al., 2015).

8.2.4 Biomass Harvesting Strategies

For the collection of algal biomass, many harvesting techniques are used such as filtration, centrifugation, flocculation, and coagulation. To collect the algae biomass, the filtration process uses a semipermeable membrane that can retain microalgae on it while letting media flow through. Various filtration techniques such as microfiltration, ultrafiltration, vacuum filtration, and pressure filtration are frequently used to harvest algal biomass (Deepa et al., 2023). Mainly, the microalgae which is more prone to shear-induced damage undergoes the filtration process as it protects the algal cells; moreover, this process is dependent upon the size of microalgae cells likewise different pore sizes of filters allow the setup to adapt according to the needs of various. Nevertheless, this process is highly susceptible to fouling as well as clogging, resulting in frequent replacement of the filter, which could raise the processing costs considerably (Tan et al., 2020). Another way is flotation which uses tiny bubbles that adhere to microalgae cells to encourage cell rise on the media surface for simple harvesting. The flotation system has several benefits, such as low expenses, substantial processing throughput, along with a relatively high harvesting efficiency (Kucmanova, et al, 2019). Flocculation is another method induced by chemical or microbe flocculants into algal media or through altering the cultivation medium to aggregate the cells while recovering biomass. Microalgal cells usually possess negative charges which keep them from clustering in suspension. Adding flocculants like multivalent cations (aluminium sulfate, ferric chloride, and calcium oxide) or cationic polymers (chitosan, polyacrylamide) to the cultures may decrease or neutralize the cell surface charge. The flocculant dose, the cultivation medium pH, and ionic strength have a significant impact on the flocculation of algal biomass (Shen et al 2013). Recently, many researchers are focusing on bio-flocculants and exploring the effects of bacteria as a bioflocculant over settling the algal biomass. It reduces the chemical treatment and costs and simultaneously improved the algal biomass due to the synergistic relationship with bacteria (Kucmanova et al., 2019). Centrifugation is the most widely used technique for harvesting especially under lab-scale cultivation. The process of centrifugation uses centrifugal force to separate cells from the surrounding media according to the density as well as the particle size of each component. Although this method has a high cell harvesting performance, it requires a lot of energy and time to complete. Furthermore, centrifugation's high gravitational force may harm cells, which makes it undesirable for some uses since fragile nutrients may be lost (Tan et al., 2020). Despite being regarded as a straightforward and inexpensive method, it is only appropriate for microalgae larger than 70 μm. Higher settling velocities for smaller microalgae require the use of flocculants. Sedimentation velocity controls additional settling features or slurry generation. Gravity sedimentation is a solid–liquid separation technique in which concentrated slurry has settled out under gravitational force, keeping above visibly transparent supernatant. Electrical methods use extremely high amounts of electrical power, and not all microalgal species can be sedimented by gravity (Kucmanova et al., 2019). Therefore, for the large-scale cultivation of microalgae biorefinery processes, it is necessary to use the most reliable method for harvesting. It is completely dependent upon the type of microalgae, its cell surface properties, cell size, and most importantly the concentration of biomass present and the cultivation system used.

8.3 HIGH VALUE-ADDED PRODUCTS

The term "metabolite" refers to small intermediate molecules formed due to the metabolic process or reactions. Metabolites perform several functions within the cell by serving as stimulants and signaling molecules, as well as providing defensive functions and structural support. Metabolites can be classified into two categories: (1) *primary metabolites*, such as carbohydrates, proteins, and lipids, which are formed as by-products of metabolic activities and are involved in physiological processes within the cell; and (2) *secondary metabolites*, which include sterols, waxes, ketones, hydrocarbons, vitamins, pigments, phycocolloids, carotenoids, lectins, mycosporine, and polyketides formed in small quantities as a by-product of primary metabolic reactions (Vuppaladadiyam et al., 2018). Since different algal metabolites exhibit distinct functional properties, the simultaneous production of multiple bioproducts can enhance economic viability at an industrial scale. Additionally, the use of algae in wastewater treatment and toxic antremediation, along with their ability to sequester carbon, makes them an excellent option for integration into the biorefinery process (Orejuela-Escobar et al., 2021). Algal metabolites have been found to have potential for numerous applications in multiple industries but their large-scale application is hindered due to various reasons as illustrated in Table 8.1.

Table 8.1 Types of metabolites and its applications.

TABLE 8.1
Types of Metabolites and Its Applications

Metabolite	Application	Properties/Benefits	Drawback
Proteins	Animal feed, human consumption, food supplements	Exhibits antimicrobial, anti-inflammatory, antioxidant, antihypertensive, anti-atherosclerotic	Significant protein content variation by species, strain, growth phase, and conditions Extraction efficiency challenges
Polysaccharides	Biostimulants for plants, biopesticides, poultry feed (antibiotic alternative), anti-obesity agents, food packaging	• Exhibits antibacterial, antifungal, and plant defense activation properties • Enhances enzyme activity that strengthens plant resistance • Retains water, maintaining soil moisture • Increases β-glucan content (immune-modulating activity in animals) • Ecological and biodegradable for food packaging	Biomass reduction during metabolite increase • Chemical complexity varies with species, making standard analysis difficult • Sensitivity to light intensity and environmental conditions impacts yield

(*Continued*)

TABLE 8.1 (*Continued*)
Types of Metabolites and Its Applications

Metabolite	Application	Properties/Benefits	Drawback
Lipids	Biofuel, biodiesel, food, feed, cosmetics, healthcare products	• Neutral lipids for biofuel and biodiesel • Polar lipids exhibit anti-inflammatory, heart and brain health, anticancer, anti-obesity, Alzheimer's treatment	Lipid content and type vary significantly by species and cultivation conditions • Nitrogen limitation may decrease the overall biomass yield • Temperature sensitivity can negatively impact lipid production in some species
Pigments	Natural colorants, food supplements, pharmaceuticals, cosmetics	Exhibits antioxidant, anticancer, anti-inflammatory, and detoxifying properties • Aids in wound healing • Secondary carotenoids help protect against reactive oxygen species (ROS) • Effective in photo-protection of photosynthetic systems	Light intensity, spectral distribution, and salinity impact production

8.3.1 Primary Metabolites

8.3.1.1 Proteins

Microalgae produce a significant amount of protein, which is typically generated during the growth phase and can account for up to 70% of their dry biomass weight. They are known to contain all essential amino acids, presenting a balanced total amino acid profile. This property has gained interest in algae as a potential source of proteins for both animal feed and human consumption. Additionally, microalgal peptides either produced natively within the algae or via enzymes have exhibited biological activities, including hormonal action, antimicrobial effects, anti-inflammatory properties, antihypertensive effects, anti-atherosclerotic properties, and antioxidant properties (Lucakova et al., 2022). These characteristics have fuelled extensive investigations into the various applications of microalgal peptides.

Species such as *Chlorella* sp. and *Scenedesmus obliquus*, and the cyanobacterium *Arthrospira* sp., have been utilized for food supplements, with *Chlorella and Arthrospira* being commercialized at the industrial level (Alam et al., 2020). The protein content produced by microalgae varies significantly due to differences in genetic traits among strains and species, along with cultivation conditions, the method of analysis, and the growth phase at the time of extraction.

To effectively introduce microalgal proteins to the consumer market, the first step is to analyze the protein production from various algal strains. Common methods for estimation are colorimetric assays, elemental analysis, summation of anhydro amino acids, and nitrogen analysis. However, these methods often yield inconsistent protein values when applied to the same sample (Wang et al., 2021). Alternatively, other methods capable of measuring the "true" concentration of protein like the anhydro amino summation method can be implied (Templeton and Laurens, 2015). Colorimetric methods are also employed but each possesses some setbacks in hands-on application. Lowry method is dependent on the presence of specific amino acids, i.e., tyrosine and tryptophan. Additionally, these reactions are affected by the presence of interferents with the reagents being reactive for a limited time and proteins used for constructing standards may not resemble the protein of interest. The major challenge with these techniques is that all methods are dependent on the successful extraction of intracellular proteins. The efficiency of cell wall disruption and protein extraction can vary significantly among different algal species. In contrast, elemental analysis does not pose these challenges, cost-effective, fast, and simple.

Protein content in microalgae also showed great variation throughout the growth cycle ranging from 7% to 40%of dry weight (Amorim et al., 2021). Since protein production is affected by cultivation conditions in addition to their genetic constitution. Numerous studies have explored the effects of environmental and nutritional factors on protein production, demonstrating that conditions such as temperature, CO_2, light spectra, and intensity, along with nutrient availability all play crucial roles (Wang et al., 2021). The effect of light on protein production is critical to understand, yet there is no consistent trend across all microalgae species. While some microalgae exhibit increased protein content with higher light intensity, others may experience a detrimental effect, and some show no change in yield at all (Wang et al., 2021). That is why, strain selection is the most important parameter while optimizing the light intensity for protein production. Nitrogen is one of the essential macronutrients involved in metabolic reactions and significantly influences the biochemical composition of algae. Microalgae can utilize many forms of nitrogen such as nitrite, nitrate, ammonium, and organic sources of nitrogen such as urea. Any form of nitrogen that is taken up by algae is converted to ammonium, which is then assimilated into amino acids. In general, protein content is increased with an increase in nitrogen concentration. But this occurs only up to a certain point, after which the increase halts. Conversely nitrogen starvation results in a decrease in protein production. Changes in cell composition, structure, and nutrient content occur during various growth phases. Protein content is primarily found higher in exponential phases and subsequently reduces in the stationary phase. Microalgae species such as *Tisochryis* sp., *Isochrysis*, *Pavlovalutheri*, *Nannochloropsis oculate, C. vulgaris*, and *N. gaditana* have exhibited enhanced protein accumulation in the exponential phase but underwent declined after different durations in the stationary phase (Wang et al., 2021). Since the growth phase of culture significantly impacts protein content that can be extracted, determining the optimal harvest time in the exponential phase by evaluating yield as well as accumulation rate is crucial for maximizing protein production.

8.3.1.2 Polysaccharides

Plants may benefit from microalgae polysaccharides as biostimulants, helping them to combat biotic and abiotic stress. The release of bioactive chemicals (vitamins, amino acids, polypeptides, antibacterial or antifungal substances, phytohormones, and polysaccharides) is linked to the enrichment of soil and plants by microalgae (Alvarez et al., 2021; Costa et al., 2021). The extracellular polysaccharides (EPS) produced by microalgae and cyanobacteria have the ability to retain water and keep soil moist and some polysaccharides may function as a biopesticide in vegetable crops (Alvarez et al., 2021). Additionally, microalgae polysaccharides, like alginate, increase the activity of defense-related enzymes, hence causing plant resistance. Producers are also looking for alternatives to antibiotics in response to regulatory restrictions and consumer preferences for antibiotic-free animal food. In poultry feed, certain microalgae polysaccharides, such as β-glucan, can be utilized as antibiotic substitutes. Microalgae can also be used as anti-obesity agents because it contains nondigestible polysaccharides. Microalgae polysaccharides are ecological and biodegradable; they hold promise for food packaging (Orejuela-Escobar et al., 2021).

Various pretreatment and extraction methods are used to convert the microalgae metabolites into valuable industrial products. Reducing starch breakdown and boosting glucan storage can both increase the amount of carbohydrates. The most popular and economical method that directly affects the biochemical composition is the manipulation of environmental factors such as pH, light, nutrition, temperature, and carbon source. However, this often results in reduced biomass production while increasing metabolites (Chen et al., 2017). Therefore, it is essential to comprehend and control the variables to increase the amount of carbohydrates without compromising the biomass production. Nutrient limitation is a practical and frequently used strategy to change the composition of metabolites (Vuppaladadiyam et al., 2018; Eltanahy et al., 2021). Restricting phosphorus and nitrogen can redirect the carbon fixed during the Calvin cycle toward the synthesis of non-nitrogenous substances such as lipids and polysaccharides, primarily starch and glycogen (de Farias Silva et al.2019a). For instance, when *Chlorella* was grown in nitrogen-starved conditions, its protein content decreased from 60% to 20%, while its carbohydrate content increased from 12% to 54% (de Farias Silva et al. 2019b). Similarly, *Chlorella* grown with phosphorus-limited conditions showed a decrease in nitrogen content from 57% to 7% and an increase in the carbohydrate content from 10% to 60% (Samiee-Zafarghandi et al. 2018). Although it comes at the expense of protein decrease, dietary constraints are a more appealing strategy than famine, even though they may lead to a decrease in protein and encourage carbohydrates accumulation. Continuous culturing or two-stage cultivation might also be effective for the integrated protein and carbohydrate accumulation. Another crucial factor is irradiance, a significant source of energy for photosynthesis, which has a direct impact on the growth rates, cell composition, and CO_2 fixation. A linear relationship between the rise in irradiance intensity and the biomass productivity and carbohydrate content has been documented, despite the fact that the effect of light intensity is species dependent. However, an excessive increase in irradiance could harm the photosystem, and biomass saturation may cause self-shading and decrease light penetration (Ho et al. 2014). Likewise, lower light intensities (less than 275 μmol/m^{2}s) can gradually decrease starch production.

Certain microalgae species have structural polysaccharides in their cell walls and can accumulate a significant amount of intracellular starch. However, the availability of multiple microalgae species with distinct cellular structures and chemical compositions poses challenges in developing standard analytical methodologies for characterizing microalgae biomass (Costa et al., 2021). Additionally, due to their significant sensitivity to cultivation conditions including temperature, salinity, and food availability, even a single species of microalgae may exhibit heterogeneity in their chemical complexity. As a result, sample preparation for determining the amount of carbohydrates in microalgae is quite difficult. Some analyses require total carbohydrate quantification, while others focus on the qualitative or quantitative composition of monosaccharides.

The study of the carbohydrate composition in algal biomass necessitates initial acid hydrolysis. Intracellular starch and structural polysaccharides can be depolymerized into their monomers using acid hydrolysis, which enables further quantification (deSouza et al., 2019). In addition to preventing excessive monosaccharide breakdown, the ideal hydrolysis conditions (acid, temperature, and time) must ensure complete hydrolysis of the polysaccharides (Schulze et al., 2017).

Another interesting approach is the selective cleavage of algal polysaccharides by enzymatic hydrolysis, provided that the breakdown products such as furfural or hydroxymethyl furfural are minimized (Schulze et al.,2017). Analytical techniques like high-performance liquid chromatography (HPLC) are frequently employed for monosaccharide quantification. Liquid chromatography mass spectrometers (LC–MS), HPLC with a refractive index detector, and HPLC with pulsed amperometric detection (high-performance anion-exchange chromatography) were used for the quantification of carbohydrates. The investigation of techniques that examine extracellular polymeric materials as well as cell wall polysaccharides is one of the developments in carbohydrate analysis. Smaller sample sizes, less reagents, and shorter analyses are the constant goals of the established procedures. Since liquid chromatography demonstrated a good separation and quantification of these substances, it has been emphasized in the determination of carbohydrates. It offers information on the makeup of individual monomeric sugars, which makes it useful for this kind of investigation even though it is a more costly analytical method.

8.3.1.3 Lipids

Lipids are a class of chemical molecules that have enormous commercial potential and can be utilized as raw materials for food, feed, cosmetics, healthcare products, and bioenergy. Lipids are categorized into neutral, polar, and nonpolar according to their polarity. One significant energy source found in microalgae is neutral lipids, which can be used to produce biofuel and biodiesel. On the other hand, polar lipids, typically show biological activities such as lower inflammation, promote heart and brain health and help in the treatment of diseases like cancer, obesity, and Alzheimer's (Chen et al., 2017). The economics of producing biofuel from microalgae biomass is generally increased by improving lipid accumulation in microalgae cell. For optimizing nutrient conditions, it is necessary to examine the uptake of nutrients and metabolic pathways and then optimize the nutrient conditions to increase biomass production and target product production (Wang et al., 2022).

Microalgae use CO_2 and $NaHCO_3$ as inorganic carbon sources to create organic metabolites. *Scenedesmus obliquus* and *Chlorella pyrenoidosa*, photosynthetic lipid content increased in response to elevated CO_2 levels (Tang et al., 2011). Under 50% CO_2 conditions, the maximal lipid content of *S. obliquus* was 24.4% of dry cell weight, which is 61% higher than at 0.03% CO_2. In heterotrophic cultivation, *Chlorella sorokiniana* showed a 2.4 times increase in lipid with 20 g/L glucose and 3.9 times with 8 g/L glucose compared to photoautotrophic cultivation (Li et al., 2014). In comparison to photoautotrophic cultivation, mixotrophic cultivation of *Chlorella vulgaris* yielded biomass productivity of 137.43 ± 13.3 mg/L/d and had a lipid content of 39% when molasses was used as an additional carbon source. These results were 1.5 and 2 times higher than photoautotrophic cultivation (Laraib et al., 2021). Regardless of whether the cultivation is photoautotrophic, heterotrophic, or mixotrophic, lipid production must be promoted by adequate carbon concentrations. One of the nutrients that have the biggest direct impact on the build-up of microalgal lipids is nitrogen (Shin et al., 2018; Feng et al., 2020). It can be supplied in various forms, including ammonium, urea, and nitrate. Nonetheless, distinct types of nitrogen sources are unique to microalgae development and lipid formation (Feng et al., 2020; Xin et al., 2024). Nitrogen stress is the most successful method for promoting lipid formation in microalgae (Feng et al., 2020). For example, Arguelles and Martinez-Goss discovered that, in the range of 1.5–0.375 g/L $NaNO_3$, there was an increasing tendency in the lipid production of both *Chlorolobion* sp. and *Chlorella* sp. under nitrogen-limited growing conditions (Arguelles and Martinez-Goss, 2021). In comparison to the 1.5 g/L $NaNO_3$ growth condition, the lipid productivity of *Chlorolobion* sp. and *Chlorella* sp. increased by 2.9 and 1.9 times, respectively, to 227.84 and 151.14 mg/L/d at 0.375 g/L $NaNO_3$ concentration. The reduction in microalgae cell division is caused by nitrogen deficiency toward neutral lipids (Feng et al., 2020). As a result, nitrogen limitation may also result in less biomass being produced, which influences the subsequent formation of lipids. One of the most important environmental factors throughout the microalgal lipid synthesis process is temperature; variations in temperature alter the nutrient's composition as well as the activity of several important enzymes involved in the microalgal cells' metabolic process. The lipid content of a *Monoraphidium* strain decreased from 32.9% to 29.6% as the ambient temperature rose from 25°C to 35°C (Wu et al., 2013). The lipid content of *C. vulgaris* decreased from 14.71% to 5.90% when the temperature was raised from 25°C to 30°C, whereas the lipid content of *Nannochloropsis oculata* increased from 7.90% to 14.92% when the temperature was raised from 25°C to 30°C (Converti et al., 2009). The environmental pH value has an impact on the dynamic forms and the relative concentration of inorganic carbon sources in the culture medium. It is also a highly significant complete indication of microalgal metabolic activities under certain conditions. As a result, pH has a significant impact on microalgae cell development and lipid accumulation. As a result, pH has a significant impact on microalgae cell development and lipid accumulation. When the pH was maintained at 7.5, the maximum lipid productivity of *T. suecica* was 92 ± 13.1 mg/L/d, while *Chlorella* sp. achieved 99 ± 17.2 mg/L/d when the pH was maintained at 7.0 (Moheimani, 2013). Zhang et al. (2014) claim that at an initial pH of 7.0, *Chlorella* sp. also produced the maximum amount of lipids (167.5 mg/L). To maximize the accumulation of microalgal lipids, systematic optimization techniques that integrate different biomass enhancement approaches with nutritional and environmental stress should be determined. More research into these synergistic techniques for large-scale microalgal lipid synthesis with financial viability is still essential.

8.3.2 Secondary Metabolites

8.3.2.1 Pigments

Chlorophylls are greenish pigments found in microalgae and cyanobacteria. They are in charge of photosynthesis, which involves the absorption and conversion of sunlight into chemical energy. There are five different types of chlorophylls, although chlorophyll A is more prevalent in microalgae. The pigment found in green microalgae with the second-highest quantity is chlorophyll b. Conversely, haptophytes, dinoflagellates, cryptophytes, and heterokonts have all been found to have chlorophyll c. However, Rhodophyta, cyanobacteria, and dinoflagellates were shown to have chlorophyll d and f. Microalgae contain between 0.5% and 1.0% chlorophyll per DW (Sun et al., 2023). Many microalgae, including *Ankistrodesmus falcatus*, *Chlorella* sp., *Chlamydomonas reinhardtii*, *Scenedesmus dimorphus*, and *Pavlova lutheri* can produce chlorophylls utilized in industry (Begum et al. 2016). Kagawa and Suetsugu (2007) state that various light spectral proportions, including red: far red, blue: red, green: red, and blue: green, influence the relative pigment content and may function as photomorphogenic signals for microalgae. Light intensity is an essential factor that influences pigment generation in microalgal cells. Microalgae cultured in high light intensity have been demonstrated to have a lower amount of chlorophyll, while microalgae cultured in low light intensity have greater chlorophyll concentration. Microalgae ability to produce pigment is greatly impacted by salinity. In the cell external environment, a higher concentration of salt results in an environment that is hypertonic. This state results in a net migration of water molecules out of the cell, which shrinks the cell and damages it and its various components (Begum et al. 2016). According to Avron (1992), salinity significantly affects the formation of β-carotene and chlorophyll. In blue green microalgae, he found that increasing salinity increased the synthesis of β-carotene while decreasing the amount of chlorophyll. Carotenes are made up of α- and β-carotene, which are made up of hydrocarbons. There are two types of carotenoids: primary and secondary. The primary carotene serves as the functional element of the cellular photosynthetic apparatus and is connected to membranes or certain proteins in thylakoid membrane. The secondary carotenoids are located in lipid vesicles and are usually generated in response to particular environmental stimuli. Carotenoids impact photosynthesis in two significant ways. First, they absorb light in visible spectrum bands where no chlorophyll is efficiently absorbed. They also provide photoprotection for photosynthetic systems. In chlorophyll, the energy state shift brought on by excessive absorption of light radiation is generally weakened by the photoprotective processes. It gives carotenoids stronger antioxidant ability and can successfully prevent the generation of reactive oxygen species (ROS). The orange-yellow hue is determined by the amount of β-carotene and other pigments present. β-carotene is the precursor of vitamin A. Because natural β-carotene has more bioactive qualities than other carotenoids, its demand is rising (Sun et al., 2023). The largest natural source of β-carotene has long been known to be the microalgae Dunaliella, which has a β-carotene content above 14% DW (Spolaore et al. 2006). Salinity has a significant impact on the synthesis of β-carotene, according to Avron (1992). They found that blue green microalgae increased salinity ad improved their production of β-carotene. Changes in pH have an impact on the solubility and bioavailability of nutrients. The optimal pH for β-carotene production in the blue-green microalgae *D. salina* was 7.5 (Garcıa-Gonzaleze et al., 2005) whereas the ideal pH for the green microalgae *Scenedesmus almeriensis* to produce the most carotenoids was 8.

8.3.2.2 Phycobiliproteins

Phycobiliproteins (PBPs) are functional proteins that exist as a pigment–protein complex consisting of a deacylated protein and phycobilins linked by covalent bonds, They are natural pigments and can serve as a safer and preferable colorant alternative in comparison to conventional synthetic options, which pose the risk of allergies and cancer. They can also find utility in microscopy due to their fluorescence property and can thus be incorporated into immunofluorescence probing. Furthermore, their anti-inflammatory and cancer-inhibitory properties on breast, lung, and liver cancer cells elevate their commercial potential in the biotechnology and pharmaceutical industry (Wang et al. 2023). To satisfy the rising market demand, PBP production on a wide scale now requires additional optimization. PBPs can be produced by various microalgal strains, such as *Arthrospira platensis, Porphyridium* sp., and *Aphanizomenon gracile* (Limrujiwat et al., 2022). The Nos. toc sp. SW02 algal strain, which is predicted to emerge as a new source of PBPs production, has a high yield of PBPs (31.9%) in challenging production conditions with inadequate light and nutrition (Wang et al. 2023). Numerous environmental conditions, including nutrients, of which nitrogen is one of the influencing elements affect the microalgae capacity to produce PBPs. Different strains of microalgae can use different sources of nitrogen, such as urea, nitrate, nitrite, and ammonium. The effects of ammonium (NH4Cl) and nitrate ($NaNO_3$ and KNO_3) were examined on various algal strains. *Pseudoscillatoria* sp. produced the most PBPs under NH_4Cl-supplemented culture conditions (19.99% ± 0.14%, dry weight), while *Arthrospira platensis* had the highest overall PBPs accumulation with $NaNO_3$ supplementation (22.27% ± 0.2%, dry weight) (Khazi et al., 2018). Furthermore, the amount of nitrogen source addition was not proportionate to the final generation of PBPs, and microalgae growth may be restricted by excessive nitrogen concentrations (Wang et al. 2023).

Therefore, optimum nitrogen supplementation is necessary to promote the production of PBP while avoiding potential growth restrictions caused by excessive nutritional concentrations.

8.3.2.3 Polyhydroxyalkanoates

Polyhydroxyalkanoates (PHAs) are aliphatic polyesters formed from hydroxyl-alkanoic acids. Because of their mechanical similarities, similar processing methods, and complete biodegradability, they are seen to be a viable alternative to petroleum-based plastics. Approximately half of the overall manufacturing expenses are currently incurred by commercial PHAs, which are made in fermenters employing bacteria, significant volumes of organic carbon sources, and salts in the growth media. The PHA increased commercial use is restricted to lowering production costs. PHAs hold immense potential for a number of uses, such as biofuels as methyl esters, agriculture for controlled pesticide release, and pharmaceuticals for drug delivery systems. They can also be utilized in chromatography as stationary phases for columns, in medicine for absorbable sutures and implants, and in disposable goods including packaging and personal care items (Costa et al., 2019). When exposed to dietary stress, *Arthrospira* biomass can accumulate large amounts of carbohydrates, which makes it a valuable feedstock for PHA production (Troschl et al., 2017). PHB levels in *Chlorella fritschii* and *Phaeodactylum tricornutum* can range from 10.6% to 17% of their dry weight. Because of their high starch or PHA content, *Calothrix scytonemicola, Neochlorisoleobundans*, and *Scenedesmus almeriensis* have been recognized as viable options for the manufacture of biopolymers (Onen Cinar et al., 2020). Optimizing the PHA content is essential to creating a commercially viable PHA production system from microalgae. Numerous researchers have tried to boost PHA content through the use of genetic engineering techniques or by optimizing operational parameters related to PHA production. Numerous factors, including temperature, pH, salinity, light (both quantity and quality), and the kind and quantity of nutrients present in the culture media, can affect the growth of microalgae (Madadi et al., 2021). Although it can influence the rise in lipids and carotenoids, salinity stress has been demonstrated to have no direct correlation with PHA synthesis (Troschl et al., 2017). Since they are assimilated into nitrogen-containing organic components like proteins, enzymes, and chlorophylls, nitrogen compounds particularly ammonium and nitrate contribute more than 10% of the biomass of microalgae (Shen et al., 2015). Another crucial component for microalgae growth is phosphorus, which is involved in the synthesis of proteins, lipids, and intermediate carbohydrate molecules (Costa et al., 2019). This allows for the manipulation of these nutrients composition to promote the production of particular by-products, like PHAs. Limiting nitrogen and phosphorus in the medium culture of certain microalgae species, for instance, has been shown to promote the accumulation of lipids, PHAs, and other carbonaceous chemicals because it inhibits the production of nitrogen molecules like proteins (Madadi et al., 2021). Nutrient deficiencies in microalgae culture media have been proposed as the primary means of encouraging PHA production (Shen et al., 2015; Madadi et al., 2021). In the end, utilizing nutrient management techniques and understanding the metabolic processes of microalgae will be essential for enhancing biorefineries targeted at producing PHA sustainably, thereby encouraging the shift to environmentally acceptable substitutes for traditional plastics.

8.4 CHALLENGES AND POTENTIAL SOLUTION FOR MICROALGAE BIOREFINERIES

Production of sustainable biomass from microalgae by using the biorefinery method has many benefits along with scaling-up issues. Microalgae are highly used for the production of biomass because their growth rate is high, and they can be cultivated in nonarable lands. However, the cost of cultivation, harvesting, and downstream processing is very high and is the main obstacle for commercialization on a large scale (Chandra et al., 2019). There is a risk of contamination in microalgae cultivation in open ponds, whereas sustainable energy and maintenance is required for closed PBRs. As filtration or centrifugation requires proper equipment; therefore, the cost for harvesting microalgae is high (Singh et al., 2024). It is difficult to yield product with good quality by aligning different biorefinery methods to extract proteins, biofuels, and pigments (Xin et al., 2024). Moreover, there are different environment factors like water, nutrient source, and light intensity, which again complicate the sustainable biomass production of microalgae. Apart from this, market acceptance is another major problem for bio-based products which are derived from microalgae, and it also requires regularity approval and consumer awareness campaigns. Although the productivity potential of microalgae is very high, the cost of overall production is challenging with biofuels or petrochemicals. Advancement in genetic engineering and optimization processes enhances the productivity and decreases the cost, by achieving proper sustainability and it will require advancement in energy-efficient technologies, waste recycling, and carbon management strategies (Sarangi et al., 2022). Investment in research is needed to build technologies and improve efficiencies. Advancement in cultivation, harvesting, and extraction technologies supported by policies and market development are important for making this approach economically and ecologically adaptable (Chandra et al., 2019; Singh et al., 2024). Therefore, the success of biomass production from the biorefinery approach depends on overcoming different interrelated technical, environmental, and economic challenges for identifying the full potential of microalgae-based biorefineries.

8.5 CONCLUSION

Industrialists are looking to use the algal biomass in multidimensional way and it can be possible by using biorefinery approaches. This could be possible with algal biomass as it is massively nutrients rich containing many high-value products, which can be commercialized and economically beneficial. Microalgae-extracted biomass can be exploited for many industry-based applications such as the wastewater industry, the pharmaceutical industry, and agri-based products. Apart from this, microalgae are a promising substitute for nonrenewable energy sources such as biofuels, biodiesel, and bioplastics. It needs an integrated approach using novel strains or a combination of strains, process designing, and successive extraction from the same biomass to reduce the cost of downstream processing. Many factors such as cultivation system, physiochemical parameters, and extraction and harvesting techniques directly impact the bio-refinery process, so it is very important to opt the best possible approach to optimize the process. That is why, it is very important to take a detailed overview of the requirements of biorefinery process and investigate thoroughly. To achieve the goals of microalgae biorefinery, a strategy is important considering the key challenges and planning the treatment processes using advanced technologies such as metabolic engineering and AI. By considering all the above-mentioned points, one can design and pursue a viable microalgae biorefinery process for the sustainable production of value-added products.

ACKNOWLEDGMENTS

We would like to thank Amity University for providing all infrastructure facilities. Financial support for this study had been provided by the AISTDF Secretariat, Science & Engineering Research Board (SERB), India [Grant number CRD/2022/000595].

REFERENCES

Alam, M. A., Xu, J. L., & Wang, Z. 2020. *Microalgae biotechnology for food, health and high value products*, 81–123. Singapore: Springer.

Alvarez, A. L., Weyers, S. L., Goemann, H. M., Peyton, B. M., & Gardner, R. D. 2021. Microalgae, soil and plants: a critical review of microalgae as renewable resources for agriculture. *Algal Research* 54: 102200.

Amorim, M. L., Soares, J., Coimbra, J., Leite, M. O., Albino, L. F. T., & Martins, M. A. 2021. Microalgae proteins: production, separation, isolation, quantification, and application in food and feed. *Critical Reviews in Food Science and Nutrition* 61: 1976–2002.

Arguelles, E. D., & Martinez-Goss, M. R. 2021. Lipid accumulation and profiling in microalgae Chlorolobion sp. (BIOTECH 4031) and chlorella sp. (BIOTECH 4026) during nitrogen starvation for biodiesel production. *Journal of Applied Phycology* 33: 1–11.

Avron, M. 1992. Osmoregulation. In *Dunaliella: Physiology, biochemistry, and biotechnology*, eds. A. Ben-Amotz, M. Avron, 1: 135–164, Boca Raton, FL: CRC Press.

Begum, H., Yusoff, F. M., Banerjee, S., Khatoon, H., & Shariff, M. 2016. Availability and utilization of pigments from microalgae. *Critical Reviews in Food Science and Nutrition* 56 (13): 2209–2222.

Boonchai, R., Kaewsuk, J., & Seo, G. 2015. Effect of nutrient starvation on nutrient uptake and extracellular polymeric substance for microalgae cultivation and separation. *Desalination and Water Treatment* 55 (2): 360–367.

Chandra, R., Iqbal, H. M., Vishal, G., Lee, H. S., & Nagra, S. 2019. Algal biorefinery: a sustainable approach to valorise algal-based biomass towards multiple product recovery. *Bioresource Technology* 278: 346–359.

Cheah, Y. T., & Chan, D. J. C. 2021. Physiology of microalgal biofilm: a review on prediction of adhesion on substrates. *Bioengineered* 12 (1): 7577–7599.

Chen, B., Wan, C., Mehmood, M. A., Chang, J.-S., Bai, F., & Zhao, X. 2017. Manipulating environmental stresses and stress tolerance of microalgae for enhanced production of lipids and value-added products – a review. *Bioresource Technology* 244 (Pt 2): 1198–1206.

Converti, A., Casazza, A. A., Ortiz, E. Y., Perego, P., & Del Borghi, M. 2009. Effect of temperature and nitrogen concentration on the growth and lipid content of *Nannochloropsis oculata* and *Chlorella vulgaris* for biodiesel production. *Chemical Engineering and Processing* 48: 1146–1151.

Costa, J. A. V., Lucas, B. F., Alvarenga, A. G. P., Moreira, J. B., & de Morais, M. G. 2021. Microalgae polysaccharides: an overview of production, characterization, and potential applications. *Polysaccharides* 2 (4): 759–772.

Costa, S. S., Miranda, A. L., de Morais, M. G., Costa, J. A. V., & Druzian, J. I. 2019. Microalgae as source of polyhydroxyalkanoates (PHAs)—a review. *International Journal of Biological Macromolecules* 131: 536–547.

de Farias Silva, C. E., Barbera, E., & Bertucco, A. 2019a. Biorefinery as a promising approach to promote ethanol industry from microalgae and cyanobacteria. In *Bioethanol production from food crops*, eds. R. C. Ray & S. Ramachandran, 343–359. London: Elsevier.

de Farias Silva, C. E., Sforza, E., & Bertucco, A. (2019b). Enhancing carbohydrate productivity in photosynthetic microorganism production: a comparison between cyanobacteria and microalgae and the effect of cultivation systems. In *Advances in feedstock conversion technologies for alternative fuels and bioproducts*, ed. M. Hossein, 37–67. London: Elsevier.

de Souza, M. P., Sanchez-Barrios, A., Rizzetti, T. M., et al. 2019. Concepts and trends for extraction and application of microalgae carbohydrates. In *Microalgae: From physiology to application*, 1. London: IntechOpen.

Deepa, P., Sowndhararajan, K., & Kim, S. 2023. A review of the harvesting techniques of microalgae. *Water* 15 (17): 3074.

Eltanahy, E., & Torky, A. 2021. Chapter 1: Microalgae as cell factories: Food and feed-grade high-value metabolites. In *Microalgal biotechnology: Recent advances, market potential, and sustainability*, eds. A. Shekh, P. Schenk, & R. Sarada, 1–35. London: The Royal Society of Chemistry.

Feng, P. Z., Xu, Z. B., Qin, L., Alam, M. A., Wang, Z. M., & Zhu, S. N. 2020. Effects of different nitrogen sources and light paths of flat plate photobioreactors on the growth and lipid accumulation of chlorella sp. GN1 outdoors. *Bioresource Technology* 301: 122762.

García-González, M., Moreno, J., Manzano, J. C., Florencio, F. J., & Guerrero, M. G. 2005. Production of Dunaliella salina biomass rich in 9-cis-β-carotene and lutein in a closed tubular photobioreactor. *Journal of Biotechnology* 115 (1): 81–90.

Ho, S.-H., Ye, X., Hasunuma, T., Chang, J.-S., & Kondo, A. 2014. Perspectives on engineering strategies for improving biofuel production from microalgae—a critical review. *Biotechnology Advances* 32 (8): 1448–1459.

Kagawa, T., & Suetsugu, N. 2007. Photometrical analysis with photosensory domains of photoreceptors in green algae. *FEBS Letters*, 581 (3): 368–374.

Khazi, M. I., Demirel, Z., & Dalay, M. C. 2018. Evaluation of growth and phycobiliprotein composition of cyanobacteria isolates cultivated in different nitrogen sources. *Journal of Applied Phycology* 30: 1513–1523.

Kucmanová, A., & Gerulová, K. 2019. Microalgae harvesting: a review. *Research Papers Faculty of Materials Science and Technology Slovak University of Technology* 27 (44): 129–143.

Laraib, N., Manzoor, M., Javid, A., et al. 2021. Mixotrophic cultivation of Chlorella vulgaris in sugarcane molasses preceding nitrogen starvation: biomass productivity, lipid content, and fatty acid analyses. *Environmental Progress & Sustainable Energy* 40: e13625.

Li, T. T., Zheng, Y. B., Yu, L., & Chen, S. L. 2014. Mixotrophic cultivation of a Chlorella sorokiniana strain for enhanced biomass and lipid production. *Biomass Bioenergy* 66: 204–213.

Limrujiwat, K., Supan, S., & Khetkorn, W. 2022. Cyanobacterial biodiversity from Thai karstic caves as a potential source for phycobiliprotein production. *Algal Research* 64: 102666.

Lucakova, S., Branyikova, I., & Hayes, M. 2022. Microalgal proteins and bioactives for food, feed, and other applications. *Applied Sciences* 12 (9): 4402.

Madadi, R., Maljaee, H., Serafim, L. S., & Ventura, S. P. 2021. Microalgae as contributors to produce biopolymers. *Marine Drugs* 19 (8): 466.

Mantzorou, A., & Ververidis, F. 2019. Microalgal biofilms: a further step over current microalgal cultivation techniques. *Science of the Total Environment* 651: 3187–3201.

Mohd-Sahib, A. A., Lim, J. W., Lam, M. K., et al. 2017. Lipid for biodiesel production from attached growth Chlorella vulgaris biomass cultivating in fluidized bed bioreactor packed with polyurethane foam material. *Bioresource Technology* 239: 127–136.

Moheimani, N. R. 2013. Inorganic carbon and pH effect on growth and lipid productivity of *Tetraselmis suecica* and *Chlorella* sp. (Chlorophyta) grown outdoors in bag photobioreactors. *Journal of Applied Phycology* 25: 387–398.

Novoveská, L., Nielsen, S. L., Eroldoğan, O. T., et al. 2023. Overview and challenges of large-scale cultivation of photosynthetic microalgae and cyanobacteria. *Marine Drugs* 21 (8): 445.

Onen Cinar, S., Chong, Z. K., Kucuker, M. A., Wieczorek, N., Cengiz, U., & Kuchta, K. 2020. Bioplastic production from microalgae: a review. *International Journal of Environmental Research and Public Health* 17 (11): 3842.

Orejuela-Escobar, L., Gualle, A., Ochoa-Herrera, V., & Philippidis, G. P. 2021. Prospects of microalgae for biomaterial production and environmental applications at biorefineries. *Sustainability* 13 (6): 3063.

Qin, S., Wang, K., Gao, F., Ge, B., Cui, H., & Li, W. 2023. Biotechnologies for bulk production of microalgal biomass: from mass cultivation to dried biomass acquisition. *Biotechnology for Biofuels and Bioproducts* 16 (1): 131.

Rosmahadi, N. A., Rwandan, H., Lim, J. W., et al. 2022. Enhancing growth environment for attached microalgae to populate onto spent coffee grounds in producing biodiesel. *Renewable and Sustainable Energy Reviews* 169: 112940.

Samiee-Zafarghandi, R., Karimi-Sabet, J., Abdoli, M. A., & Karbassi, A. 2018. Increasing microalgal carbohydrate content for hydrothermal gasification purposes. *Renewable Energy* 116: 710–719.

Sarangi, P. K., Singh, T. A., Singh, N. J., et al. 2022. Sustainable utilization of pineapple wastes for production of bioenergy, biochemicals and value-added products: a review. *Bioresource Technology* 351: 127085.

Schnurr, P. J., & Allen, D. G. 2015. Factors affecting algae biofilm growth and lipid production: a review. *Renewable and Sustainable Energy Reviews* 52: 418–429.

Schulze, C., Strehle, A., Merdivan, S., & Mundt, S., 2017. Carbohydrates in microalgae: comparative determination by TLC, LC-MS without derivatization, and the photometric thymol-sulfuric acid method. *Algal Research* 25: 372–380.

Shen, X. F., Chu, F. F., Lam, P. K., & Zeng, R. J. 2015. Biosynthesis of high yield fatty acids from Chlorella vulgaris NIES-227 under nitrogen starvation stress during heterotrophic cultivation. *Water Research* 81: 294–300.

Shen, Y., Cui, Y., & Yuan, W. 2013. Flocculation optimization of microalgae *Nannochloropsis oculata*. *Applied Biochemistry and Biotechnology* 169: 2049–2063.

Shen, Y., Xu, X., Zhao, Y., et al. 2014. Influence of algae species, substrata and culture conditions on attached microalgal culture. *Bioprocess and Biosystems Engineering* 37 (3): 441–445.

Shin, Y. S., Choi, H. I., Choi, J. W., Lee, J. S., Sung, Y. J., & Sim, S. J. 2018. Multilateral approach on enhancing economic viability of lipid production from microalgae: a review. *Bioresource Technology* 258: 335–344.

Singh, A. K., Srivastava, R. K., Pal, P., et al. 2024. Microalgal biorefinery as a sustainable and cost-effective platform for co-production of high-value-added products/metabolites: an insight into emerging trends, challenges, and opportunities. *Biocatalysis and Agricultural Biotechnology* 58: 103192.

Spolaore, P., Joannis-Cassan, C., Duran, E., & Isambert, A. 2006. Commercial applications of microalgae. *Journal of Bioscience and Bioengineering* 101 (2): 87–96.

Sun, H., Wang, Y., He, Y., Liu, B., Mou, H., Chen, F., & Yang, S. 2023. Microalgae-derived pigments for the food industry. *Marine Drugs* 21 (2): 82.

Tan, J. S., Lee, S. Y., Chew, K. W., et al., 2020. A review on microalgae cultivation and harvesting, and their biomass extraction processing using ionic liquids. *Bioengineered* 11 (1): 116–129.

Tang, D. H., Han, W., Li, P. L., Miao, X. L., & Zhong, J. J. 2011. CO_2 biofixation and fatty acid composition of *Scenedesmus obliquus* and *Chlorella pyrenoidosa* in response to different CO_2 levels. *Bioresource Technology* 102: 3071–3076.

Templeton, D., & Laurens, L. M. 2015. Nitrogen-to-protein conversion factors revisited for applications of microalgal biomass conversion to food, feed and fuel. *Algal Research* 11: 359–367.

Troschl, C., Meixner, K., & Drosg, B. 2017. Cyanobacterial PHA production—review of recent advances and a summary of three years' working experience running a pilot plant. *Bioengineering* 4 (2): 26.

Trovão, M., Schüler, L. M., Machado, A., et al. 2022. Random mutagenesis as a promising tool for microalgal strain improvement towards industrial production. *Marine Drugs* 20 (7): 440.

Vuppaladadiyam, A. K., Prinsen, P., Raheem, A., Luque, R., & Zhao, M. 2018. Microalgae cultivation and metabolites production: a comprehensive review. *Biofuels, Bioproducts and Biorefining* 12 (2): 304–324.

Wang, J., Qin, S., Lin, J., Wang, Q., Li, W., & Gao, Y. 2023. Phycobiliproteins from microalgae: research progress in sustainable production and extraction processes. *Biotechnology for Biofuels and Bioproducts* 16 (1): 170.

Wang, J., Singer, S. D., Souto, B. A., et al., 2022. Current progress in lipid-based biofuels: feedstocks and production technologies. *Bioresource Technology* 351: 127020.

Wang, Y., Tibbetts, S. M., & McGinn, P. J. 2021. Microalgae as sources of high-quality protein for human food and protein supplements. *Foods* 10 (12): 3002.

Wu, L. F., Chen, P. C., & Lee, C. M. 2013. The effects of nitrogen sources and temperature on cell growth and lipid accumulation of microalgae. *International Biodeterioration and Biodegradation* 85: 506–510.

Xin, Y., Wu, S., Miao, C., Xu, T., & Lu, Y. 2024. Towards lipid from microalgae: products, biosynthesis, and genetic engineering. *Life* 14 (4): 447.

Zhang, Q., Wang, T., & Hong, Y. 2014. Investigation of initial pH effects on growth of an oleaginous microalgae *Chlorella* sp. HQ for lipid production and nutrient uptake. *Water Science and Technology* 70: 712–719.

9 Linking Microalgae-Based Processes and Conversion of Water Pollutants for Obtaining Valuable Compounds

Simone Kubeneck, Aline Frumi Camargo, Júlia Pieper Nerling, Loisleini Fontoura Saldanha, Thamarys Scapini, Natalia Klanovicz, and Helen Treichel

9.1 INTRODUCTION

The bioremediation processes are an environmentally viable alternative for removing contaminants in water. The use of microalgae as a bioremediator of polluting compounds has been discussed to reduce environmental impacts due to their metabolic versatility and adaptability to different environments. Emerging contaminants in water bodies have increasingly been reported due to anthropogenic actions involving improper disposal of effluents, inappropriate application of chemical inputs, lack of monitoring, and poor sanitation conditions. Among the primary sources of pollution are the agricultural sector and wastewater treatment plants. The inappropriate application of chemical inputs causes the leaching of these compounds and reaches groundwater. At the same time, wastewater treatment plants discharge water into aquatic environments with high levels of chemical and biochemical demand for oxygen, suspended solids, oils and greases, and xenobiotic substances (Bhatt et al. 2022).

Xenobiotic compounds are chemical substances found in the environment or organisms from industrial processes and the use of chemical compounds that can cause interference in ecosystems, especially pharmaceutical-based compounds, agricultural pesticides, and insecticides, already considered toxic to animals and human health (Goswami, Agrawal, and Verma 2022a; Štefanac, Grgas, and Dragičević 2021). Pesticides and insecticides such as diazinon and trichlorfon are applied directly in the soil. These compounds end up contaminating rivers and surface waters due to the transport of rain. These contaminants are also considered Class II chemicals according to the World Health Organization, and therefore toxic

 DOI: 10.1201/9781003536277-9

to organisms, had their metabolites detected in wastewater and urban water. The pharmaceuticals are introduced into water bodies through the disposal of domestic effluents because they are not able to metabolize entirely and, like chemical inputs, are converted into several metabolites that are more toxic than their initial formulation (Štefanac, Grgas, and Dragičević 2021). The presence of these contaminants in water bodies raises nitrogen and phosphorus levels in water, causing eutrophication, bioaccumulation, and biomagnification in invertebrate organisms (Bhatt et al. 2022; Ray and Shaju 2023).

Microalgae are chosen as an environmentally friendly alternative for removing these contaminants from effluents due to their characteristics. Known as cell factories, microalgae use photosynthetic processes for their growth and biomass production, fixing CO_2 in the environment and generating lipids, carbohydrates, and proteins in their composition (Bhatt et al. 2022). When exposed to wastewater containing pollutants, they use carbon, nitrogen, and phosphorus available in high quantities, such as nutrients for growth and based on processes of accumulation, precipitation, adsorption, complexation, and bioaccumulation, microalgae remove contaminants (Bhatt et al. 2022; Goswami, Agrawal, and Verma 2022a).

In addition, microalgae continue to be a preference in wastewater treatment due to the possibility of using biomass produced post-treatment. In a circular economy context that aims at reduction, reuse, and recycling, the use of microalgal biomass from xenobiotic bioremediation meets the objectives by obtaining valuable compounds through bioconversion such as fuels, fertilizers, biopolymers, and animal feed supplements more sustainable (Nagarajan et al. 2020). In this sense, this chapter aims to address the state-of-the-art remediation processes based on the use of microalgae of xenobiotics pollutants such as pharmaceutical-based and agricultural-based present in wastewater and to explore the compounds and value-added products that can be obtained from the bioconversion of this biomass, addressing the aspects described in Figure 9.1.

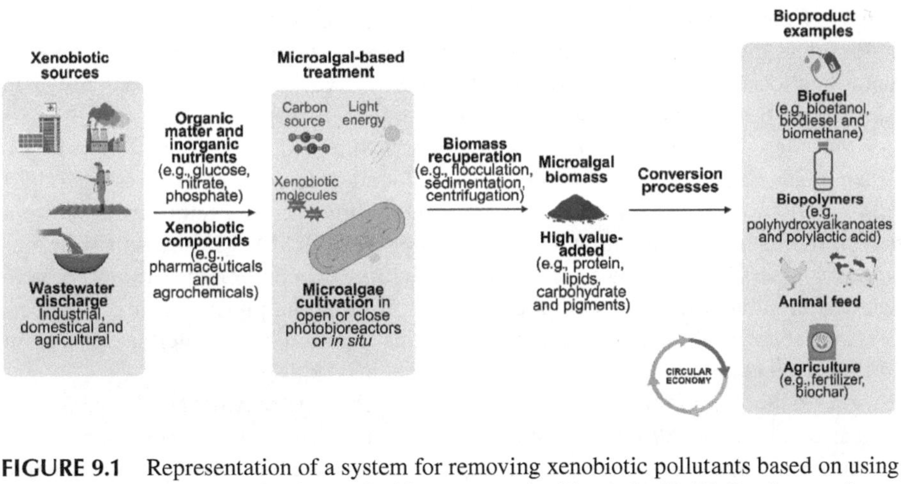

FIGURE 9.1 Representation of a system for removing xenobiotic pollutants based on using microalgae and bioconversion into valuable compounds. (Created with BioRender.com.)

9.2 MAIN SOURCES OF XENOBIOTIC POLLUTION IN THE AQUATIC ENVIRONMENT

The occurrence of xenobiotics in environmental matrices, such as surface waters, groundwater, and sediments, can be attributed to society's increasing chemical production and consumption. Wastewater is the primary pollution source for other aqueous matrices, specifically when generated by industries, households, hospitals, and agriculture (Burri et al. 2019; Islam et al. 2022). This is due to the higher concentrations of chemicals that these activities release compared to decentralized sources such as municipal wastewater. Because of the low efficiency of conventional treatment technologies to degrade and remove these xenobiotics from wastewater, these chemicals reach natural aquatic environments (Islam et al. 2022; Goswami, Agrawal, and Verma 2022a).

With this emerges the necessity of constantly and systematically monitoring these compounds, aiming to understand the extension of such environmental pollution. For this, financial support and creating research networks can be a way to exchange information and collect data. The information has been used to create lists of prioritization compounds, which are valuable as a guideline for future research, monitoring campaigns, and mitigation actions (Islam et al. 2022; Burri et al. 2019; Wilkinson et al. 2022).

An example is the NORMAN network formed by reference laboratories, research centers, and related organizations in Europe. This network has been monitoring emerging environmental substances since 2005 with the financial support by the European Commission. In addition, they maintain an open-access database with geo-referenced monitoring data on emerging substances on samples from several countries and ecosystems. The network has monitored more than four thousand compounds, mainly in aquatic environments (98% of the samples) (https://www.norman-network.net/).

When searching at NORMAN Chemical Occurrence Database (https://www.norman-network.com/nds/empodat/chemicalSearch.php?s=new) for wastewater samples from 2019 to 2024, it was found 580 xenobiotics (8396 samples). The country that had more number (n) of samples at concentrations higher than the limit of quantification (LoQ) was Switzerland (n = 1296), followed by Germany (n = 820), Austria (n = 631), Slovenia (n = 552), and Romania (n = 550).

From these 580 xenobiotics, 23 compounds had n ≥ 50, totalizing 1283 samples. When analyzing these compounds, it was found that four of them (benzothiazole-2-sulfonic acid, ensulizole, N-acetylaminoantipyrine, and telmisartan) are not part of the list of prioritization, determined by the NORMAN Prioritisation framework for emerging substances. Thus, the remaining 19 compounds (n = 1069) were detailed in Table 9.1, which compiles general information, physicochemical properties, and concentration data of these xenobiotics found in wastewater samples from 14 European countries.

In a first look at Table 9.1, it is possible to observe the prevalence of compounds used as pharmaceuticals (10 substances) and agrochemicals (5 substances). These two categories of substances can reach the aquatic environment through different routes. Still, the ones with significant concern are those related to anthropogenic activities,

TABLE 9.1
General Information and Concentration Data of Xenobiotics Most Found in Urban Wastewater Samples Collected During 2019 in Europe

Compound Name (Acronym)	CAS Number	Chemical Formula	Use Category[a]	Physicochemical Properties[b]					Concentration (µg/L)[a]		
				Mass (g/mol)	pK_a	Log K_{ow}	K_{oc} (L/kg)	Biodeg. Half-Time (Days)	Ave ± SD (*n*)	Min	Max
1,2,3-Benzotriazole (BTA)	95-14-7	$C_6H_5N_3$	Industrial chemical	119.13	8.38	1.37	55.7	3.98	3.03 ± 4.91 (63)	0.02	25.92
5-Methyl-1H-benzotriazole (MBT)	136-85-6	$C_7H_7N_3$	Industrial chemical	133.15	8.90	1.56	63.1	2.51	1.79 ± 4.06 (53)	0.02	22.21
Carbamazepine (CBZ)	298-46-4	$C_{15}H_{12}N_2O$	Pharmaceutical	236.27	5.07	2.41	550.0	6.61	0.29 ± 0.33 (67)	5.0×10^{-4}	1.27
Carbendazim (CAR)	10605-21-7	$C_9H_9N_3O_2$	PPP[c]/Biocide	191.19	7.12	1.55	172.0	4.47	1.21 ± 4.45 (52)	0.01	22.02
Diclofenac (DIC)	15307-86-5	$C_{14}H_{11}C_{12}NO_2$	Pharmaceutical	296.15	0.91	4.27	5750.0	4.68	1.21 ± 1.14 (67)	0.01	6.46
Hydrochlorothiazide (HCTZ)	58-93-5	$C_7H_8ClN_3O_4S_2$	Pharmaceutical	297.73	9.25	−0.06	52.5	7.24	1.06 ± 1.20 (54)	4.2×10^{-3}	4.91
Imidacloprid (IMI)	138261-41-3	$C_9H_{10}ClN_5O_2$	PPP/Biocide	255.66	1.10	0.43	74.2	3.55	0.07 ± 0.15 (52)	2.4×10^{-3}	0.97
Lamotrigine (LTG)	84057-84-1	$C_9H_7C_{12}N_5$	Pharmaceutical	256.09	5.22	1.26	178.0	6.17	0.39 ± 0.42 (55)	0.02	2.60
Metformin (MET)	657-24-9	$C_4H_{11}N_5$	Pharmaceutical	129.17	13.3	−1.42	9.5	4.79	15.83 ± 33.93 (56)	4.1×10^{-3}	139.96
N,N-Diethyl-meta-toluamide (DEET)	134-62-3	$C_{12}H_{17}NO$	PCP[d]/Biocide	191.27	2.14	2.16	191.0	3.39	0.71 ± 1.69 (50)	1.7×10^{-3}	10.17
Sucralose (SCL)	56038-13-2	$C_{12}H_{19}Cl_3O_8$	Food additive	397.63	7.31	−0.29	28.2	4.37	9.75 ± 11.22 (56)	0.02	47.08

(Continued)

TABLE 9.1 (*Continued*)
General Information and Concentration Data of Xenobiotics Most Found in Urban Wastewater Samples Collected During 2019 in Europe

Compound Name (Acronym)	CAS Number	Chemical Formula	Use Category[a]	Physicochemical Properties[b] Mass (g/mol)	pK_a	Log K_{ow}	K_{oc} (L/kg)	Biodeg. Half-Time (Days)	Concentration (µg/L)[a] Ave ± SD (*n*)	Min	Max
Sulfamethoxazole (SMX)	723-46-6	$C_{10}H_{11}N_3O_3S$	Pharmaceutical	253.28	6.18	0.73	91.2	3.31	0.25 ± 0.35 (67)	9.7×10^{-4}	1.64
Sulfapyridine (SPD)	144-83-2	$C_{11}H_{11}N_3O_2S$	Pharmaceutical	249.29	5.37	0.35	148.0	3.39	0.16 ± 0.32 (53)	1.0×10^{-3}	2.07
Tebuconazole (TEB)	107534-96-3	$C_{16}H_{22}ClN_3O$	PPP/Biocide	307.82	7.03	3.73	715.0	3.16	2.78 ± 9.71 (53)	8.0×10^{-4}	45.89
Terbutryn (TER)	886-50-0	$C_{10}H_{19}N_5S$	PPP/Biocide	241.36	4.83	3.53	657.0	3.39	1.40 ± 5.94 (62)	4.1×10^{-4}	37.22
Tramadol (TRA)	27203-92-5	$C_{16}H_{25}NO_2$	Pharmaceutical	263.38	8.69	2.52	562.0	3.39	0.46 ± 0.56 (50)	0.01	2.59
Triethyl phosphate (TEP)	78-40-0	$C_6H_{15}O_4P$	Flame retardant	182.16	–	0.94	60.3	4.90	0.53 ± 0.88 (52)	1.2×10^{-3}	4.74
Trimethoprim (TMP)	738-70-5	$C_{14}H_{18}N_4O_3$	Pharmaceutical	290.32	7.11	0.89	115.0	4.27	0.12 ± 0.12 (51)	0.01	0.41
Valsartan (VAL)	137862-53-4	$C_{24}H_{29}N_5O_3$	Pharmaceutical	435.53	1.71	3.18	589.0	7.41	2.13 ± 7.24 (56)	3.0×10^{-3}	49.40

Ave, Average; SD, standard deviation; Max, maximum; Min, minimum concentrations. The number of analyzed samples (n) is given in parentheses. pK_a; Dissociation constant; Log K_{ow}; Octanol-water partition coefficient. K_{oc}, soil adsorption coefficient.

[a] NORMAN EMPODAT Database (https://www.norman-network.com/nds/empodat/chemicalSearch.php?s=new), restricting the search for wastewater samples, period 2019–2024, concentrations >LOQ, and samples per compound was ≥50.

[b] CompTox predicted average (https://comptox.epa.gov/dashboard/).

[c] PPP, plant protection product.

[d] PCP, personal care product.

such as industries, urbanization, and agriculture. Once in contact with water, these chemicals can migrate or accumulate because of natural water cycle processes, such as infiltration, surface and subsurface runoff, percolation, and leaching (Burri et al. 2019; Islam et al. 2022).

Pharmaceuticals have been considered xenobiotics only in recent decades, and with the advances in the medical field, the abundance of new compounds to treat human and animal diseases has been growing. They are mainly found in anthropogenically impacted water bodies, and the serious concern regarding these chemicals is that they are designed to be bioavailable to living organisms. Thus, reaching aquatic environments can potentially negatively affect the life cycle of untargeted organisms (Burri et al. 2019).

Agrochemicals, on the other hand, were recognized as environmental pollutants in the early 1960s. These substances have a crucial role in providing food security for society, which is one reason for a continued increase in their production and consumption. Meanwhile, the increasing awareness of their environmental impacts induced the creation of regulations regarding their use over the past two decades. The restrictions are advancing, particularly in Europe and North America, where several agrochemicals have been banned; however, they continue to be measured in aquatic environments, including groundwater.

Although the scientific community does not fully understand why, their physical and chemical properties and environmental parameters seem to drive the persistence of these substances. In addition, they have the potential to be transported over long distances, being found far away from the pollution source (Burri et al. 2019; Islam et al. 2022; Ray and Shaju 2023). On the other hand, several pharmaceuticals, including carbamazepine (CBZ) and diclofenac (DIC) reported in Table 9.1, do not have regulations and restrictions in leading chemical-producing countries (Europe, USA, China, Canada, and Switzerland) (Burri et al. 2019).

The compounds reported in Table 9.1 present a wide variation of chemical composition and physicochemical properties. As a consequence, these chemicals will behave differently when reaching aquatic environments. Indeed, some are more likely to achieve and persist in these environments because of their characteristics. For example, the pK_a values vary from 9.25 (hydrochlorothiazide (HCTZ)) to 0.91 (DIC). Both compounds belong to the pharmaceutical class, but the first one is more likely to dissociate completely in aqueous solution. The pK_a is related to the pH of the environment, and thus, when a compound has a high pK_a value, it will adsorb better when in acidic conditions (National Research Council 2014; Jjemba 2018).

Log K_{ow}'s values are related to a compound's mobility in aquatic environments. Negative values, such as for HCTZ (−0.06), metformin (MET) (−1.42), and sucralose (SCL) (−0.29), indicate that these substances are more likely to remain dissolved in water rather than adsorbing onto soil, sediment, and living organisms. Compared to these compounds, the xenobiotics DIC, tebuconazole (TEB), and terbutryn (TER) have higher log K_{ow} values (see Table 9.1), thus presenting hydrophobic behavior and lower bioaccumulation and toxicity potential (National Research Council 2014).

The K_{oc} is also related to mobility, but it has a particular relationship with soil adsorption in this case. Higher K_{oc} values, such as 5,750 for DIC, 715 for TEB, and 657 for TER, indicate a high affinity for soil particles, which is a good characteristic when considering water contamination. When the compound has a low K_{oc}, as is the case of MET, SCL, and HCTZ (see Table 9.1), it increases its mobility in soil and raises concerns about surface and subsurface water contamination.

Note that the physicochemical properties above are related and corroborate synergistically for a compound to migrate to the aqueous phase rather than adsorb on surfaces and pores. Besides, the environmental conditions (e.g., temperature, pH, microbial community composition, and nutrient availability) are essential for understanding the behavior of a compound once it reaches aquatic environments (National Research Council 2014).

The implications of this behavior in terms of human and environmental health will be discussed in detail in the next section of this chapter. For instance, we can highlight (1) the biodegradation half-time and (2) the concentration data on environmental samples. These two factors are dependent, i.e., the rate at which the concentration of a compound decreases by half due to biological processes depends on the initial concentration of the compound; in the same way, the initial concentration will drive the decay rate and can induce or inhibit biological activity (Mattoli et al. 2022; Hammershøj et al. 2020).

As presented in Table 9.1, biodegradation half-time values of the 19 compounds can vary from 2.51 (5-methyl-1H-benzotriazole) to 7.41 (valsartan) days, indicating that natural biological processes are inefficient in degrading these xenobiotics. It is important to note that these chemicals are often released into aquatic environments in a complex chemical mixture, i.e., the active ingredient of a pharmaceutical or biocide product is part of a formulation combining several chemicals. This fact is often ignored when estimating the biodegradation half-time, which results in the wrong output, as the global result is not the sum of the parts (Mattoli et al. 2022).

Regarding the concentration data on wastewater samples, the data presented in Table 9.1 reinforces that conventional wastewater treatment technologies have not been efficient in degrading these persistent xenobiotics. The average concentration ranged from 0.07 μg/L for the biocide imidacloprid (IMI, $n = 52$ samples) to 15.83 μg/L for the pharmaceutical MET ($n = 56$ samples). In the case of MET, there is a high standard deviation between the minimum (4.1×10^{-3} μg/L) and maximum (139.96 μg/L) concentrations.

By searching for the location of the wastewater sample that presented the higher concentration of MET, using the geo-referenced data, it was found that it is from a WWTP that collects and treats municipal wastewater from Hodonín District, Czech Republic (https://vak-hod.cz/). The plant comprises a pretreatment step, a primary sedimentation tank, and a sequencing batch reactor with activated sludge and operates at a daily flow of 6549 m^3/day. When analyzing the complete dataset of this treatment plant ($n = 8396$), it was found that the sample that contained a high concentration of MET (collected in August 2019) also had 167 other compounds, which resulted in a summed concentration of 1015.62 μg/L of mixed xenobiotics.

A similar case was reported for a municipal WWTP located in Giurgiu District (Romania), which comprises the exact treatment steps previously mentioned and operates at a daily flow of 9600 m^3/day. The sample collected in August 2019 contained low concentrations of the compounds listed in Table 9.1, but it had a total of 152 compounds, resulting in a summed concentration of 3441.13 μg/L. The case of these two WWTPs mentioned above highlights the importance of considering that the environmental pollutants are found as a complex mixture that sometimes reaches bulk concentrations. This fact creates a cumulative effect on the environment (Burri et al. 2019).

From a socioeconomic perspective, it is essential to highlight the differences between low-, middle-, and high-income countries. While monitoring campaigns, regulations, and awareness are moving in the right direction in regions such as North America and Europe, in countries from sub-Saharan Africa, south Asia, and South America, the scenarios of wastewater management and availability of safe drinking water are alarming. The study by Wilkinson et al. (2022) of pharmaceutical occurrence along 258 rivers in 104 countries of all continents (1052 sampling sites) found that the most contaminated samples were from low- and middle-income countries (e.g., Ethiopia, Tunisia, Pakistan, and India). Besides the reality of poor sanitation conditions, these regions had received limited or no previous monitoring campaigns of the aquatic environment up to this study (conducted between 2018 and 2020).

9.3 THE CONSEQUENCES FOR AQUATIC ECOSYSTEMS AND HUMAN HEALTH DUE TO EXPOSURE TO XENOBIOTICS

Aquatic ecosystems can be divided into two large groups: freshwater and marine environments. Both are home to many plant and animal species and are also an essential part of the life cycle of some species. On the other hand, aquatic environments suffer a series of effects due to exposure to different substances from different sources of pollution (AbuQamar et al. 2024).

More specifically, when it comes to runoff pollution, pesticides used in agricultural management increase the concentration of nitrogen and phosphorus, which in aquatic environments, in addition to creating dead zones, affect the biochemical processes of organisms that inhabit these environments. More specifically, these chemical compounds can break the cell membranes of aquatic organisms, hindering their reproduction and survival, and may even contribute to the extinction of species. As for human health, studies provide information on the relationship between contact with pesticides and the occurrence of cancer, reproductive problems, and cardiovascular and neurological diseases (long-term consequences) (Cavalier, Trasande, and Porta 2023).

The stability and solubility of these contaminants will guide their behavior and bioavailability in the aquatic environment. As a rule, when the marine environment is contaminated with pesticides, these compounds enter the human body through skin contact or ingesting contaminated food (via irrigation with contaminated water or direct application). After contact, poisoning may present acute symptoms, such as burning eyes, skin rashes, nausea, and dizziness. On the other hand, in long-term manifestations, the list of associations with chronic diseases is extensive, some of which are leukemia, autism, childhood cancer, Alzheimer's disease, Parkinson's disease, endocrine system disruption, neurological and developmental toxicity, among others (Xu et al. 2023). Still focusing on the effects of pesticides, studies have demonstrated their impact on the male reproductive system, including damage to sperm DNA, directly affecting the number, mobility, density, and viability of sperm (Islam et al. 2022).

As for aquatic life, the improper disposal of pesticides mainly affects mammals, corals, and fish. The main effects include DNA damage, rupture of cellular components, thinning of eggshells, reproductive failure, suppression of the immune system, excessive production of slime on scales and gills, and development of tumors. In addition to the organisms themselves, pesticides strongly impact the entire cycle of the aquatic ecosystem, limiting the flow of the food chain, causing ecological balance, and, unfortunately, extinction (AbuQamar et al. 2024).

When the aquatic environment is exposed to these compounds, the main mechanisms involved are bioaccumulation and biomagnification. Initially, contaminants can adhere to suspended particles (adsorption) and form accumulation layers, directly affecting, for example, benthic invertebrates that usually inhabit these places. The adsorption process can occur directly through contact with water, when the contaminants dissolve in the water and are absorbed by the gills and skin of fish, amphibians, and primary producers such as algae and aquatic plants. In a more complex way, these contaminants can be transmitted via the food chain, with some organisms able to metabolize and eliminate the pollutants, while others, more limited, favor bioaccumulation over time. As contaminants move up the food chain, their concentrations also rise, accumulating pollutants in the bodies of organisms. This phenomenon is called biomagnification, which is highly harmful to the health and biodiversity of the aquatic ecosystem and can have implications for humans through the consumption of fish and other contaminated organisms (Ray and Shaju 2023).

9.4 MICROALGAE AS BIOREMEDIATORS OF XENOBIOTIC POLLUTANTS

Microalgae are considered unique bioremediators because they present metabolic diversity and their ability to change in response to the environment (Kurade et al. 2016). Based on these characteristics and the need to establish more sustainable treatments, systems using microalgae have been implemented to remove various pollutants, such as pesticides, pharmaceuticals, personal care products, heavy metals, and oils, in different environments (Goswami, Agrawal, and Verma 2022a).

Some reports showing the efficiency of using microalgae to remove these pollutants in different effluents have already been published. Removal efficiency changes according to the concentration of compounds present in the effluent, the wastewater's origin and components, and the microalgae species used (Goswami, Agrawal, and Verma 2022a).

Pollutant remediation by microalgae occurs through different pathways, mainly bioadsorption, bioaccumulation, and biodegradation. Nitrogen, phosphorus, and carbon, from both organic and inorganic molecules, are assimilated and serve as nutrients for microalgae, being used by cells in the construction of carbohydrates, proteins, and lipids that contribute to the growth of these microalgae (Sutherland and Ralph 2019; Goswami, Agrawal, and Verma 2022). Extracellular degradation of pollutants through photodegradation, hydroxyl ion reaction, and enzymatic reactions can also occur (Goswami, Agrawal, and Verma 2022a).

Bioadsorption occurs through a nonmetabolic association between the surface of microalgae and pollutants. The cell wall of microalgae has negatively charged functional groups, such as carboxyl, phosphoryl, and amine, while pollutants have cationic groups, thus being attracted to the cell wall of microalgae. Among the chemical processes observed in this pathway are adsorption, complexation, ion exchange, precipitation, and chelation (Bhatt et al. 2022). Bioadsorption is carried out even by dead microalgae cells, with reports of adsorption of up to 16.7% for some drugs, such as ibuprofen and paracetamol (Goswami, Agrawal, and Verma 2022a).

In bioaccumulation, pollutants are transported from the cell surface into the cell. The bio-concentration factor indicates the accumulation of these compounds, which varies according to factors such as the ionization state, bioavailability of pollutants, metabolic rate of microalgae, and environmental parameters. This pathway is only observed in living microalgae cells and can occur through passive, passive-facilitated, or active transport. In passive transport, pollutants cross the cell wall from a region with high concentration to an area with low concentration of these compounds. In contrast, these pollutants diffuse through the cell membrane using transport proteins in passive-facilitated transport. Active transport is done with the application of energy (Bhatt et al. 2022). Reactive oxygen species, important in signaling different cellular activities of microalgae, can be generated in this process. When in very high concentrations, these free radicals can damage proteins, lipids, and even DNA, increasing the rate of mutagenesis (Goswami, Agrawal, and Verma 2022a).

Unlike bioadsorption and bioaccumulation, which filter and remove pollutants from the effluent, biodegradation transforms these pollutants into simpler, nontoxic molecules (Wu et al. 2021). In this type of remediation, the pollutant is degraded by the microalgae's metabolism. It serves as a source of nutrients and electrons, producing algal biomass free of contaminants compared to those generated in the above-mentioned mechanisms. Pollutant degradation can occur outside the cell, inside the cell, or both through catalytic reactions divided into two phases. In the first phase, pollutants are degraded into more polar compounds through hydroxylation, carboxylation, decarboxylation, methylation, demethylation, hydrogenation, ring cleavage, and oxidation-reduction. This hydrophilic nature of the metabolites formed facilitates their removal from the cells. In the second phase, the compounds formed in the first phase are transformed into larger polar molecules, increasing their solubilization in water (Bhatt et al. 2022).

Pollutants can also undergo photodegradation and photo-oxidation, both assisted by microalgae. In these cases, microalgae secrete dissolved organic matter, contributing to increased degradation of xenobiotics. In photodegradation, the compounds absorb light and undergo chemical changes, resulting in their decomposition. Already in photo-oxidation, the pollutants react with hydroxyl radicals (Bhatt et al. 2022; Goswami, Agrawal, and Verma 2022a).

Therefore, some examples of microalgae considered for the removal of contaminants are *Desmodesmus serratus* and *Parachlorella kessleri*, which are capable of degrading pharmaceutical metabolites present in ibuprofen, paracetamol, and DIC (Jiménez-Bambague et al. 2020), as well as *Scenedesmus* sp. which are effective in removing xenobiotics contaminants including CBZ and 4-octylphenol (Gojkovic et al. 2019), and *Chlamydomonas reinhardtii* which are capable for biodegradation of trichlorfon (Wan et al. 2020).

Still, other microalgae that deserve to be highlighted when removing contaminants are *Chlorella* sp., *Neochloris* sp., and *Chlorococcum* sp. as they have high efficiency in BOD removal. The biomass of *Neochloris* sp. stands out in the production of the highest amount of lipids, as well as in the removal of BOD (<90%) (Singh and Ummalyma 2020). *Chlorococcum* sp. is also known for its versatility and resilience, standing out in the field of biodegradation mitigating damage from organic compounds. In addition, it stands out for its ability to form biofilms that can aid in treatment efficiency, increasing the surface area available for pollutant removal (Singh and Ummalyma 2020).

The use of microalgae in wastewater remediation is a sustainable and efficient strategy. In addition to purifying water, they can produce biomass and biofuels, among other bioproducts with high-value-added. This strategy helps reduce pollution and promotes the circular economy and environmental sustainability (Singh and Ummalyma 2020).

9.5 BIOCONVERSION OF POLLUTANT-REMEDIATING MICROALGAE INTO VALUE-ADDED PRODUCTS

In addition to acting in the bioremediation of xenobiotic pollutants, microalgae from these processes can be exploited to obtain value-added products. In a circular bioeconomy, the use of microalgal biomass from the treatment of contaminated water can be explored to remove pollutants and recover bioproducts, such as carbohydrates, proteins, lipids, and pigments (Nagarajan et al. 2020).

It is worth mentioning that, for biomass from the bioremediation of xenobiotic pollutants to be used for bioconversion into value-added products, biomass recovery methods such as flocculation, sedimentation, centrifugation, filtration, and flotation are necessary. The flocculation process consists of the accumulation of microalgae utilizing charge neutralization, bridging, and net capture, followed by solid–liquid separation of the biomass from the medium. This method involves flocculants that can be chemical, physical, or biological and can be used as a single recovery method or as a pre-step for physicochemical recovery processes such as sedimentation (Morais Junior et al. 2020; Liu and Hong 2021). Gravitational sedimentation is among the most used methods of microalgal biomass recovery due to its low operating cost; however, to be efficient in the process, its use is only recommended for microalgae with particles larger than 70 μm and high cell density (Liu and Hong 2021).

Following sedimentation, centrifugation is another widely used recovery method due to its speed and recovery yields of up to 98%. However, even if there is a significant recovery yield, some peculiarities arise from the process that must be considered, such as the operating cost, which ends up becoming high compared to sedimentation and cell damage caused by shear forces (Morais Junior et al. 2020). Filtration, another method of biomass recovery, consists of separating solid biomass from the liquid phase employing filters or membranes, which can be performed in various ways such as vacuum filtration, pressure filtration, microfiltration, ultrafiltration, tangential flow filtration, and dead-end filtration. However, even though it is an effective process, using filter membranes increases its cost since it is often necessary to replace them due to their clogging (Morais Junior et al. 2020). On the other hand, to the recovery methods already mentioned, flotation is a simple process in which microbubbles of air are dispersed in the aqueous medium, causing the adsorbed cells to move to the surface for recovery (Morais Junior et al. 2020; Liu and Hong 2021).

Once the biomass recovery is carried out through one of the methods mentioned above and based on its composition, these bioproducts can be converted into biofuels, biofertilizers, and biopolymers, which will be addressed below (Bhatt et al. 2022; Shahid et al. 2020).

9.5.1 Biofuels

Obtaining biofuels from microalgal biomass is a widely explored field, especially for converting carbohydrates into bioethanol and lipids into biodiesel. Biomass for biofuels is an attractive alternative since it is cultivated and obtained from alternative media, fixing CO_2, and can be converted into bioenergy, such as biodiesel, bioethanol, and biomethane (Bhatt et al. 2022).

Conversion to biodiesel is of interest since most microalgae have a high content of lipids such as triglycerides, glycolipids, and phospholipids in their composition, which, after being extracted, are subjected to the transesterification process, in which they are converted into methyl esters and fatty acids for the production of biodiesel (Bhatt et al. 2022). Biomass from cultivation in contaminated water media tends to have a higher lipid content due to the presence of chemical compounds that cause oxidative stress, increasing the accumulation of lipids for cellular protection and, as a consequence, increasing the yield in biodiesel production (Okeke et al. 2022; Goswami, Agrawal, and Verma 2022a).

This behavior was observed in the removal of phenol by *Tetradesmus obliquus*, with a 1.6-fold increase in lipid content, which, when converted to biodiesel, showed characteristics by international standards (Gomaa et al. 2022). Even so, it should be noted that oxidative stress needs to be controlled to be beneficial in the accumulation of lipids and prevent cell death. Strategies for the cell to be protected from oxidative damage with high removal of xenobiotics are reported, such as the use of exogenous phytohormones, such as gibberellin, which promoted the protection of cells from *Chlorella vulgaris* that removed 91.8% of sulfamethoxazole (Yang et al. 2023).

The carbohydrate fraction of microalgae can be converted to ethanol by fermentation processes, being considered one of the most sustainable ways to obtain the chemical and produce energy. The carbohydrates in the microalgae's cell wall are broken down by hydrolysis processes to monosaccharides, followed by conversion by fermentation, usually by yeasts (Serrà et al. 2020).

Microalgal biomass can also produce biogas through anaerobic digestion that converts the carbon in the biomass into methane and CO_2 by the activity of bacteria and archaea methanogenic (Bhatt et al. 2022). Biogas production also allows the evaluation of integrated biofuel production approaches since residual biomass can be obtained sequentially. For example, biodiesel and biogas were produced from *Chlorella* sp. after the microalgae were used to treat wastewater from agricultural runoff, demonstrating a vital potential to make the technology viable (Rana et al. 2024).

Another integrated strategy was using *Arthrospira* (*Spirulina*) *platensis* biomass for ethanol production, followed by pelleting and combustion of the residual biomass for thermal energy. In this scenario, the generated ash was used to treat effluent, which was returned as a culture medium for the cyanobacterium *A. platensis* (Serrà et al. 2020).

9.5.2 Products Aimed at the Agricultural Sector

In addition to its use to obtain biofuels, microalgal biomass from the bioremediation processes can be used to get products aimed at agriculture, consolidating its advantages as a raw material in a context of circular bioeconomy, since microalgae are mainly used for the removal of chemical compounds such as pesticides from use in agriculture.

Among the main products obtained from microalgal biomass are biochar, biofertilizers, and biomass as supplementation in animal feed. Biochar is one of the by-products obtained from microalgal biomass and is widely applied in the environment. Its production occurs through pyrolysis, a process in which biomass is thermally decomposed by limiting the presence of oxygen. It can be used as an adsorbent catalyst to improve soil conditions and create a CO_2 reservoir in carbon capture. However, the use of biomass from the bioremediation process to obtain this bioproduct is still little reported in the literature, and biomass of the genus *Chlorella* is a potential for securing biochar, as it is the most used in the removal of pollutants (Nie et al. 2020; Amaral et al. 2023).

Microalgal biomass from bioremediation processes can be an alternative to the use of chemical fertilizers; due to the presence of vitamins, minerals, peptides, and amino acids in its composition, biomass can act as a biostimulant in plants, improving germination, nutrient absorption, and growth rates (Moreira et al. 2023). The preparation of microalgae-based biofertilizers takes place through aerobic or anaerobic digestion processes. Aerobic digestion can be carried out using solid or liquid compounds, whereas in solid composting, biomass, CO_2, and water are responsible for the generation of biofertilizers. In liquid composting, heterotrophic microorganisms degrade organic matter. In anaerobic digestion through hydrolysis, acidogenesis, acetogenesis, and methanogenesis processes carried out by bacteria that use microalgal biomass as a substrate, biofertilizer and biogas are produced (Magro et al. 2021). *Monoraphidium* sp. is one of the microalgae that have the potential to be used as a biofertilizer after being used in bioremediation processes. Jimenez et al. (2020) found through their study the removal of approximately 65% of PO_4 from liquids from the digestate of a biogas plant through these microalgae, a compound present in the vast majority of chemical inputs. When applied to tomato seedlings, the authors found its biostimulant potential with the improvement of plant growth and number of leaves.

Also, among the applications aimed at the agricultural sector, microalgal biomass is used as a supplement in animal feed. Studies evaluating this application began in 1950, and according to Kusmayadi et al. (2021), using this biomass as feed brings advantages to animal physiology, such as improved immune response, antibacterial and antiviral action, and resistance to diseases. Microalgae of the genera *Spirulina* and *Chlorella* are reported in the literature for their use as feed; however, when it comes to the use of biomass from pollutant bioremediation processes, the toxicity of the compound must be evaluated since there may be an accumulation of these substances in the biomass (Kusmayadi et al. 2021).

9.5.3 Biopolymers

Another application for microalgal biomass grown in alternative media, such as contaminants in water, is in the production of biopolymers. It is possible to obtain oil-based polymers such as polyhydroxyalkanoates and polylactic acid through fermentation. Therefore, the stages of conversion of biomass to polymers consist of (1) plasticization, organic and nonvolatile molecules are added to the medium to improve characteristics such as flexibility and durability; (2) mixture: compatible polymers such as polyethylene or vinyl alcohol are added to the mixture; and finally (3) compatibility: stage in which the bioplastic is stabilized to be then molded and prepared for application in the market. In this stage, the interfacial properties of the polymers added in stage (2) are modified (Morais Junior et al. 2020). Among the microalgae with potential for use in the production of biopolymers are *Chlorella* sp. and *Spirulina* sp. due to the proteins present in their composition and the carbohydrate content (Morais Junior et al. 2020; Zeller et al. 2013).

9.6 CHALLENGES AND LIMITATIONS

The removal of contaminants through the use of microalgae and the use of biomass from these processes for conversion into valuable compounds are promising and environmentally sustainable; however, they still present some challenges and limitations. Bioremediation with microalgae has been applied, and its efficiency in the process has been demonstrated. Still, for this to be possible, it is necessary to know the composition of the wastewater to which it will be exposed to choose the microalgae to be used because some compounds can inhibit growth, and therefore, there is no removal of contaminants or even the intercropping of microalgae with other microorganisms is necessary to be efficient in bioremediation (Yadav et al. 2021; Goswami, Agrawal, and Verma 2022).

Furthermore, most of the studies found in the literature are processes carried out on a laboratory scale due to the high cost of obtaining photobioreactors on an industrial scale (Bhatt et al. 2022) and having, as a consequence, the low yield in obtaining biomass for conversion into valuable compounds. Another factor that must be considered is the toxicity of the compounds obtained from the bioconversion of biomass from bioremediation processes due to the lack of studies investigating how these contaminants are stored in microalgae cells (Yadav et al. 2021).

As seen, there are still many factors to be investigated when it comes to the removal of xenobiotic contaminants based on the use of microalgae, from their application as a bioremediation agent to the pollutants that are evaluated for the use of this process due to the presence of several compounds that have not yet been investigated for their removal using microalgae.

9.7 CONCLUSION

The presence of xenobiotic contaminants in water is one of the many consequences of anthropogenic actions that affect aquatic ecosystems and human health. The search for sustainable alternatives, such as using microalgae to remove these contaminants from polluting sources, such as wastewater treatment plants, is a way to reduce the negative impacts caused by these actions. However, as observed, there are still many limitations when employing bioremediation, mainly aiming at the bioconversion of biomass into valuable compounds, and studies are still needed to fill these gaps.

REFERENCES

AbuQamar, Synan F., Mohamed T. El-Saadony, Samar S. Alkafaas, et al. 2024. "Ecological Impacts and Management Strategies of Pesticide Pollution on Aquatic Life and Human Beings." *Marine Pollution Bulletin* 206, 116613. doi:10.1016/j.marpolbul.2024.116613

Amaral, Eduarda Torres, Luana Bertoldo Y.Castro Bender, Tiele Medianeira Rizzetti, and Rosana de Cassia de Souza Schneider. 2023. "Removal of Organic Contaminants in Water Bodies or Wastewater by Microalgae of the Genus Chlorella: A Review." *Case Studies in Chemical and Environmental Engineering* 8, (December): 100476. doi:10.1016/j.cscee.2023.100476.

Bhatt, Pankaj, Geeta Bhandari, Kalpana Bhatt, and Halis Simsek. 2022. "Microalgae-Based Removal of Pollutants from Wastewaters: Occurrence, Toxicity and Circular Economy." *Chemosphere* 306, (November): 135576. doi:10.1016/j.chemosphere.2022.135576.

Burri, Nicole M., Robin Weatherl, Christian Moeck, and Mario Schirmer. 2019. "A Review of Threats to Groundwater Quality in the Anthropocene." *Science of the Total Environment* 684, 136–154. doi:10.1016/j.scitotenv.2019.05.236.

Cavalier, Haleigh, Leonardo Trasande, and Miquel Porta. 2023. "Exposures to Pesticides and Risk of Cancer: Evaluation of Recent Epidemiological Evidence in Humans and Paths Forward." *International Journal of Cancer* 152, (5): 879–912. doi:10.1002/ijc.34300.

Gojkovic, Zivan, Richard H. Lindberg, Mats Tysklind, and Christiane Funk. 2019. "Northern Green Algae Have the Capacity to Remove Active Pharmaceutical Ingredients." *Ecotoxicology and Environmental Safety* 170 (April): 644–656. doi:10.1016/j.ecoenv.2018.12.032.

Gomaa, Mohamed, Eman H. El-Naeb, Awatief F. Hifney, Mahmoud S. Adam, and Mustafa A. Fawzy. 2022. "Coupling Phenol Bioremediation and Biodiesel Production by Tetradesmus Obliquus: Optimization of Phenol Removal, Biomass Productivity and Lipid Content." *South African Journal of Botany* 151, (December): 604–613. doi:10.1016/j.sajb.2022.10.044.

Goswami, Rahul K., Komal Agrawal, and Pradeep Verma. 2022. "Phycoremediation of Nitrogen and Phosphate from Wastewater Using Picochlorum Sp.: A Tenable Approach." *Journal of Basic Microbiology* 62, 279–295. doi:10.1002/jobm.202100277.

Goswami, Rahul Kumar, Komal Agrawal, and Pradeep Verma. 2022a. "An Exploration of Natural Synergy Using Microalgae for the Remediation of Pharmaceuticals and Xenobiotics in Wastewater." *Algal Research* 64, (May): 102703. doi:10.1016/j.algal.2022.102703.

Hammershøj, Rikke, Karina K. Sjøholm, Heidi Birch, Kristian K. Brandt, and Philipp Mayer. 2020. "Biodegradation Kinetics Testing of Two Hydrophobic UVCBs-Potential for Substrate Toxicity Supports Testing at Low Concentrations." *Environmental Science: Processes and Impacts* 22, (11): 2172–2180. doi:10.1039/d0em00288g.

Islam, Md Ariful, S. M. Nurul Amin, M. Aminur Rahman, et al. 2022. "Chronic Effects of Organic Pesticides on the Aquatic Environment and Human Health: A Review."

Environmental Nanotechnology, Monitoring and Management 18, 100740. doi:10.1016/j.enmm.2022.100740.

Jimenez, Raquel, Giorgos Markou, Saida Tayibi, Abdellatif Barakat, Camille Chapsal, and Florian Monlau. 2020. "Production of Microalgal Slow-Release Fertilizer by Valorizing Liquid Agricultural Digestate: Growth Experiments with Tomatoes." *Applied Sciences* 10, (11): 3890. doi:10.3390/app10113890.

Jiménez-Bambague, Eliana M., Carlos A. Madera-Parra, Aura C. Ortiz-Escobar, Paola A. Morales-Acosta, Enrique J. Peña-Salamanca, and Fiderman Machuca-Martínez. 2020. "High-Rate Algal Pond for Removal of Pharmaceutical Compounds from Urban Domestic Wastewater under Tropical Conditions. Case Study: Santiago de Cali, Colombia." *Water Science and Technology* 82, (6): 1031–1043. doi:10.2166/wst.2020.362.

Jjemba, Patrick K. 2018. *Pharma-Ecology: The Occurrence and Fate of Pharmaceuticals and Personal Care Products in the Environment*. Wiley.

Kurade, Mayur B., Jung Rae Kim, Sanjay P. Govindwar, and Byong Hun Jeon. 2016. "Insights into Microalgae Mediated Biodegradation of Diazinon by Chlorella Vulgaris: Microalgal Tolerance to Xenobiotic Pollutants and Metabolism." *Algal Research* 20, (December): 126–134. doi:10.1016/j.algal.2016.10.003.

Kusmayadi, Adi, Yoong Kit Leong, Hong Wei Yen, Chi Yu Huang, and Jo Shu Chang. 2021. "Microalgae as Sustainable Food and Feed Sources for Animals and Humans – Biotechnological and Environmental Aspects." *Chemosphere* 271, (May): 129800. doi:10.1016/j.chemosphere.2021.129800.

Liu, Xiao-Ya, and Yu Hong. 2021. "Microalgae-Based Wastewater Treatment and Recovery with Biomass and Value-Added Products: A Brief Review." *Current Pollution Reports*, (7): 227–45. doi:10.1007/s40726-021-00184-6.

Magro, F. G., J. F. Freitag, A. Bergoli, V. A.F. Cavanhi, and L. M. Colla. 2021. "Microalgae Consortia Cultivation Using Effluents for Bioproduct Manufacture." *Reviews in Environmental Science and Biotechnology* 20, 865–886. doi:10.1007/s11157-021-09587-9.

Mattoli, Luisa, Giacomo Proietti, Claudio Marzio Quintiero, et al. 2022. "New Insight into the Evaluation of Complex Mixture Biodegradability: An UHPLC-QToF' All-Ion MS/MS' Acquisition Technique for the Untargeted and Targeted Analysis of Pharmaceutical Formulation Biodegradation." *Environmental Science: Advances* 1, (5): 725–735. doi:10.1039/d1va00038a.

Morais Junior, Wilson G., Malihe Gorgich, Priscila S. Corrêa, António A. Martins, Teresa M. Mata, and Nídia S. Caetano. 2020. "Microalgae for Biotechnological Applications: Cultivation, Harvesting and Biomass Processing." *Aquaculture* 528, 735562. doi:10.1016/j.aquaculture.2020.735562.

Moreira, Juliana Botelho, Thaisa Duarte Santos, Jessica Hartwig Duarte, Priscilla Quenia Muniz Bezerra, Michele Greque de Morais, and Jorge Alberto Vieira Costa. 2023. "Role of Microalgae in Circular Bioeconomy: From Waste Treatment to Biofuel Production." *Clean Technologies and Environmental Policy* 25, (2): 427–437. doi:10.1007/s10098-021-02149-1.

Nagarajan, Dillirani, Duu Jong Lee, Chun Yen Chen, and Jo Shu Chang. 2020. "Resource Recovery from Wastewaters Using Microalgae-Based Approaches: A Circular Bioeconomy Perspective." *Bioresource Technology* 302, 122817. doi:10.1016/j.biortech.2020.122817.

National Research Council. 2014. *A Framework to Guide Selection of Chemical Alternatives*. The National Academies Press.

Nie, Jing, Yuqing Sun, Yaoyu Zhou, et al. 2020. "Bioremediation of Water Containing Pesticides by Microalgae: Mechanisms, Methods, and Prospects for Future Research." *Science of the Total Environment* 707, (March): 136080. doi:10.1016/j.scitotenv.2019.136080.

Okeke, Emmanuel Sunday, Onome Ejeromedoghene, Charles Obinwanne Okoye, et al. 2022. "Microalgae Biorefinery: An Integrated Route for the Sustainable Production of High-Value-Added Products." *Energy Conversion and Management: X* 16, 100323. doi:10.1016/j.ecmx.2022.100323.

Rana, Qurrat ul ain, Saira Latif, Saleha Perveen, et al. 2024. "Utilization of Microalgae for Agricultural Runoff Remediation and Sustainable Biofuel Production through an Integrated Biorefinery Approach." *Bioresources and Bioprocessing* 11, (1): 8. doi:10.1186/s40643-023-00720-w.

Ray, Suryapratap, and Sanjana Thanjan Shaju. 2023a. "Bioaccumulation of Pesticides in Fish Resulting Toxicities in Humans through Food Chain and Forensic Aspects." *Environmental Analysis Health and Toxicology* 38, (3): e2023017. doi:10.5620/eaht.2023017.

Serrà, Albert, Raul Artal, Jaume García-Amorós, Elvira Gómez, and Laetitia Philippe. 2020. "Circular Zero-Residue Process Using Microalgae for Efficient Water Decontamination, Biofuel Production, and Carbon Dioxide Fixation." *Chemical Engineering Journal* 388, (May): 124278. doi:10.1016/j.cej.2020.124278.

Shahid, Ayesha, Sana Malik, Hui Zhu, et al. 2020. "Cultivating Microalgae in Wastewater for Biomass Production, Pollutant Removal, and Atmospheric Carbon Mitigation; a Review." *Science of the Total Environment* 704, (February): 135303. doi:10.1016/j.scitotenv.2019.135303.

Singh, Anamika, and Sabeela Beevi Ummalyma. 2020. "Bioremediation and Biomass Production of Microalgae Cultivation in River Water Contaminated with Pharmaceutical Effluent." *Bioresource Technology* 307, (July): 123233. doi:10.1016/j.biortech.2020.123233.

Štefanac, Tea, Dijana Grgas, and Tibela Landeka Dragičević. 2021. "Xenobiotics—Division and Methods of Detection: A Review." *Journal of Xenobiotics* 11, (4): 131–141. doi:10.3390/jox11040009.

Sutherland, Donna L., and Peter J. Ralph. 2019. "Microalgal Bioremediation of Emerging Contaminants – Opportunities and Challenges." *Water Research* 164, 114921. doi:10.1016/j.watres.2019.114921.

Wan, Liang, Yixiao Wu, Huijun Ding, and Weihao Zhang. 2020. "Toxicity, Biodegradation, and Metabolic Fate of Organophosphorus Pesticide Trichlorfon on the Freshwater Algae Chlamydomonas reinhardtii." *Journal of Agricultural and Food Chemistry* 68, (6): 1643–1653. doi:10.1021/acs.jafc.9b05765.

Wilkinson, John L., Alistair B. A. Boxall, Dana W. Kolpin, et al. 2022. "Pharmaceutical Pollution of the World's Rivers." *Environmental Sciences* 119: e2113947119. doi:10.1073/pnas.2113947119.

Wu, Xiaozhen, Jiayi Li, Zhe Zhou, et al. 2021. "Environmental Occurrence, Toxicity Concerns, and Degradation of Diazinon Using a Microbial System." *Frontiers in Microbiology* 12, 717286. doi:10.3389/fmicb.2021.717286.

Xu, Yifan, Xu Yang, Danrong Chen, et al. 2023. "Maternal Exposure to Pesticides and Autism or Attention-Deficit/Hyperactivity Disorders in Offspring: A Meta-Analysis." *Chemosphere* 313, (February): 137459. doi:10.1016/j.chemosphere.2022.137459.

Yadav, Geetanjali, Sabarathinam Shanmugam, Ramachandran Sivaramakrishnan, et al. 2021. "Mechanism and Challenges behind Algae as a Wastewater Treatment Choice for Bioenergy Production and Beyond." *Fuel* 285, (February): 119093. doi:10.1016/j.fuel.2020.119093.

Yang, Lei, Ashiwin Vadiveloo, Ai Jie Chen, Wen Zhu Liu, Dong Zhi Chen, and Feng Gao. 2023. "Supplementation of Exogenous Phytohormones for Enhancing the Removal of Sulfamethoxazole and the Simultaneous Accumulation of Lipid by *Chlorella Vulgaris*." *Bioresource Technology* 378, (June): 129002. doi:10.1016/j.biortech.2023.129002.

Zeller, Mark Ashton, Ryan Hunt, Alexander Jones, and Suraj Sharma. 2013. "Bioplastics and Their Thermoplastic Blends from Spirulina and Chlorella Microalgae." *Journal of Applied Polymer Science* 130, (5): 3263–3275. doi:10.1002/app.39559.

10 Advances in the Use of Membrane Technology for Efficient Removal of Pollutants from Water Bodies

Tesfaye Kassaw Bedru, Workisa Bacha, R. Baskaran, Mani Jayakumar, and Gurunathan Baskar

10.1 INTRODUCTION

Water pollution remains one of the most critical environmental challenges of the 21st century, presenting significant threats to ecosystems, human health, and the availability of clean water resources. The rapid industrialization and urbanization of the last century have led to the continuous discharge of various contaminants, including heavy metals, organic compounds, pharmaceuticals, and microplastics into water bodies worldwide (Jayakumar et al., 2024; Egli, 2014). These pollutants, often persistent and toxic, undermine the quality of water in both surface and groundwater systems, complicating efforts to provide safe drinking water and maintain ecological balance (Sureshvarr et al., 2010; Rajasimman et al., 2007). Traditional water treatment methods, such as sedimentation, coagulation, and activated carbon filtration, are increasingly insufficient in addressing the complexity and variety of these emerging pollutants (Bolong et al., 2009). This growing challenge has driven the search for advanced technologies that can efficiently remove pollutants and ensure the sustainability of water resources.

Among the available technologies, membrane-based filtration has emerged as one of the most promising solutions for treating polluted water due to its high separation efficiency, versatility, and adaptability (Jatoi et al., 2024; Osman et al., 2024). Membrane technology is based on physical separation processes that allow selective removal of contaminants at the molecular or ionic level, making it highly effective for a wide range of pollutants, including dissolved salts, organic compounds, pathogens, and even micropollutants (Shannon et al., 2008). The versatility of membrane technology, combined with its potential for integration into existing treatment systems, has made it a focal point in the advancement of water purification strategies (Bringas et al., 2010).

DOI: 10.1201/9781003536277-10

The early use of membrane technology in water treatment dates back to the 1960s, when reverse osmosis (RO) was first employed for desalination (El-aassar, 2012). Early membranes were primarily made from polymers such as cellulose acetate, which, while effective, suffered from limitations in terms of fouling, chemical resistance, and permeability (Strathmann, 2001). Research over the past few decades has focused on overcoming these limitations by developing new membrane materials and surface modifications that improve performance. For instance, the incorporation of nanomaterials, such as carbon nanotubes (CNTs), graphene oxide (GO), and metal–organic frameworks (MOFs), has significantly enhanced the selectivity, permeability, and fouling resistance of membranes (Liu et al., 2015).

Recent advances in membrane technology have focused not only on improving material properties but also on optimizing operational processes to increase energy efficiency and reduce operational costs. One of the major challenges in membrane-based water treatment is fouling, which occurs when contaminants accumulate on the membrane surface, leading to reduced permeability and increased energy consumption (Jepsen et al., 2018). To address this issue, researchers have developed various antifouling strategies, including surface modifications, the use of hydrophilic coatings, and the integration of antimicrobial agents into membrane structures (Thamaraiselvan and Noel, 2015). These innovations have helped extend membrane lifespan and improve the overall sustainability of membrane processes, making them more suitable for large-scale applications.

In addition to fouling mitigation, significant efforts have been directed toward enhancing the energy efficiency of membrane processes, particularly in high-pressure systems such as RO ("Fouling and Chemical Cleaning of Microfiltration Membranes", 2021). These developments, coupled with the continuous evolution of membrane materials and designs, have expanded the applicability of membrane technology beyond conventional water treatment to more complex applications, such as the treatment of industrial wastewater, desalination of brackish water, and the removal of micropollutants and emerging contaminants (Qin et al., 2009).

The objective of this chapter is to deliver a comprehensive review of the recent advances in membrane technology for the efficient removal of pollutants from water bodies. By examining the progress in membrane material development, operational improvements, and emerging applications, this chapter highlights the key breakthroughs that have enhanced the performance of membrane-based water treatment systems. Additionally, it discusses the remaining challenges and future directions for research, with a focus on developing more sustainable, energy-efficient, and scalable membrane technologies to address the growing global demand for clean water.

10.1.1 Overview of Water Pollution and Its Sources

Water pollution has become a significant environmental issue, exacerbated by rapid industrialization, agricultural intensification, and urban expansion. Water bodies worldwide are increasingly contaminated with harmful substances, ranging from chemicals to pathogens, which compromise water quality and the biodiversity of aquatic ecosystems. The pollutants that enter water bodies can be broadly categorized into chemical, biological, and physical contaminants. Chemical pollutants include heavy metals, pesticides, and pharmaceuticals, which often persist in the environment and accumulate in the food chain (Schwarzenbach et al., 2010). Biological pollutants, such as bacteria and viruses, pose serious public health risks, while physical pollutants, including sediments and plastics, disrupt aquatic habitats. The complexity and diversity of these pollutants highlight the urgent need for effective management and remediation strategies.

Point source pollution originates from identifiable, discrete locations, such as industrial facilities, wastewater treatment plants, and sewage outfalls. These sources are generally regulated through permits and monitoring systems, making them easier to manage and control (Seleman et al., 2023). Industrial wastewater, for instance, often contains high levels of toxic chemicals, including heavy metals, solvents, and hazardous organic compounds. If untreated or inadequately treated, these pollutants can enter nearby rivers and lakes, leading to long-term contamination of water resources (Wang et al., 2021). On the other hand, nonpoint source pollution is diffuse and arises from multiple, often widespread, activities. Agricultural runoff, urban stormwater, and atmospheric deposition are common examples. Nonpoint sources are harder to regulate because pollutants are transported over large areas by rainfall, snowmelt, or irrigation water, making their entry into water bodies less predictable and more difficult to control (Bernhardt et al., 2008).

Agricultural activities are particularly notable contributors to nonpoint source water pollution. Fertilizers, pesticides, and animal waste applied to farmlands are washed into streams and rivers during rainfall events, introducing excess nutrients, such as phosphorus and nitrogen, into aquatic ecosystems. This nutrient enrichment often leads to eutrophication, a process where excessive nutrient loads stimulate algal blooms that deplete oxygen levels in water, causing dead zones and severely impacting aquatic life. Additionally, the widespread use of pesticides in agriculture introduces toxic compounds into water bodies, which can harm nontarget organisms and persist in the environment for long periods. Industrial activities also contribute significantly to water pollution, releasing contaminants such as petroleum products, heavy metals, and synthetic chemicals, which are often resistant to natural degradation.

Urbanization and industrialization have compounded the issue of water pollution, especially in developing countries where regulatory frameworks are often insufficient or poorly enforced. Rapid urban growth intensifies the discharge of untreated sewage and stormwater into rivers and lakes, resulting in the proliferation of harmful pathogens and chemical contaminants. Moreover, the increasing prevalence of emerging contaminants, such as pharmaceuticals, personal care products, and microplastics, has added a new dimension to the water pollution crisis (Sharma, n.d.). These pollutants are not fully removed by conventional wastewater treatment processes and are now frequently detected in aquatic environments. Addressing these challenges requires a holistic approach that integrates improved regulatory mechanisms, advanced treatment technologies, and sustainable landuse practices to mitigate both point and nonpoint sources of pollution.

10.1.2 Importance of Clean Water for Environmental and Human Health

Clean water is fundamental to both environmental sustainability and human health. As an essential resource, water supports a wide range of ecological processes, sustains biodiversity, and is a critical component of human survival. Clean water is necessary for drinking, agriculture, sanitation, and industrial processes, making it a key driver of social and economic development (Hellegers and Halsema, 2019). However, the availability of clean water is increasingly threatened by pollution, overextraction, and climate change, leading to a global water crisis that affects millions of people and ecosystems. Ensuring the quality and availability of clean water is not only vital for maintaining healthy ecosystems but also for preventing waterborne diseases, supporting food production, and fostering economic growth.

From an environmental perspective, clean water is crucial for maintaining the balance of aquatic ecosystems. Aquatic habitats such as rivers, lakes, wetlands, and oceans depend on unpolluted water to sustain biodiversity and ecological integrity. These ecosystems provide essential services, including nutrient cycling, flood regulation, and carbon sequestration, which are vital for the overall health of the planet (Everard and Powell, 2002). Polluted water bodies, on the other hand, disrupt these functions, leading to the degradation of habitats, loss of biodiversity, and reduced resilience of ecosystems to environmental stressors. For example, excessive nutrient pollution from agricultural runoff can cause eutrophication in freshwater systems, resulting in harmful algal blooms that deplete oxygen levels and create dead zones, severely impacting aquatic life.

Clean water is equally important for human health. The World Health Organization approximates that around 2 billion people absence access to safely managed drinking water leading to a significant burden of waterborne diseases such as diarrhea, cholera, and dysentery (Jayawardena, 2021). Contaminated water also contributes to long-term health issues, including exposure to toxic chemicals like arsenic, lead, and mercury, which can cause neurological damage, cancer, and other chronic conditions. Access to clean water is not only a matter of public health but also a human right recognized by the United Nations (Salman, 2015). The challenges of maintaining clean water supplies are multifaceted and require coordinated efforts across various sectors. Water pollution from industrial discharges, agricultural runoff, and untreated sewage continues to degrade water quality, while overextraction of freshwater resources for agriculture and industry creates water scarcity in many regions (Naidoo, 2022). Addressing these issues requires a holistic approach that includes stricter regulations on water pollution, investment in sustainable water infrastructure, and the promotion of water conservation practices. Protecting clean water resources is essential not only for environmental sustainability but also for ensuring the health and well-being of future generations.

10.2 MEMBRANE TECHNOLOGY: PRINCIPLES AND IMPORTANCE

Membrane technology encompasses a range of separation processes that employ semipermeable membranes to selectively separate components from mixtures based on their physical and chemical properties. These membranes serve as barriers that allow certain molecules or ions to pass while retaining others, enabling efficient separation and purification. The application of membrane technology spans various industries, including water treatment, food processing, pharmaceuticals, and energy production, making it a crucial element in addressing contemporary challenges such as resource scarcity and environmental sustainability (San Román et al., 2010).

The membranes used in these processes can be categorized into different groups based on their pore size: microfiltration (MF), ultrafiltration (UF), nanofiltration (NF), and RO, each suited for specific applications. For example, MF membranes typically remove larger particles and microorganisms, while RO membranes can eliminate dissolved salts and small organic molecules. The growing emphasis on sustainable practices has driven innovations in membrane materials and configurations, enhancing their performance and efficiency (Li et al., 2011). As a result, membrane technology is increasingly recognized as a vital tool for achieving cleaner production processes and conserving resources.

The core principle of membrane technology is selective permeability, which allows membranes to differentiate between various components of a mixture. This selectivity arises from the physical and chemical characteristics of the membrane material, including pore size, charge, and affinity for specific solutes. For instance, in MF processes, membranes with larger pore sizes can effectively retain bacteria and suspended solids while allowing smaller solutes like water and ions to pass through (Baker and Lokhandwala, 2008). Conversely, RO employs dense membranes that prevent even small ions from passing, making it highly effective for desalination applications.

Membrane fouling represents a significant challenge in membrane technology that can adversely affect system efficiency and longevity. Fouling occurs when unwanted materials accumulate on the membrane surface or within its pores, leading to reduced permeate flux and increased operational costs (Gkotsis et al., 2014). Addressing fouling is essential for maintaining the efficacy of membrane processes and ensuring their long-term viability.

Another critical aspect of membrane technology is the choice of membrane materials. The development of advanced membrane materials, such as polymeric, ceramic, and composite membranes, has significantly enhanced performance characteristics such as permeability, selectivity, and resistance to fouling (Osman et al., 2024).

The configuration of membrane systems also plays a vital role in determining their efficiency. Common configurations include spiral-wound modules, flat-sheet modules, and hollow-fiber membranes. Each configuration offers distinct advantages regarding space utilization, ease of cleaning, and scalability (Kaur et al., 2022). Selecting the appropriate configuration based on the application requirements is crucial for optimizing system performance and minimizing operational costs.

Finally, the integration of membrane technology with other separation processes can enhance overall efficiency. Hybrid systems that combine membrane processes with conventional methods such as adsorption or distillation can lead to improved separation performance and lower energy consumption (Obotey Ezugbe and Rathilal, 2020). This synergistic approach leverages the strengths of different technologies to achieve optimal outcomes in various applications.

The significance of membrane technology is particularly evident in its part in addressing global water scarcity issues. As freshwater resources become increasingly limited due to population growth and climate change, membrane-based desalination processes offer a viable solution for providing potable water from seawater or brackish sources. RO has emerged as a standard method for desalination, supplying millions with safe drinking water worldwide (Baker and Lokhandwala, 2008). The continuous advancements in membrane materials and designs have further enhanced the efficiency and cost-effectiveness of these processes.

In addition to water treatment, membrane technology plays a crucial role in food processing. Techniques such as MF and UF are employed for clarifying juices and concentrating dairy products while preserving their nutritional quality. These methods not only improve product quality but also reduce energy consumption compared to traditional thermal methods. The ability to achieve high separation efficiencies with minimal energy input makes membrane technology an attractive option for sustainable food production.

The pharmaceutical industry also benefits significantly from membrane technology. Membranes are utilized in various applications, including drug formulation, purification of active pharmaceutical ingredients, and controlled drug delivery systems. The precision offered by membrane processes ensures high purity levels and consistent product quality, which are critical in pharmaceutical manufacturing (Y. Zhang et al., 2023). As regulatory requirements become more stringent, the role of membrane technology in ensuring compliance and enhancing product safety will continue to grow.

Membranes designed for gas separation can selectively remove carbon dioxide from industrial emissions, thereby reducing greenhouse gas concentrations in the atmosphere. This application not only aids in compliance with environmental regulations but also contributes to global efforts aimed at reducing carbon footprints. As industries increasingly focus on sustainability, the development of efficient membrane systems for CCS will be vital in achieving climate goals.

Finally, the versatility of membrane technology allows for its application across diverse sectors beyond those mentioned. Industries such as textiles, electronics manufacturing, and wastewater treatment increasingly rely on membrane processes for efficient separation and purification (Pendergast and Hoek, 2011). The ongoing research into novel membranes and hybrid systems ensures that this technology will continue to evolve and adapt to meet emerging challenges in various fields.

10.3 TYPES OF MEMBRANE TECHNOLOGIES

Membrane technology plays an important role in various industries by enabling the separation and purification of substances through semipermeable membranes. The main types of the membrane include MF, UF, NF, and RO. Each type has different filtration abilities depending on the particle size. MF is used for larger particles, such as bacteria. UF targets macromolecules, such as proteins. NF removes dissolved organic matter and polyvalent ions. While RO is the most advanced, filtering out even dissolved salts. In addition to these established processes, emerging membrane technologies (EMTs) is developing new materials and designs to improve performance, selectivity, and durability in applications such as water purification and gas separation (Obotey Ezugbe and Rathilal, 2020). Table 10.1 shows the working principle, pros, and cons of different membrane technologies.

10.3.1 Microfiltration

MF is a widely used membrane technology. It is mainly used to separate suspended particles, microorganisms, and large contaminants from liquids or gases. Working with membranes with pore sizes of 0.1–10 μm, MF is ideal for applications that

TABLE 10.1
Various Membrane Technology, Working Principle, Pros and Cons

Technology	Working Principle	Pros	Cons	Reference
Microfiltration (MF)	Using a membrane with a pore size between 0.1 and 10 μm works by retaining particles such as bacteria and suspended solids and allowing small molecules to flow through.	High flow rate, effective in removing large particles and saving energy.	Cannot remove dissolved salts or viruses and is limited to larger contaminants.	Sohn et al. (2011)
Ultrafiltration (UF)	It works similar to microfiltration but has a smaller pore size (0.01–0.1 μm). It retains macromolecules, proteins, and some viruses.	High efficiency in removing bacteria, proteins, and colloids, lower energy consumption compared to RO.	Unable to remove dissolved ions, gets fouling easily, and needs to be cleaned every day.	Castro and Abej (2024)
Nanofiltration (NF)	The membrane has a pore size between UF and RO (1–10 nm). It can reject small organic molecules and divalent ions. However, monovalent salts are allowed to pass through.	Effective in removing polyvalent ions and organic substances, the operating pressure is lower than RO.	Cannot effectively remove monovalent salts and limited use in seawater desalination.	Yangali-Quintanilla et al. (2011)
Reverse osmosis (RO)	Semipermeable membranes are used to remove ions, molecules, and large particles by applying pressures greater than osmotic pressure and only pure water will pass.	The highest level of purification removes almost all dissolved salts, organic substances, and microorganisms.	High energy consumption tends to scale and fouling and requires extensive pretreatment.	Skuse et al. (2021)
Emerging membrane technologies (EMTs)	Include forward osmosis, membrane distillation, and other new hybrid systems. These systems use osmotic pressure or phase change to separate contaminants.	Has the potential to use less energy compared to RO and has promising applications in challenging feed water treatment.	Operating costs are high, long-term efficacy data are limited, fouling and scalability issues,	Castro and Abej (2024)

require the elimination of larger particles while allowing small particles and molecules can pass through. The process works at low pressure, typically 1–3 bar, making it energy efficient compared to higher pressure methods such as RO. MF is often used in water purification by eliminating bacteria, suspended solids, and certain viruses and serves as an important step in the production of drinking water. It is also widely used in the food and beverage industry, especially in the dairy and brewing industries, to remove microbial contaminants without affecting the quality or taste of the product. In addition, in the field of pharmaceuticals and biotechnology, MF membranes are important for sterilizing injectable drugs and other liquids such as vaccines and other injectable drugs ensure that bacteria and large particles are removed, at the same time maintaining the effectiveness of the biologically active components (Fatima et al., 2019; Nasir et al., 2022).

The key advantage of MF is high throughput due to the larger pore size. This allows for higher flow and efficient processing in industrial applications. But the main limitation of MF is its inability to remove dissolved substances such as salts or small organic molecules. For this reason, it is often used as a pretreatment before more complex filtration processes such as UF or RO. Another challenge faced by MF systems is membrane fouling in which particles accumulate on the membrane surface over time decreased efficiency and the membrane needs to be frequently cleaned or replaced to maintain optimum performance. Despite these limitations, MF is still an indispensable technology in many industries due to its value and versatility (Le-clech et al., 2006).

10.3.2 Ultrafiltration

UF is a pressurized membrane technology used to separate macromolecules, colloids, and small particles from liquids. It works with a membrane with a pore size of 0.01–0.1 μm, which is smaller than that used in MF but the larger NF membrane, and is particularly effective at removing protein, polysaccharide, and other high molecular weight materials. This makes it widely applicable in various industries such as food and beverage technology, and wastewater treatment. This technology operates at medium pressures, typically 1–5 bar, making it an energy-efficient option for large-scale processes (Nasir et al., 2022).

Typical applications of UF are water and wastewater treatment, which is used to remove suspended solids, bacteria, viruses, and organic matter and is important for various processes including separating milk protein from other components such as lactose and water (5,6). UF is also used to purify vaccines and enzymes in pharmaceutical manufacturing. This ensures that unwanted large molecules can be eliminated while maintaining the desired bioactive compounds.

The key advantage of UF membranes is their ability to maintain high selectivity and permeability. It provides a balance between efficient filtering and throughput, but like other membrane technologies UF is prone to membrane fouling. This occurs when particles accumulate on the membrane surface, resulting in decreased efficiency. This is still a limitation of UF membrane technology but can be reduced with periodic cleaning and proper system design. Despite these challenges, advances in membrane materials, such as ceramic polymer membranes, have improved the durability and chemical resistance of UF systems, making them more durable for industrial applications (S. Al et al., 2023; J. Zhang et al., 2023). Overall, UF is a versatile and efficient membrane technology that plays an important role in purification and separation.

10.3.3 Nanofiltration

NF is a membrane technology that operates between UF and RO in terms of pore size. They typically range in size from 1 to 10 nm. They are especially effective at removing polyvalent ions, small organic molecules, and larger solutes including sodium chloride and other monovalent ions that pass through, making NF highly selective. It provides a balance between filtration and permeability. This is useful in various industries such as water purification, food processing, and pharmaceutical production. NF operates at moderate pressures, typically 4–30 bar, which offers greater energy efficiency than RO while also achieving high levels of purification and separation (Guo et al., 2022; Hilal et al., 2005).

The major applications of NF membranes are water softening and desalination. It selectively removes divalent ions such as calcium and magnesium to reduce water hardness. It is also widely used in wastewater treatment, where NF effectively removes organic pollutants, dyes, and heavy metals from industrial wastewater (García et al., 2020). NF is used in the food and beverage industry for the concentration of sugars and the removal of unwanted salts in dairy and juice production. NF is also used in the pharmaceutical industry to purify antibiotics and other small-molecule drugs, as it helps separate the active ingredient from impurities while maintaining the integrity of the compound (Mallakpour and Azadi, 2022).

Despite its advantages, NF faces challenges such as membrane fouling and limited removal of monovalent ions. When organic or inorganic substances accumulate on the membrane, fouling will occur and reduce efficiency. However, advances in membrane materials such as thin-film ceramic composite membranes improve fouling resistance of the NF membrane and increased service life. New developments in NF membrane design have focused on improving the trade-off between selectivity and permeability. This makes the filtration process more efficient in various industrial applications (Xing et al., 2024). NF is a versatile and efficient technology that plays an important role in demanding applications requiring the selective removal of specific contaminants from liquids.

10.3.4 Reverse Osmosis

RO is a high-performance membrane technology used to separate solutes from water and other solvents through a semipermeable membrane. It works using high pressure, typically 15–80 bar, and forces water molecules through the membrane while rejecting dissolved salts, organic molecules, and other impurities. The pore size of RO is very small, typically less than 0.001 μm, making it one of the most efficient filtration technologies to produce high purity water. RO is widely used in separating salt from seawater to turn seawater into drinking water as well as in the water treatment industry, pharmaceutical industry, and food processing industry (Energy et al., 2011).

The main advantage of RO is that it provides ultrapure water that removes almost all dissolved salts and contaminants. This makes desalination plants to produce fresh drinking water in regions with limited water resources. RO is also used in the electronics and pharmaceutical industries, which require ultrapure water in the production process. In food and beverage production, RO is used to concentrate liquids, such as juices, and to remove unwanted components from products (Greenlee et al., 2009). But the high energy consumption of RO caused that this technique is expensive and have some other drawbacks. Efforts are being made to improve the energy efficiency of RO systems through advances in materials and systems (Subramani and Jacangelo, 2015).

One of the major challenges associated with RO technology is membrane fouling. This occurs when particles and organic substances accumulate on the membrane surface. Fouling can reduce membrane permeability and performance. This requires frequent cleaning and maintenance. Despite these challenges, improvements in membrane technology, such as the use of thin-film composite membranes and improved pretreatment, have significantly improved RO efficiency and prolonged membrane life; additionally, innovations aimed at reducing the energy requirements of RO and membranes are helping to make the process more sustainable and cost-effective (Jay R Werber et al., 2016a).

10.3.5 Emerging Membrane Technologies

EMT is at the cutting edge of membrane science. It aims to address the limitations of traditional membrane processes such as MF, UF, NF, and RO. These new technologies focus on improving efficiency, reducing energy consumption, and increasing selectivity and durability. Important innovations in this field include the development of advanced membrane materials, such as graphene-based membranes, mixed matrix membranes, and forward osmosis (FO) systems. GO membrane provides exceptional water permeability and selectivity, with the potential to revolutionize seawater desalination and water treatment processes due to their mechanical strength (Tousley and Elimelech, 2014). Hybrid matrix membranes, which combine polymers with inorganic materials such as zeolites or metal–organic frameworks (MOFs), are gaining attraction for their capabilities to improve efficiency, reduce fouling, and increase chemical resistance (Siddique and Dutta, 2021). FO is another promising EMT. It offers an energy-efficient alternative to RO. Instead of using high pressure, FO relies on natural osmotic gradients to move water across a semipermeable membrane. This process is being explored for use in seawater desalination, wastewater treatment, and energy production. This is especially true of hybrid systems that combine FO with other technologies for more sustainable water management (Shaffer et al., 2014). EMT also includes the development of membrane distillation which uses a thermal gradient for water purification and an electrochemical membrane system that combines an electric field to increase filtration efficiency.

The main advantage of these emerging technologies lies in their ability to overcome challenges such as membrane contamination, high energy costs, and limited selectivity found in conventional membranes, but these technologies are still in the development stage meeting the challenge of scaling up for industrial applications especially in terms of material costs and long-term stability. Despite these hurdles, ongoing research and technological advancements suggest that EMTs play an important role in solving global water scarcity, environmental pollution, and the demand for sustainable energy (Roy and Smruti, 2018). As these technologies continue to develop, they are expected to provide more efficient, sustainable, and energy-efficient solutions for a wide range of industrial and environmental applications.

10.4 MECHANISMS OF POLLUTANT REMOVAL

Pollutant removal is a critical aspect of environmental management, aiming to reduce contaminants in water, air, and soil. Various mechanisms play a role in the removal of pollutants, including physical separation, chemical interactions, and biological processes. Understanding these mechanisms is essential for developing effective treatment strategies and technologies. This chapter delves into each of these mechanisms, highlighting their principles, applications, and effectiveness in pollutant removal.

10.4.1 Physical Separation

Physical separation involves the use of physical forces to remove pollutants from mixtures without altering their chemical composition. Common techniques include filtration, sedimentation, and flotation. Filtration processes utilize membranes or porous materials to separate suspended solids from liquids based on size exclusion. MF and UF are particularly effective in removing bacteria and larger particles, making them suitable for water treatment applications (Baker, 2023).

Sedimentation relies on gravitational forces to settle heavier particles out of a liquid medium. This process is widely employed in wastewater treatment plants to remove suspended solids before further treatment (Eddy et al., 2014).

Flotation techniques, including dissolved air flotation (DAF), leverage the attachment of air bubbles to suspended particles, causing them to rise to the surface for removal. DAF is particularly effective in treating oily wastewater and can achieve high removal efficiencies for various contaminants (Muñoz-Alegría et al., 2021).

Another physical separation method is adsorption, where pollutants adhere to solid surfaces. Activated carbon is commonly used due to its high surface area and porosity, enabling the removal of organic compounds and certain heavy metals from water (Wong et al., 2018). Membrane filtration technologies have gained prominence due to their ability to achieve high levels of purity with low energy consumption. RO membranes can effectively remove dissolved salts and small organic molecules, making them suitable for desalination and advanced water treatment (Ibrahim et al., 2023). Overall, physical separation mechanisms are vital for initial pollutant removal stages and often serve as precursors to more complex treatment processes.

10.4.2 Chemical Interactions

Chemical interactions involve the transformation of pollutants through various chemical reactions. These processes can include oxidation–reduction reactions, precipitation, and neutralization. Oxidation–reduction (redox) reactions are particularly significant in the degradation of organic pollutants (Garcia-Costa and Casas, 2022).

Precipitation is another chemical interaction mechanism where soluble pollutants are converted into insoluble forms that can be removed from the solution. This method is widely used for removing heavy metals from wastewater through the addition of reagents that form insoluble metal salts (Tang et al., 2016). For example, lime is often added to precipitate calcium carbonate, effectively removing lead and cadmium ions.

Neutralization processes are crucial for treating acidic or alkaline waste streams. By adding a reagent that counteracts the pH imbalance, harmful substances can be converted into less toxic forms. For instance, sulfuric acid can be neutralized with sodium hydroxide to form sodium sulfate, significantly reducing toxicity (Tyagi et al., 2011). In addition to these methods, ion exchange is a powerful chemical interaction technique used to remove ionic pollutants from water. Ion exchange resins can selectively capture specific ions while releasing others, making this method effective for softening hard water and removing heavy metals (Silva, 2023).

Chemical interactions also play a role in bioremediation strategies that employ microorganisms to degrade pollutants. In this context, chemical reactions facilitate the breakdown of contaminants into less harmful by-products (Gupta et al., 2015). Overall, chemical interactions are fundamental in enhancing pollutant removal efficiency and transforming hazardous substances into more manageable forms.

10.4.3 Biological Processes

Biological processes harness the metabolic capabilities of microorganisms to degrade or transform pollutants into less harmful substances. Bioremediation is a prominent application of this mechanism, where bacteria, fungi, or plants are utilized to clean contaminated environments. Microbial degradation can effectively remove organic pollutants such as hydrocarbons through enzymatic pathways (Gupta et al., 2021).

Phytoremediation involves the use of plants to absorb, accumulate, or degrade pollutants from soil and water. Certain plant species possess the ability to uptake heavy metals and organic contaminants through their root systems, subsequently transforming or storing these substances in their tissues (Liang et al., 2017). This method offers an eco-friendly approach to remediating contaminated sites.

Composting is another biological process that utilizes microbial activity to degrade organic matter and pollutants in soil (Hargreaves et al., 2008). This process not only mitigates pollution but also enhances soil fertility. Anaerobic digestion is a biological process that breaks down organic matter in the absence of oxygen. This method is particularly effective in treating wastewater rich in organic content. The end products include biogas—composed mainly of methane—and digestate, which can be used as a fertilizer (Omotoso and Owolabi, 2015).

This approach has been successful in accelerating the breakdown of complex organic compounds in various environments (Goswami et al., 2018). Finally, constructed wetlands utilize natural biological processes to treat wastewater. These systems mimic natural wetlands' functions by using plants and microorganisms to filter and degrade contaminants effectively (Zhao et al., 2020). Constructed wetlands offer a sustainable solution for treating municipal and industrial wastewater while providing habitat for wildlife.

10.5 RECENT ADVANCES IN THE USE OF MEMBRANE TECHNOLOGY

10.5.1 Material Innovations

Material innovation in membrane technology has made great strides in this field. This makes it possible to use it more efficiently and in various ways in water treatment, seawater desalination, gas separation, and other industrial processes. Traditional membrane materials including polymers have been enhanced by developing advanced nanomaterials like graphene, CNTs, and metal-organic structures (MOF) to provide selectivity and resistance to fouling (Gindaba et al., 2023; Goh et al., 2016). These innovations address the limitations of conventional membranes such as limited chemical stability and sensitivity to decomposition by combining materials with improved thermal and mechanical properties. Integrating nanotechnology also paves the way for more sustainable and energy-efficient processes because nano-based membranes require less energy while maintaining high filtration efficiency (Lee et al., 2016). As research continues to push the boundaries of material science and engineering, the innovations therefore have the potential to expand the scope of membrane technology to solve global challenges related to water scarcity and environmental pollution.

10.5.1.1 Development of New Membrane Materials (e.g., Graphene, Nanocomposites)

Recent advances in membrane technology have been greatly influenced by the development of new materials such as graphene, GO, and polymer-based nanocomposites. These materials have amazing properties including higher mechanical strength, chemical stability, and increased permeability, which addresses some important limitations. Compared to traditional membrane materials, graphene-based membranes have exceptional water permeability and selectivity due to their atomic structure, which makes them suitable for water desalination and filtration processes (Jiang et al., 2016). Similarly, polymer/graphene nanocomposite membranes combine the flexibility and low-cost polymer with the excellent performance of nanomaterials (Kausar and Bocchetta, 2022).

Furthermore, integrating MOFs and CNTs with graphene in nanocomposites has opened new avenues for enhancing membrane performance including selective gas separation and removal of specific contaminants from water (Bhol et al., 2021). These developments have led to membrane technology being used in advanced applications ranging from wastewater treatment to energy-efficient gas separation, providing more sustainable solutions to various industrial and environmental challenges (Ingole, 2019).

10.5.1.2 Improved Durability, Permeability, and Selectivity

Recent advances in membrane technology have led to significant improvements in durability, permeability, and selectivity. This is an important factor in increasing membrane performance in applications such as water desalination, gas separation, and wastewater treatment. The development of new materials like graphene-based membranes and MOFs has enhanced increased strength and chemical resistance leading to a longer service life (Jiang et al., 2016). These materials also allow for higher permeability without compromising selectivity (Jay R. Werber et al., 2016b). Polymers bonded to nanoparticles improve membrane fouling resistance (Kausar and Bocchetta, 2022). In addition, innovations in surface modification and structural design are helping in improving membrane selectivity. This allows for the precise separation of ions or molecules. This is important for high-efficiency applications such as carbon capture and water purification (Kausar and Bocchetta, 2022).

10.5.2 Surface Modification Techniques

10.5.2.1 Antifouling and Anti-Biofouling Coatings

Recent advances in membrane technology have led to the development of antifouling and bio-protective coatings that greatly increase the efficiency and lifespan of the filtration system. The use of fouling coatings, such as layers using aqueous polymers and nanomaterials like GO, is shown to reduce fouling adhesion to the membrane surface (Kalajahi et al., 2024). These coatings increase the wettability of the membrane, making it easier for water to flow through to prevent the build-up of unwanted materials (Lu et al., 2017).

In the context of anti-biofouling, recent innovations include the incorporation of antimicrobial agents and nanoparticles such as silver and copper into the membrane coating having antibacterial properties that prevent biofilms' growth, which is an important factor causing biological contamination (Song et al., 2018). Zwitterionic materials are known to be charge-neutral and have a high resistance to protein adsorption. It has emerged as an effective solution for controlling biological contamination. This is especially true in applications such as water purification and medical equipment (Zhang et al., 2021).

10.5.3 Hybrid Membrane Systems

10.5.3.1 Integration with Other Treatment Technologies (e.g., Adsorption and Photocatalysis)

Recent advances in membrane technology show great potential. This is especially true when integrated with other treatment methods such as adsorption and photocatalysis. The combination of these technologies increases functionality for treatment efficiency. Membrane technology alone is highly effective at filtering contaminants due to its selective permeability but membrane fouling reduced lifespan and limited removal of specific pollutants, such as dissolved organic compounds or heavy metals. To overcome these limitations integrating with adsorption techniques that capture contaminants through surface reactions or photocatalysis which uses light to decompose pollutants received a lot of attention (Gindaba et al., 2023; Nan et al., 2010; Ridhwan et al., 2022).

For example, a connected membrane with an adsorption system allows pollutants to be adsorbed on a specific material, such as activated carbon, before filtering through the membrane (Jeyapaul and Ganesapillai, 2020, 2022, 2024). This combination not only improves the efficiency of contaminant removal but also reduces the rate of membrane fouling which extends the membrane's operating life (Loganathan et al., 2023). In the same way, integrating photocatalysis, which often uses semiconductor materials such as titanium dioxide (TiO_2), allows the degradation of organic pollutants under ultraviolet light. It allows the degradation of organic pollutants under UV light. This process creates reactive particles that break down persistent contaminants that would otherwise pass through conventional membranes (M. A. Al et al., 2023). Recent studies have focused on the development of hybrid membrane photocatalytic systems wherein photocatalytic nanoparticles are immobilized on the membrane surface. This configuration that contaminants are degraded directly onto the membrane surface brings clean water to use at the same time. The result of the interaction between membrane filtration and photocatalysis leads to a system that not only physically separates pollutants but also changes chemically. This leads to a higher overall elimination rate.

In addition, combining adsorption and photocatalysis with membrane technology is being investigated to explore its potential to treat emerging contaminants including pharmaceuticals and endocrine-disrupting chemicals (EDCs), which are often resistant to conventional treatment methods. These integrated systems have been shown to have improved durability and provide a versatile solution to deal with a wide range of contaminants in municipal and industrial wastewater (Katibi et al., 2021). Recent advances highlight the need for further optimization to increase the scalability, efficiency, and cost-effectiveness of these hybrid systems. Ongoing research is exploring new materials, configurations, and operating parameters to ensure that membrane-based integrated systems continue to evolve and meet the increasing demand for effective water treatment solutions.

10.5.3.2 Benefits of Combined Approaches for Pollutant Removal

Latest advances in membrane technology especially integrating with other treatment methods have significantly increased efficiency in removing pollutants. One of the important advantages of combining membrane technology with complementary approaches such as adsorption, photocatalysis, and biological treatment is the ability to overcome limitations commonly associated with standalone membrane processes. These hybrid systems improve the removal of a wide range of contaminants, reduce membrane fouling, and increase the overall sustainability and cost-effectiveness of water and wastewater treatment systems.

Membrane filtration is highly effective in removing particles, bacteria, and some dissolved compounds often face challenges related to membrane fouling which reduces performance over time. The adsorption method was integrated into the membrane system to solve this problem. Adsorption traps contaminants at the surface of materials such as activated carbon or zeolite. This significantly improves the removal of dissolved organic substances, heavy metals, and pharmaceuticals. For example, research has shown that combining adsorption with membrane filtration not only increases efficiency but also helps delay membrane fouling, thus extending the operational lifespan of the membrane system. This is especially useful in treating industrial wastewater and complex mixtures of pollutants, which are difficult to remove by membrane filtration alone (Stoquart et al., 2012).

Another promising integration method is the integration of photocatalysis with membrane technology. Photocatalysis involves the use of light-activated carbon, commonly titanium dioxide (TiO_2), which is used to degrade organic pollutants into less harmful by-products. When used in conjunction with membrane filtration, photocatalysis breaks down persistent pollutants such as drugs and pesticides that may pass through cell membranes. Hybrid photocatalytic membrane systems had significantly higher organic contaminant removal rates. It provides a more efficient solution for wastewater treatment. Photocatalytic activity on the membrane surface helps prevent the accumulation of pollution. This reduces fouling and the permeability of the membrane increases over time (Khan et al., 2024; Nugraha et al., 2022).

Combining biological treatments with membrane technology has also gained attention in recent years. Membrane bioreactor (MBR) combines biodegradation with membrane filtration allowing for the simultaneous physical separation of organic pollutants and the separation of suspended solids. This method is especially effective for treating municipal wastewater and provides better pollutant removal compared to traditional activated sludge processes. MBR also offers the advantage of producing high-quality effluent that can be directly recycled for industrial application or released into sensitive environments (Wu and Sakdaronnarong, 2024).

Overall, the benefits of the integrated membrane technology pollutant removal approach are evident in increased removal efficiency, longer membrane life, and improved operational stability. Research and development of future hybrid membrane systems integrates processes such as adsorption, photocatalysis, and biological treatment. These combined approaches reduce costs and improve energy efficiency focused on expanding application to both industrial and municipal water treatment systems.

10.5.4 Smart and Responsive Membranes

10.5.4.1 Development of Membranes with Stimuli-Responsive Properties

Recent advances in membrane technology have led to the development of stimulus-responsive membranes which are often called "smart membrane" that dynamically responds to external stimuli such as pH, temperature, light, and magnetic or electric fields. These membranes provide control over the separation process, making applications in water purification, drug delivery, and biochemical separation highly desirable (Mills et al., 2024).

pH-responsive membranes are among the most studied including polymers such as poly(acrylic acid) or poly(methacrylic acid) that change structure in response to different pH levels. This behavior causes a selective permeability change that makes these membranes for wastewater treatment or control drug release. Temperature-responsive membranes using polymers such as poly(N-isopropyl acrylamide) adjust permeability according to temperature changes. At temperatures below the critical solution temperature (LCST), the polymer swells and increases permeability, while above the lower critical solution temperature (LCST), the polymer collapses and decreases permeability. These membranes are particularly useful in biotechnological processes such as protein or enzyme separation (Sato et al., 2023).

Light-responsive membranes use photochromic compounds, such as azobenzene, which change their structure when exposed to light. It allows precise and noninvasive control of the separation process, with applications in photodynamic therapy and controlled substance release. Similarly, membranes that respond to magnetic and electric fields use magnetic nanoparticles or conductive materials to adjust their permeability or structure when external fields are applied, offering potential in targeted drug delivery or electro-driven separations (Nicoletta et al., 2012). The development of these smart membranes offers important advantages such as tunable selectivity and the ability to adapt to changing conditions. However, challenges remain in scaling up production, ensuring long-term stability and improved cost-effectiveness for wider use.

10.5.4.2 Applications in Adaptive and Selective Pollutant Removal

Recent advances in membrane technology have greatly expanded their applications in the enhancement and selective removal of pollutants, making it a highly versatile tool for environmental protection. The ability to target and remove specific contaminants including heavy metals, organic pollution, and emerging contaminants, such as pharmaceuticals, is critical to meeting increasingly stringent water quality standards. Modern membranes are engineered with special functions such as surface modification and composites to improve adaptability in response to various types of pollutants and concentrations (Castro and Abej, 2024).

One of the most important advances has been the development of membranes that can selectively remove heavy metals from wastewater. These membranes are often modified with chelating agents or metal-binding ligands that have a high affinity for specific metal ions, such as lead (Pb^{2+}), cadmium (Cd^{2+}), or mercury (Hg^{2+}). The -SH group functionalized membrane showed a remarkable ability to selectively adsorb mercury from industrial wastewater. It provides a more efficient and environmentally friendly solution compared to conventional methods (Nickisch et al., 2023).

Another promising application of adaptive membrane technology is the removal of organic pollutants, including dyes, pesticides, and EDCs. Selective membranes designed with molecular imprinting techniques can create recognition sites that specifically bind to certain organic molecules (Kaczorowska et al., 2023). For example, the use of molecularly printed polymer membranes for the selective removal of pesticide residues from agricultural wastewater. These membranes have high selectivity and adaptability. It allows efficient adsorption of pollutants in the complex water matrix while reducing the removal of nontarget molecules (Metwally et al., 2021).

Membrane technology also advances the removal of pharmaceuticals and personal care products, an emerging contaminant that poses a significant environmental risk. Advanced hybrid membranes that combine adsorption or photocatalysis with filtration have been developed to target these micropollutants effectively. The aggregation of photocatalytic nanoparticles within a membrane selects degradative drugs in wastewater. This hybrid method not only removes the target contaminant but also reduces membrane fouling and improves the overall performance and longevity of the membrane system (Wang et al., 2018).

10.6 APPLICATIONS OF MEMBRANE TECHNOLOGY

10.6.1 Industrial Wastewater Treatment

Membrane technology plays a key role in industrial wastewater treatment by providing effective solutions for removing contaminants and facilitating water reuse. Membrane-based processes such as MF, UF, NF, and RO are widely used in various industrial sectors including petrochemicals, textiles, food, and pharmaceuticals (Tomaszewska, 2007). These processes are highly efficient in removing dissolved solids organic compounds and heavy metals, making them ideal for treating complex industrial wastewater.

MBRs are particularly important for their efficiency in combining biological treatment with membrane filtration. This results in a high-quality effluent with minimal sludge production (Al-Asheh et al., 2021). This combination helps remove more organic and inorganic pollutants, which is important in an industry with strict environmental regulations. In addition, ceramic membranes due to their durability and high chemical resistance are increasingly used in industries that have high temperatures or harsh chemicals. This results in a long service life and low maintenance costs (Zheng et al., 2015).

Although there are many advantages, the main challenges facing membrane technology are membrane fouling and energy consumption. However, advances in membrane materials and cleaning techniques continue to improve the durability and operational efficiency of these systems continuously to ensure durability in large-scale industrial applications (Obotey Ezugbe and Rathilal, 2020).

10.6.2 Municipal Water Treatment

Membrane technology has revolutionized municipal water treatment to remove contaminants and provide high performance in membrane processes for clean drinking water, such as MF, UF, NF, and RO, which are widely used in various steps of water treatment in municipal areas. These technologies provide solutions for removing suspended solids, bacteria, and viruses and provide solutions to eliminate dissolved pollutants. It is a versatile approach that can deal with both surface and underground water sources. A key advantage of membrane technology is its ability to meet strict water quality standards and requires fewer chemicals compared to traditional treatment (Othman et al., 2022). In addition, the membrane process is highly effective in treating municipal wastewater. It facilitates water reuse by converting wastewater into a valuable resource for nonpotable uses (Yang et al., 2020). The integration of membrane systems and other treatments such as MBRs has increased the efficiency and sustainability of municipal water treatment processes (Alzahrani and Wahab, 2014).

Despite the clear benefits, membrane technology faces challenges such as membrane fouling and high operating costs. This is especially true in terms of energy consumption. However, continued advances in membrane materials and cleaning technologies can overcome these problems. This makes the membrane system more robust and cost-effective for large-scale applications (Bilal and Zhang, 2021).

The increasing demand for reliable and sustainable water treatment solutions has led to the widespread adoption of membrane technology especially in water shortage areas, and this emphasizes the importance of these technologies in municipal water management (Zheng et al., 2012).

10.6.3 Agricultural Runoff and Contaminant Management

Membrane technology has shown potential in managing agricultural runoff and solving contamination problems associated with agricultural practices. The implementation of membrane-based processes has shown significant efficiency in removing nutrients, pesticides, and other pollutants. The wastewater often includes fertilizers and chemicals used in agriculture (Hamda et al., 2023; Kumar et al., 2022). This approach is especially useful in areas where there is an excess of nutrients in water sources because it helps remove excess nitrogen and phosphorus, which is an important factor that causes eutrophication.

MBRs are also increasingly being used to treat wastewater from agricultural runoff combining biological treatment with advanced membrane filtration. This technology improves water quality, promotes water reuse, and supports sustainable agricultural practices. The removal of emerging contaminants such as pesticides and pharmaceuticals through membrane technology is an important development, which deals with human and environmental health issues (Rahman et al., 2023).

Challenges such as membrane fouling and high operating costs remain; however, innovations in membrane materials and cleaning techniques have made these systems more efficient and economical (Nicolaisen, 2002). The future of membrane technology in agricultural runoff management is promising as research continues to optimize these systems for large-scale applications, and this is especially true in areas facing water shortages and pollution from agricultural activities.

10.7 CONCLUSION

This chapter explored advances in membrane technology for efficiently removing pollutants from water sources. With increasing concerns about water pollution due to pollutants such as chemical residues, heavy metals, dyes, and pharmaceuticals, this chapter emphasized the need for innovative and effective water treatment solutions. This chapter also highlighted the physical, chemical, and biological mechanisms of pollution removal and recent developments including new materials surface modification and hybrid systems that optimize membrane performance. The durability of the membrane and the ability to select and adapt are improved due to these innovations. This makes it an important tool in treating industrial, municipal, and agricultural wastewater.

REFERENCES

Al-Asheh, S., Bagheri, M., Aidan, A., 2021. Membrane bioreactor for wastewater treatment: A review. *Case Stud. Chem. Environ. Eng.* 4, 100109. https://doi.org/10.1016/j.cscee.2021.100109

Al, M.A., Asawer, N., Zainab, A.A., 2023. The photocatalytic process in the treatment of polluted water. *Chem. Pap.* 77, 677–701. https://doi.org/10.1007/s11696-022-02468-7
Al, S., Mustafa, T.N., Hilal, N., 2023. Ultra filtration membranes for wastewater and water process engineering : A comprehensive statistical review over the past decade. *J. Water Process Eng.* 35, 101241. https://doi.org/10.1016/j.jwpe.2020.101241
Alzahrani, S., Wahab, A., 2014. Challenges and trends in membrane technology implementation for produced water treatment : A review. *J. Water Process Eng.* 4, 107–133. https://doi.org/10.1016/j.jwpe.2014.09.007
Baker, R.W., 2023. *Membrane Technology and Applications*. John Wiley & Sons.
Baker, R.W., Lokhandwala, K., 2008. Natural gas processing with membranes: An overview. *Ind. Eng. Chem. Res.* 47, 2109–2121.
Bernhardt, E.S., Band, L.E., Walsh, C.J., Berke, P.E., 2008. Understanding, managing, and minimizing urban impacts on surface water nitrogen loading. *Ann. New York Acad. Sci.* 96, 61–96. https://doi.org/10.1196/annals.1439.014
Bhol, P., Yadav, S., Altaee, A., Saxena, M., Misra, P.K., Samal, A.K., 2021. Graphene-based membranes for water and wastewater treatment: A review. *ACS Appl. Nano Mater.* 4, 3274–3293. https://doi.org/10.1021/acsanm.0c03439
Bilal, M., Zhang, Z., 2021. Ceramic membrane technology for water and wastewater treatment : A critical review of performance, full-scale applications, membrane fouling and prospects. *Chem. Eng. J.* 418, 129481. https://doi.org/10.1016/j.cej.2021.129481
Bolong, N., Ismail, A.F., Salim, M.R., Matsuura, T., 2009. A review of the effects of emerging contaminants in wastewater and options for their removal. *DES* 239, 229–246. https://doi.org/10.1016/j.desal.2008.03.020
Bringas, E., Iba, R., Ortiz, I., Rom, M.F.S., 2010. Liquid membrane technology : Fundamentals and review of its applications, *J. Chem. Technol. Biotechnol.* 85, 2–10. https://doi.org/10.1002/jctb.2252
Castro, K., Abej, R., 2024. Removal of heavy metals from wastewaters and other aqueous streams by pressure-driven membrane technologies: An outlook on reverse osmosis, nanofiltration, ultrafiltration and microfiltration potential from a bibliometric analysis. *Membranes (Basel).* 14, 180. https://doi.org/10.3390/membranes14080180
Eddy, M., Abu-Orf, M., Bowden, G., Burton, F.L., Pfrang, W., Stensel, H.D., Tchobanoglous, G., Tsuchihashi, R., (Firm), A., 2014. *Wastewater Engineering: Treatment and Resource Recovery*. McGraw Hill Education.
Egli, T., 2014. Global water pollution and human health. *Ann. Rev. Environ. Resour.* 35, 109–136. https://doi.org/10.1146/annurev-environ-100809-125342
El-aassar, A.M.A., 2012. Polyamide thin film composite membranes using interfacial polymerization : Synthesis, characterization and reverse osmosis performance for water desalination. *Aust. J. Basic Appl. Sci.* 6, 382–391.
Energy, D., Elimelech, M., Phillip, W.A., 2011. The future of seawater and the environment. *Science* 333, 712–718. https://doi.org/10.1126/science.1200488
Everard, M., Powell, A., 2002. Rivers as living systems. *Aquat. Conserv.: Mar. Freshw. Ecosyst.* 337, 329–337. https://doi.org/10.1002/aqc.533
Fatima, S., Hashaikeh, R., Hilal, N., 2019. Micro filtration membrane processes : A review of research trends over the past decade. *J. Water Process Eng.* 32, 100941. https://doi.org/10.1016/j.jwpe.2019.100941
Garcia-Costa, A.L., Casas, J.A., 2022. Intensification strategies for thermal H_2O_2-based advanced oxidation processes: Current trends and future perspectives. *Chem. Eng. J. Adv.* 9, 100228.
García, N., Purcell-milton, F., Gun, Y.K., 2020. Recent progress and future prospects in development of advanced materials for nano filtration. *Mater. Today Commun.* 23, 100888. https://doi.org/10.1016/j.mtcomm.2019.100888

Gindaba, G.T., Demsash, H.D., Jayakumar, M., 2023. Green synthesis, characterization, and application of metal oxide nanoparticles for mercury removal from aqueous solution. *Environ. Monit. Assess.* 195, 9.

Gkotsis, P.K., Banti, D.C., Peleka, E.N., Zouboulis, A.I., Samaras, P.E., 2014. Fouling issues in membrane bioreactors (MBRs) for wastewater treatment: Major mechanisms, prevention and control strategies. *Processes* 2, 795–866.

Goh, P.S., Ismail, A.F., Hilal, N., 2016. Nano-enabled membranes technology: Sustainable and revolutionary solutions for membrane desalination? *Desalination* 380, 100–104. https://doi.org/10.1016/j.desal.2015.06.002

Goswami, M., Chakraborty, P., Mukherjee, K., Mitra, G., Bhattacharyya, P., Dey, S., Tribedi, P., 2018. Bioaugmentation and biostimulation: A potential strategy for environmental remediation. *J. Microbiol Exp.* 6, 223–231.

Greenlee, L.F., Lawler, D.F., Freeman, B.D., Marrot, B., Moulin, P., Ce, P., 2009. Reverse osmosis desalination : Water sources, technology, and today's challenges. *Water Res.* 43, 2317–2348. https://doi.org/10.1016/j.watres.2009.03.010

Gul, A., Hruza, J., Yalcinkaya, F., (2021). Fouling and chemical cleaning of microfiltration membranes . *Polymers* 13(6), 846.

Guo, H., Li, X., Yang, W., Yao, Z., Mei, Y., Peng, L.E., Yang, Z., Shao, S., 2022. Nano filtration for drinking water treatment : A review. *Front. Chem. Sci. Eng.*16, 681–698. https://doi.org/10.1007/s11705-021-2103-5

Gupta, S., Ivvala, J., Grewal, H.S., 2021. Development of natural wax based durable superhydrophobic coatings. *Ind. Crops Prod.* 171, 113871. https://doi.org/10.1016/j.indcrop.2021.113871

Gupta, S., Pathak, B., Fulekar, M.H., 2015. Molecular approaches for biodegradation of polycyclic aromatic hydrocarbon compounds: A review. *Rev. Environ. Sci. Bio/Technol.* 14, 241–269.

Hamda, A.S., Abo, L.D., Jayakumar, M., Selvakumar, K.V., Periyasamy, S., Emana, A.N., 2023. Introduction: Growth of Agricultural Waste, Its Disposal, and Related Environmental Issues, in: *Agricultural Waste to Value-Added Products: Technical, Economic and Sustainable Aspects*. Springer, pp. 1–19.

Hargreaves, J.C., Adl, M.S., Warman, P.R., 2008. A review of the use of composted municipal solid waste in agriculture. *Agric. Ecosyst. Environ.* 123, 1–14.

Hellegers, P., Halsema, G. Van, 2019. Weighing economic values against societal needs : Questioning the roles of valuing water in practice. *Water Policy* 21, 514–525. https://doi.org/10.2166/wp.2019.048

Hilal, N., Al-zoubi, H., Mohammad, A.W., Darwish, N.A., 2005. Nanofiltration of highly concentrated salt solutions up to seawater salinity. *Desalination* 184, 315–326. https://doi.org/10.1016/j.desal.2005.02.062

Ibrahim, M., Nawaz, M.H., Rout, P.R., Lim, J.-W., Mainali, B., Shahid, M.K., 2023. Advances in produced water treatment technologies: An in-depth exploration with an emphasis on membrane-based systems and future perspectives. *Water* 15, 2980.

Ingole, P.G., 2019. Application of Sustainable Nanocomposites in Membrane Technology, in: Inamuddin, Thomas, S., Kumar Mishra, R., Asiri, A. (eds.), *Sustainable Polymer Composites and Nanocomposites*. Springer, pp. 935–960. https://doi.org/10.1007/978-3-030-05399-4_32

Jatoi, A.S., Ahmed, J., Bhutto, A.A., Jeyapaul, A.S., 2024. Recent advances and future perspectives of carbon-based nanomaterials for environmental remediation. *Brazilian J. Chem. Eng.* 42, 151–169. https://doi.org/10.1007/s43153-024-00439-x

Jayakumar, M., Sundramurthy, V.P., Gebeyehu, K.B., Selvakumar, K.V., Emana, A.N., Manivannan, S., Mohanasundaram, S., Sagadevan, S., Baskar, G., 2024. Artificial neural network guided optimization of limiting factors for enhancing photocatalytic treatment of textile wastewater using UV/TiO_2 and kinetic studies. *Desalin. Water Treat.* 320, 100828.

Jayawardena, A.W., 2021. An inconvenient truth about access to safe drinking water. *Int. J. Environ. Clim. Chang.* 11, 158–168. https://doi.org/10.9734/ijecc/2021/v11i1030503

Jepsen, K.L., Bram, M.V., Pedersen, S., 2018. Membrane fouling for produced water treatment: A review study from a process control perspective. *Water* 10, 847. https://doi.org/10.3390/w10070847

Jeyapaul, A.S., Ganesapillai, M., 2020. Dual packed bed adsorption of sulphur dioxide from surface modified haematite/III-ferric oxide: Characterization of the mass transfer zone. *South African J. Chem. Eng.* 33, 95–102. https://doi.org/10.1016/j.sajce.2020.06.008

Jeyapaul, A.S., Ganesapillai, M., 2022. Reduction of sulphur dioxide emission into the environment by adsorption on enhanced α-hematite surface. *Environ. Monit. Assess.* 194, 871. https://doi.org/10.1007/s10661-022-10531-9

Jeyapaul, A.S., Ganesapillai, M., 2024. A sustainable solution: Mitigating sulphur dioxide emissions through adsorption on chemically modified iron oxide nanoparticles. *Clean Technol. Environ. Policy*. https://doi.org/10.1007/s10098-024-02807-0

Jiang, Y., Biswas, P., Fortner, J.D., 2016. A review of recent developments in graphene-enabled membranes for water treatment. *Environ. Sci. Water Res. Technol.* 2, 915–922. https://doi.org/10.1039/c6ew00187d

Kaczorowska, M.A., Bozejewicz, D., Witt, K., 2023. The application of polymer inclusion membranes for the removal of emerging contaminants and synthetic dyes from aqueous solutions — A mini review. *Membranes (Basel).* 13, 132. https://doi.org/10.3390/membranes13020132

Kalajahi, S.T., Misra, A., Koerdt, A., 2024. Nanotechnology to mitigate microbiologically in fluenced corrosion (MIC). *Front. Nanotechnol.* 6, 1340352. https://doi.org/10.3389/fnano.2024.1340352

Katibi, K.K., Yunos, K.F., Man, H.C., Aris, A.Z., Zuhair, M., Nor, M., Azis, R.S., Umar, A.M., 2021. Contemporary techniques for remediating endocrine-disrupting compounds in various water sources: Advances in treatment methods and their limitations. *Polymers (Basel).* 13, 3229. https://doi.org/10.3390/polym13193229

Kaur, R., Goyat, R., Singh, J., Umar, A., Chaudhry, V., Akbar, S., 2022. An overview of membrane distillation technology: One of the perfect fighters for desalination. *Eng. Sci.* 21, 771.

Kausar, A., Bocchetta, P., 2022. Polymer/graphene nanocomposite membranes: Status and emerging prospects. *J. Compos. Sci.* 6, 76. https://doi.org/10.3390/jcs6030076

Khan, S., Noor, T., Iqbal, N., Yaqoob, L., 2024. photocatalytic dye degradation from textile wastewater: A review. *A Rev. ACS Omega*. 9, 21751–21767. https://doi.org/10.1021/acsomega.4c00887

Kumar, P., Sharma, N., Ranjan, R., Kumar, S., Bhat, Z.F., Jeong, D.K., 2013. Perspective of membrane technology in dairy industry : A review. *Asian Australas. J. Anim. Sci.* 26, 1347–1358.

Kumar, R., Qureshi, M., Kumar, D., Al-ansari, N., 2022. Case studies in chemical and environmental engineering case report: A review on emerging water contaminants and the application of sustainable removal technologies. *Case Stud. Chem. Environ. Eng.* 6, 100219. https://doi.org/10.1016/j.cscee.2022.100219

Le-clech, P., Chen, V., Fane, T.A.G., 2006. Fouling in membrane bioreactors used in wastewater treatment. *J. Memb. Sci.* 284, 17–53. https://doi.org/10.1016/j.memsci.2006.08.019

Lee, A., Elam, J.W., Darling, S.B., 2016. Membrane materials for water purification: Design, development, and application. *Environ. Sci. Water Res. Technol.* 2, 17–42. https://doi.org/10.1039/c5ew00159e

Li, N.N., Fane, A.G., Ho, W.S.W., Matsuura, T., 2011. *Advanced Membrane Technology and Applications*. John Wiley & Sons.

Liang, L., Liu, W., Sun, Y., Huo, X., Li, S., Zhou, Q., 2017. Phytoremediation of heavy metal contaminated saline soils using halophytes: Current progress and future perspectives. *Environ. Rev.* 25, 269–281.

Liu, G., Jin, W., Xu, N., 2015. Graphene-based membranes. *Chem. Soc. Rev.* 44, 5016–5030. https://doi.org/10.1039/c4cs00423j

Loganathan, P., Kandasamy, J., Ratnaweera, H., Vigneswaran, S., 2023. Submerged membrane/adsorption hybrid process in water reclamation and concentrate management — A mini review. *Environ. Sci. Pollut. Res.* 30, 42738–42752. https://doi.org/10.1007/s11356-022-23229-9

Lu, X., Peng, Y., Qiu, H., Liu, X., Ge, L., 2017. Anti-fouling membranes by manipulating surface wettability and their anti-fouling mechanism. *Desalination* 413, 127–135. https://doi.org/10.1016/j.desal.2017.02.022

Mallakpour, S., Azadi, E., 2022. Nanofiltration membranes for food and pharmaceutical industries. *Emergent Mater.* 5, 1329–1343. https://doi.org/10.1007/s42247-021-00290-7

Metwally, M.G., Benhawy, A.H., Khalifa, R.M., Nashar, R.M. El, Trojanowicz, M., 2021. Application of molecularly imprinted polymers in the analysis of waters and wastewaters. *Molecules* 26, 6515. https://doi.org/10.3390/molecules26216515

Mills, R., Baldridge, K.C., Bernard, M., Bhattacharyya, D., 2024. Recent advances in responsive membrane functionalization approaches and applications. *Sep. Sci. Technol.* 58, 1202–1236. https://doi.org/10.1080/01496395.2022.2145222

Muñoz-Alegría, J.A., Muñoz-España, E., Flórez-Marulanda, J.F., 2021. Dissolved air flotation: A review from the perspective of system parameters and uses in wastewater treatment. *TecnoLógicas* 24, 281–303.

Naidoo, S., 2022. Commentary on the contribution of Working Group III to the Sixth Assessment Report of the Intergovernmental Panel on Climate Change. *S. Afr. J. Sci.* 118, 1–4.

Nan, M., Jin, B., Chow, C.W.K., Saint, C., 2010. Recent developments in photocatalytic water treatment technology: A review. *Water Res.* 44, 2997–3027. https://doi.org/10.1016/j.watres.2010.02.039

Nasir, A.M., Ridhwan, M., Nur, S., Aqmar, E., Kamal, M., 2022. A review of the potential of conventional and advanced membrane technology in the removal of pathogens from wastewater. *Sep. Purif. Technol.* 286, 120454.

Nickisch, R., Vos, W.M. De, Meier, M.A.R., Baig, M.I., 2023. Removal of transition-metal ions by metal-complexing polythiosemicarbazone membranes. *ACS Appl. Polym. Mater.* 5, 7240–7251. https://doi.org/10.1021/acsapm.3c01192

Nicolaisen, B., 2002. Developments in membrane technology for water treatment. *Desalination* 153, 355–360.

Nicoletta, F.P., Cupelli, D., Formoso, P., Filpo, G. De, Colella, V., Gugliuzza, A., 2012. Light responsive polymer membranes: A review. *Membranes (Basel).* 2, 134–197. https://doi.org/10.3390/membranes2010134

Nugraha, H., Dharma, C., Jaafar, J., Widiastuti, N., Matsuyama, H., Rajabsadeh, S., Hafiz, M., Othman, D., Rahman, M.A., Natasha, N., Jafri, M., Suhaimin, N.S., Nasir, A.M., Alias, N.H., 2022. A review of titanium dioxide (TiO_2) based photocatalyst for oil-field-produced water treatment. *Membranes (Basel).* 12, 345. https://doi.org/10.3390/membranes12030345

Obotey Ezugbe, E., Rathilal, S., 2020. Membrane technologies in wastewater treatment: A review. *Membranes (Basel).* 10, 89. https://doi.org/10.3390/membranes10050089

Omotoso, M., Owolabi, A., 2015. Pulp and paper evaluation of solid wastes from agricultural produce. *Int. J. Chem.* 7, 113. https://doi.org/10.5539/ijc.v7n2p113

Osman, A.I., Chen, Z., Elgarahy, A.M., Farghali, M., Mohamed, I.M.A., Priya, A.K., Hawash, H.B., Yap, P., 2024. Membrane technology for energy saving: Principles, techniques, applications, challenges, and prospects. *Adv. Energy Sustain. Res.* 5, 2400011.

Othman, N.H., Alias, N.H., Fuzil, N.S., Marpani, F., 2022. A review on the use of membrane technology systems in developing countries. *Membranes (Basel).* 12, 30. https://doi.org/10.3390/membranes12010030

Pendergast, M.M., Hoek, E.M.V, 2011. A review of water treatment membrane nanotechnologies. *Energy Environ. Sci.* 4, 1946–1971.

Peyravi, M., Jahanshahi, M., Banafti, S., 2020. *Application of Membrane Technology in Beverage Production and Safety, Safety Issues in Beverage Production.* Elsevier Inc. https://doi.org/10.1016/B978-0-12-816679-6.00008-5

Qin, J., Liberman, B., Kekre, K.A., 2009. Direct osmosis for reverse osmosis fouling control: Principles, applications and recent developments. *Open Chem. Eng. J.* 3, 8–16.

Rahman, T.U., Roy, H., Islam, R., Tahmid, M., Fariha, A., Mazumder, A., Tasnim, N., Pervez, N., Cai, Y., Naddeo, V., 2023. The advancement in membrane bioreactor (MBR) technology toward sustainable industrial wastewater management. *Membranes (Basel).* 13, 181. https://doi.org/10.3390/ membranes13020181

Rajasimman, M., Jayakumar, M., Ravindranath, E., Chitra, K., 2007. Treatment of solid and liquid wastes from tanneries in an UASB reactor. *Proc. 60th Annu. Sess. Indian Inst. Chem. Eng. CHEMCON-2007*, Kolkata, India.

Ridhwan, M., Hafiz, M., Othman, D., Agustiono, T., Hafiz, M., Ismail, A.F., Khongnakorn, W., Rahman, M.A., Jaafar, J., 2022. Advances in adsorptive membrane technology for water treatment and resource recovery applications : A critical review. *J. Environ. Chem. Eng.* 10, 107633.

Roy, S., Smruti, R., 2018. Emerging membrane technologies for water and energy sustainability: Future prospects, constraints and challenges. *Energies* 11, 2997. https://doi.org/10.3390/en11112997

Salman, S.M.A., 2015. The human right to water and sanitation : Is the obligation deliverable ? *Water Int.* 39, 969–982. https://doi.org/10.1080/02508060.2015.986616

San Román, M.F., Bringas, E., Ibañez, R., Ortiz, I., 2010. Liquid membrane technology: Fundamentals and review of its applications. *J. Chem. Technol. Biotechnol.* 85, 2–10.

Sato, T., Makino, K., Tamesue, S., Ishiura, G., Itoh, N., 2023. Preparation and permeation properties of a pH-responsive polyacrylic acid coated porous alumina membrane. *Membranes (Basel).* 13, 82. https://doi.org/10.3390/membranes13010082

Schwarzenbach, P., Egli, T., Hofstetter, T.B., Gunten, U. Von, Wehrli, B., 2010. Global water pollution and human health. *Annu. Rev. Environ. Resour.* 35, 109–136. https://doi.org/10.1146/annurev-environ-100809-125342

Seleman, M., Sime, T., Ayele, A., Sergawie, A., Nkambule, T., Fito, J., 2023. Isotherms and kinetic studies of copper removal from textile wastewater and aqueous solution using powdered banana peel waste as an adsorbent in batch adsorption systems. *Int. J. Biomater.* 2023, 10.

Shaffer, D.L., Werber, J.R., Jaramillo, H., Lin, S., Elimelech, M., 2014. Forward osmosis: Where are we now ? *Desalination* 356, 271–284. https://doi.org/10.1016/j.desal.2014.10.031

Shannon, M.A., Bohn, P.W., Elimelech, M., Georgiadis, J.G., Marin, B.J., Mayes, A.M., 2008. Science and technology for water purification in the coming decades. *Nature* 452, 301–310. https://doi.org/10.1038/nature06599

Sharma, S.K., n.d. *SpringerBriefs in Molecular Science Green Chemistry for Sustainability.* Springer.

Siddique, T., Dutta, N.K., 2021. Mixed-matrix membrane fabrication for water treatment. *Membranes (Basel).* 11, 557. https://doi.org/10.3390/membranes11080557

Silva, J.A., 2023. Water supply and wastewater treatment and reuse in future cities: A systematic literature review. *Water (Switzerland)* 15, 3064. https://doi.org/10.3390/w15173064

Skuse, C., Gallego-Schmid, A., Azapagic, A., Gorgojo, P., 2021. Can emerging membrane-based desalination technologies replace reverse osmosis? *Desalination* 500, 114844. https://doi.org/10.1016/j.desal.2020.114844

Sohn, J., Valavala, R., Han, J., Her, N., Yoon, Y., 2011. Pretreatment in reverse osmosis seawater desalination: A short review. *Environ. Eng. Res.* 16, 205–212. https://doi.org/10.4491/eer.2011.16.4.205

Song, W., Li, Z., Li, Y., You, H., Qi, P., Liu, F., Loy, D.A., 2018. Facile sol-gel coating process for anti-biofouling modification of poly (vinylidene fluoride) micro filtration membrane based on novel zwitterionic organosilica. *J. Memb. Sci.* 550, 266–277. https://doi.org/10.1016/j.memsci.2017.12.076

Stoquart, C., Servais, P., Bérubé, P.R., Barbeau, B., 2012. Hybrid membrane processes using activated carbon treatment for drinking water: A review. *J. Memb. Sci.* 411–412, 1–12. https://doi.org/10.1016/j.memsci.2012.04.012

Strathmann, H., 2001. Membrane separation processes: Current relevance and future opportunities. *AIChe* 47, 1077–1087.

Subramani, A., Jacangelo, J.G., 2015. Emerging desalination technologies for water treatment : A critical review. *Water Res.* 75, 164–187. https://doi.org/10.1016/j.watres.2015.02.032

Sureshvarr, K., Bharathiraja, B., Jayakumar, M., Jayamuthunagai, J., Balaji, L., 2010. Removal of azo dye compounds from paper industries wastes using phytoremediation methodology. *Int. J. Chem. Sci.* 8, 687–700.

Tang, X., Zheng, H., Teng, H., Sun, Y., Guo, J., Xie, W., Yang, Q., Chen, W., 2016. Chemical coagulation process for the removal of heavy metals from water: A review. *Desalin. Water Treat.* 57, 1733–1748.

Thamaraiselvan, C., Noel, M., 2015. Membrane processes for dye wastewater treatment : Recent progress in fouling control. *Crit. Rev. Environ. Sci. Technol.* 45, 3389. https://doi.org/10.1080/10643389.2014.900242

Tomaszewska, M., 2007. Industrial wastewater treatment by means of membrane techniques. *Polish J. Chem. Technol.* 9, 138–142. https://doi.org/10.2478/v10026-007-0074-z

Tousley, M.E., Elimelech, M., 2014. Thin-film composite polyamide membranes functionalized with biocidal graphene oxide nanosheets. *Environ. Sci. Technol. Lett.* 1, 71–76.

Tyagi, M., da Fonseca, M.M.R., de Carvalho, C.C.C.R., 2011. Bioaugmentation and biostimulation strategies to improve the effectiveness of bioremediation processes. *Biodegradation* 22, 231–241.

Wang, M., Liu, X., Yang, B., Fei, Y., Yu, J., An, R., Duan, L., 2021. Heavy metal contamination in surface sediments from lakes and their surrounding topsoils of China. *Environ. Sci. Pollut. Res. Int.* 28, 29118–29130.

Wang, Y., Wang, X., Li, M., Dong, J., Sun, C., Chen, G., 2018. Removal of pharmaceutical and personal care products (PPCPs) from municipal waste water with integrated membrane systems, MBR-RO/NF. *Int. J. Environ. Res. Public Heal.* 15, 269. https://doi.org/10.3390/ijerph15020269

Werber, Jay R., Deshmukh, A., Elimelech, M., 2016a. The critical need for increased selectivity, not increased water permeability, for desalination membranes. *Environ. Sci. Technol. Lett.* 3, 112–120. https://doi.org/10.1021/acs.estlett.6b00050

Werber, Jay R., Osuji, C.O., Elimelech, M., 2016b. Materials for next-generation desalination and water purification membranes. *Nat. Rev. Mater.* 1, 16018. https://doi.org/10.1038/natrevmats.2016.18

Wong, S., Ngadi, N., Inuwa, I.M., Hassan, O., 2018. Recent advances in applications of activated carbon from biowaste for wastewater treatment: A short review. *J. Clean. Prod.* 175, 361–375.

Wu, K.C., Sakdaronnarong, C., 2024. Wastewater treatment using membrane bioreactor technologies : Removal of phenolic contaminants from oil and coal refineries and pharmaceutical industries. *Polymers (Basel).* 16, 443. https://doi.org/10.3390/polym16030443

Xing, S., Du, S., Huang, Y., Qi, X., Sui, M., 2024. Advances in high-performance nanofiltration membranes facilitated by two-dimensional materials. *Water* 16, 988. https://doi.org/10.3390/w16070988

Yang, J., Monnot, M., Ercolei, L., Moulin, P., 2020. Membrane-based processes used in municipal wastewater treatment for water reuse: State-of-the-art and performance analysis. *Membranes (Basel).* 10, 131. https://doi.org/10.3390/membranes10060131

Yangali-Quintanilla, V., Maeng, S.K., Fujioka, T., Kennedy, M., Li, Z., Amy, G., 2011. Nanofiltration vs. reverse osmosis for the removal of emerging organic contaminants in water reuse. *Desalin. Water Treat.* 34, 50–56. https://doi.org/10.5004/dwt.2011.2860

Zhang, J., Li, G., Yuan, X., Li, P., Yu, Y., Yang, W., Zhao, S., 2023. Reduction of ultrafiltration membrane fouling by the pretreatment removal of emerging pollutants: A review. *Membranes (Basel).* 13, 77. https://doi.org/10.3390/membranes13010077

Zhang, Juan, Wang, G., Zhang, Jianhua, Xu, Z., Zhao, Y., Wang, Y., She, F., Gray, S., Kong, L., 2021. Substrate-independent, regenerable anti-biofouling coating for polymeric membranes. *Membranes (Basel).* 11, 205.

Zhang, Y., Tan, Y., Sun, R., Zhang, W., 2023. Preparation of ceramic membranes and their application in wastewater and water treatment. *Water* 15, 3344.

Zhao, Y., Ji, B., Liu, R., Ren, B., Wei, T., 2020. Constructed treatment wetland: Glance of development and future perspectives. *Water Cycle* 1, 104–112.

Zheng, X., Yu, M., Liang, H., Qi, L., Zheng, H., Exler, H., Schier, W., Frechen, F., 2012. Membrane technology for municipal drinking water plants in China: Progress and prospect. *Desalin. Water Treat.* 49, 281–295. https://doi.org/10.1080/19443994.2012.719337

Zheng, X., Zhang, Z., Yu, D., Chen, X., Cheng, R., Min, S., Wang, J., Xiao, Q., Wang, J., 2015. Resources, conservation and recycling overview of membrane technology applications for industrial wastewater treatment in China to increase water supply. *Resour. Conserv. Recycl.* 105, 1–10. https://doi.org/10.1016/j.resconrec.2015.09.012

11 Application and Research Progress of Hydrochar in Nutrient Recovery from Livestock Manure Wastewater

Xiaowei Zhang, Zhuoya Lv, and Tao Zhang

11.1 INTRODUCTION

With the rapid development of the global livestock industry, the scale of livestock and poultry farming has been expanding, and the emissions of livestock and poultry manure have increased significantly (Hu et al., 2020; Springmann et al., 2018). According to the Ministry of Agriculture and Rural Affairs of China, China produces 4.2 billion tons of manure annually (Kong et al., 2022). With the rising demand for meat and animal products, manure is concentrated in intensive agricultural systems and has become a major source of agricultural pollution (Jin et al., 2021). Livestock manure wastewater is rich in organic matter and nutrients, such as nitrogen and phosphorus, but also contains pollutants, including suspended solids, heavy metals, and pathogenic microorganisms. Improper treatment and direct discharge may lead to eutrophication of water bodies, jeopardize the safety of aquatic ecosystems, and even trigger algal blooms and water quality deterioration (Xu et al., 2024a). In addition, long-term wastewater accumulation in soil can lead to soil crusting, acidification, and heavy metal pollution, affecting soil health and crop growth. The widespread use of antibiotics in intensive farming increases the risk of antibiotic residues, about 54% of which enter the soil and farmland and may persist in the environment for extended periods (Spielmeyer et al., 2017). The decomposition process of untreated fecal wastewater also releases greenhouse gases, such as methane and nitrous oxide, which contribute to climate change. Therefore, efficient treatment and resourceful utilization of livestock manure wastewater have become pressing concerns (Shehata et al., 2023).

In recent years, researchers have increasingly focused on innovative wastewater treatment technologies aimed at improving nutrient recovery and promoting environmentally friendly treatment methods. Hydrothermal carbonization (HTC) technology has gained significant attention due to its advantages in wastewater treatment and resource recovery (Xie et al., 2024). Hydrochar, a carbon-based material prepared by HTC under relatively mild conditions (typically 150°C–250°C and

DOI: 10.1201/9781003536277-11

2–10 MPa pressure), offers promise for sustainable wastewater treatment (Liu et al., 2024). Unlike traditional carbon materials, hydrochar is prepared in a humid environment, requires no high-temperature pyrolysis, and involves lower energy consumption. Moreover, HTC technology can process wet biomass, such as agricultural waste, forestry by-products, and livestock and poultry manure, as raw materials. These raw materials, when heated and pressurized in a closed hydrothermal reactor, yield hydrochar with a porous structure and abundant surface functional groups.

Hydrochar exhibits a high specific surface area, porous structure, and abundant surface oxygenated functional groups, such as alcohol, phenol, and carboxyl groups, which confer advantages for adsorption and nutrient immobilization (Zhang et al., 2020). The pore structure enhances physical adsorption capacity, while oxygen-containing functional groups effectively adsorb and immobilize nitrogen and phosphorus ions in wastewater through chemical bonding (Zhang et al., 2019). The hydrochar production process is environment-friendly, and porosity and functional groups can be optimized by adjusting temperature, duration, and raw material type, thereby enhancing adsorption performance in wastewater treatment. Current studies indicate that hydrochar shows promising results for nitrogen and phosphorus recovery in laboratory settings, though practical applications still face challenges related to adsorption performance, production cost, and post-regeneration stability.

Using VOSviewer 1.6.20, a term co-occurrence cluster map was created after a literature search was carried out for the particular subjects of hydrothermal charcoal and nutrient recovery using the Web of Science core ensemble database as the literature data source. According to the search results, 2247 publications were published in the subject of hydrothermal charcoal throughout the course of 5 years (2019–2024). Figure 11.1 displays the hot topics that were covered by the publications that were categorized by Web of Science. As hydrothermal charcoal has been used gradually in the medium, the number of papers has remained consistent, demonstrating the academic community's interest in using hydrothermal charcoal to recycle nitrogen, phosphorus, and other nutrients.

Therefore, the aim of this study is to systematically summarize the progress of hydrochar application in nutrient recovery from livestock and poultry manure wastewater and to thoroughly explore its preparation process, physicochemical properties, and mechanism of nitrogen and phosphorus adsorption and conversion in wastewater treatment. Meanwhile, this chapter will also analyze the bottlenecks in the existing research and propose corresponding optimization strategies and solutions, with a view to providing theoretical and technical support for the promotion of hydrochar in industrial applications in the future. This study not only helps expand the application scope of hydrochar in the field of wastewater treatment but also provides new ideas for realizing the resource utilization of livestock and poultry manure wastewater and environmental protection.

11.2 PREPARATION AND CHARACTERIZATION OF HYDROCHAR

Hydrochar is a carbon-based material produced through HTC under relatively mild temperature and high-pressure conditions. The preparation and characterization of hydrochar are essential for its application in wastewater treatment and

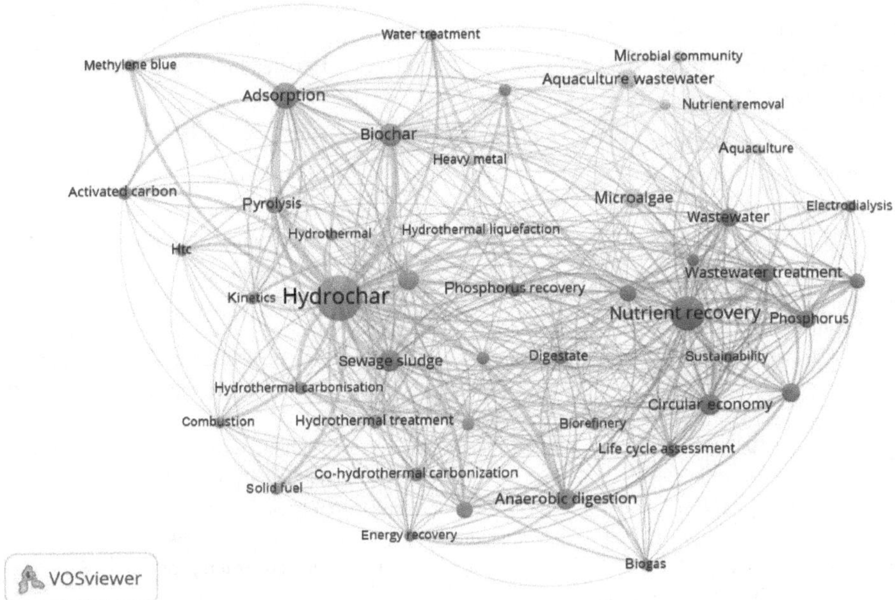

FIGURE 11.1 Hop map of keywords for articles on the use of hydrochar using VOSviewer 1.6.20 software.

nutrient recovery. This section provides a detailed discussion of the preparation process, physicochemical properties of hydrochar, and adsorption mechanisms for nutrients, including nitrogen and phosphorus, in wastewater treatment.

11.2.1 Preparation Process of Hydrochar

Hydrochar is produced by HTC of biomass feedstocks (e.g., animal manure and plant residues) in a hydrothermal reactor under high temperature and pressure conditions. The reaction occurs in a closed system by heating water in a subcritical state (150°C–250°C, 2–10 MPa). The process begins by dispersing the water-soluble components from the biomass into the water at 100°C. Hydrolysis occurs when the temperature exceeds 150°C, partially breaking down cellulose and hemicellulose into monomer chains, which form a slurry at 200°C. Thus, the reaction induces the decomposition and carbonization of biomass in the presence of water (Xie et al., 2024). Throughout the process, water functions as both a reaction medium and a reactant in the decomposition, dehydration, and carbonization of biomass. Table 11.1 presents several research findings on the effect of different factors on the nutrient recovery process in the hydrochar synthesis process.

11.2.1.1 Effect of Reaction Temperature and Time

Temperature is one of the most important parameters in the hydrochar preparation process. Typically, the HTC process takes place between 150°C and 250°C. Different temperature conditions can significantly affect the physical and chemical properties

TABLE 11.1
Effect of Different Factors on Nutrient Recovery Process in Hydrochar Preparation Process

Factors	Important Parameters	Raw Materials	Main Results	References
Temperature	180°C–240°C	Dairy cow manure	Phosphorus concentration in manure-derived aqueous carbon increased at elevated temperatures.	Belete et al. (2021)
Temperature	180°C–260°C	Livestock manure digestate (LMD)	Ca and Fe concentrations of aqueous char increased with increasing temperature, highest at 240°C.	Wang et al. (2024)
Reaction time	10, 30, 60,120, 180 min	Swine manure	The rate of solid reduction decreased gradually at stages less than 60 min.	Xie et al. (2024)
Temperature, reaction time	250°C, 275°C, 300°C 30, 60, 120, 240 min	Cow manure	The effect of reaction time was not as significant as temperature.	Liu et al. (2023)
Raw material type	Mixing ratio 1:1	Livestock manure (SM) and corn cob (CC)	The addition of CC can effectively reduce the ash content and improve the nitrogen recovery.	Li et al. (2022)
Raw material type	Pepper branches and swine manure mass ratio (1:0, 7:3, 5:5, 3:7, and 0:1)	Pepper branches (ZB), swine manure (SM)	H/C and O/C decreased with increasing ZB:SM ratio.	Xiong et al. (2024)
pH	0.1, 0.3, 0.5 M citric and hydrochloric acids	Cow manure	Low pH increased phosphorus recovery in the liquid phase with no significant effect on nitrogen.	Qaramaleki et al. (2020)
Chemical oxidation methods	Hydrogen peroxide oxidation, Walkley–Black oxidation, pyrolysis	Cow manure	The two-step pyrolysis process produces better char stability than the process alone.	Bever et al. (2024)
Additives	CaO	Swine manure	Through CaO-assisted HTC, almost all phosphorus in swine manure is recovered as AP in aqueous carbon.	Lang et al. (2018)
Additives	Formic acid, oxalic acid, citric acid, sulfuric acid, NH_4Cl, $MgCl_2$	Dairy cow manure	Addition of acid extracted more than 90% of phosphorus from manure and ammonia and magnesium salt precipitation significantly recovered phosphorus.	Qaramaleki et al. (2023)

of hydrochar. For example, Song et al. (2020) found that when the HTC temperature was increased from 170°C to 190°C, the soluble organic matter content of hydrochar from swine manure (SM) decreased dramatically from 12.0% to 3.3%, while its aromaticity, molecular weight, and types of organic functional groups also changed with the increase in temperature. Low temperatures (150°C–180°C) help preserve the organic content of the feedstock, and the resulting hydrochar contains more oxygenated functional groups and organic matter. Belete et al. (2021) found experimentally that the oxygen concentration decreases significantly from 18.8% to 3.2% as the HTC temperature increases from 180°C to 240°C and that high temperatures (200°C–250°C) help further dehydration and aromatization, leading to the generation of hydrochar with greater specific surface area and porosity, along with relatively abundant micro- and mesoporous structures (Wang et al., 2024). In general, high temperatures increase the densification and ash content, but slightly decrease the total energy yield. This temperature variation directly affects the adsorption properties and pore structure of hydrochar, so controlling the appropriate temperature is crucial for preparing hydrochar with specific properties.

Reaction time is also a critical factor. Shorter reaction times (e.g., 1–3 hours) are generally associated with hydrochar containing higher organic content, likely due to the decreased dielectric constant of subcritical water under increased pressure, which weakens hydrogen bonding and cleaves oxygen-containing functional groups, ultimately leading to lower H and O content (Wang et al., 2024). Although extending reaction time (e.g., 5–10 hours) may further enhance carbonization and pore structure development in hydrochar, its effect is generally less significant than that of temperature. Liu et al. (2023) reported that hydrochar yield decreased by 13.30% when the temperature was increased from 250°C to 275°C with a reaction time of 30 minutes, whereas the yield reduction was only 2.97% with extended reaction times. Additionally, product distributions at 250°C with reaction times of 120 or 240 minutes closely resembled those achieved at 300°C. Extending reaction time during the HTC process allows for sufficient reaction to support hydrochar formation at lower temperatures. However, excessively long reaction times can lead to higher energy consumption and reduced preparation efficiency. Therefore, in practical applications, reaction time should be optimized to meet the specific characteristic requirements of the target hydrochar.

11.2.1.2 Selection of Feedstock Type

The type of biomass feedstock plays a decisive role in the preparation of hydrochar. Different raw materials (e.g., wood chips, straw, and livestock manure) contain different organic components and minerals, which will form different pore structures and surface functional groups during the HTC process. For example, due to the high content of cellulose and hemicellulose in plant residues, the hydrochar formed after carbonization usually has a high specific surface area and good pore structure; while in livestock and poultry manure, due to the rich organic nitrogen and phosphorus elements, the hydrochar generated after carbonization tends to have a high nitrogen content and reactive functional groups, which are important for the adsorption of nitrogen and phosphorus in wastewater. Although HTC can realize the conversion of livestock manure into organic fertilizer, there are some problems such as high ash content and low nutrient recovery rate. Also, higher levels of furans, PAHs, and carboxylic acids produced during hydrolysis, dehydration, and decarboxylation of hydrochar using carbon-rich lignocellulosic materials may be acutely toxic to plants and soil microorganisms (Karatas et al., 2022). Since the properties of hydrochar are highly dependent on the type of feedstock, cohydrothermal carbonization (co-HTC) of two or more feedstocks has been widely recognized as an effective strategy to produce hydrochar with better physicochemical properties (Zhang et al., 2017). Li et al. (2023) found that corn cobs (CC) have excellent nitrogen fixation during HTC, which can lead to a reduction in the ash content, and proposed that the co-HTC of two typical agroforestry and livestock wastes, SM and CC. Xiong et al. (2024) used SM and peppercorn branches (ZB) for co-HTC, and the H/C and O/C ratios decreased with an increase in the ZB:SM ratio, highlighting the significant effect of feedstock composition on the properties of hydrochar.

In addition, the mineral content (e.g., calcium, magnesium, and iron) in different feedstocks also affects the properties of hydrochar. Minerals can promote the formation of stable carbon structures, while the combined effect of high temperature and minerals promotes the formation of highly aromatic structures. These mineral elements have the potential to be embedded in the skeletal matrix of hydrochar, thus enhancing its application in specific environmental areas, for example, iron and aluminum promote the adsorption of nutrients in the anionic form (sulfates and phosphates) (Oumabady et al., 2024). These minerals can act as catalysts during HTC to promote hydrothermal reactions and may also act as adsorption sites to increase the capacity of hydrochar to adsorb nutrients from wastewater.

11.2.1.3 Selection and Regulation of Hydrothermal Media

The preparation of hydrochar also involves the selection and regulation of hydrothermal medium. As a reaction medium, the physicochemical properties of water (e.g., pH value and redox property) will affect the reaction mechanism and product properties during the HTC process. By adjusting the pH of the hydrothermal medium, the acidity and alkalinity of the reaction environment can be changed, thus affecting the formation of functional groups in hydrochar. Previous studies have found that acid addition facilitates the extraction of phosphorus into the aqueous phase (AP), presumably because the hydrothermal treatment of biomass feedstocks converts all phosphorus species to orthophosphates, and the large amounts of Ca and Fe in feces can induce complexation of orthophosphate anions leading to the formation of insoluble minerals (Qaramaleki et al., 2020). The addition of acidic additives lowered the pH, altered the solubility of Ca and Fe-conjugated phosphate minerals, and enhanced the recovery of nutrients from fecal wastewater. In the alkaline-biased hydrothermal environment, the generation of oxygen-containing functional groups on the surface of hydrochar can be promoted, which improves its adsorption performance of phosphorus.

In addition, by adding external additives, such as metal salts or alkaline compounds, the reaction kinetics during HTC can be altered to accelerate the decomposition and aromatization of organic matter, and even the formation of metal oxides on the surface of hydrochar. A study using CaO-assisted HTC for the preparation of hydrochar from SM promoted the conversion of phosphorus from NAIP to AP, and almost all of the phosphorus in SM was recovered by CaO-assisted HTC as AP in hydrochar (Lang et al., 2019). Qaramaleki et al. (2023) dissolved 90.4%, 89.0%, 94.0%, and 99.6% of phosphorus in fecal sludge as phosphate to HTC by using citric, formic, oxalic, and sulfuric acids, respectively, after which ammonia and magnesium salts were added to hydrothermal aqueous products for precipitation to achieve phosphorus recovery. These modifications in the process water were effective in enhancing the adsorption and immobilization capacity of the hydrochar for specific nutrients (e.g., nitrogen and phosphorus).

11.2.2 Physicochemical Properties of Hydrochar

The physicochemical properties of hydrochar have a direct impact on its adsorption and immobilization of nutrients in wastewater treatment. The following are some of the main characteristics of hydrochar:

11.2.2.1 Pore Structure and Specific Surface Area

Hydrochar usually has a porous structure and these pores provide physical space for nutrient adsorption. The size and structure of pores can affect their adsorption properties; for example, micropores are expected to provide adsorption sites for volatile organic compounds (VOCs), mesopores can accelerate the transport of VOCs, and macropores can promote their diffusion (Zhou et al., 2020). During HTC, the size and distribution of pores can be adjusted by controlling the temperature and reaction time, thus optimizing the specific surface area. The increase in specific surface area helps increase the adsorption capacity of hydrochar. Meanwhile, it has been shown that the proportion of water has an effect on the specific surface area and pore structure of hydrochar, and the results of toluene cyclic adsorption suggest that the hierarchical porous structure produced by adjusting the amount of water during the hydrothermal process can improve the adsorption capacity of hydrochar (Fan et al., 2023). However, the optimization of the pore structure does not only depend on the physical conditions but also needs to consider the selection of raw materials and modification techniques. Stelgen et al. (2024) found that the co-HTC of blunt-topped *Spirulina platensis* with cellulose/lignin derivatives significantly increased the specific surface area and the number of well-developed pore structures of hydrochars, which provided sufficient sites for adsorption and minimize the spatial site resistance for mass transfer through the abundant mesopores. For different types of nutrients in wastewater (e.g., ammonia and phosphate), different pore size ranges can be selected to enhance the adsorption effect. For example, small pore size hydrochar is more suitable for adsorption of small molecules of nutrients, while large pore size is suitable for capturing larger molecules or particulate matter.

11.2.2.2 Distribution of Surface Functional Groups

Oxygen-containing functional groups (e.g., carboxyl, hydroxyl, carboxyl, and ester groups) enriched on the surface of hydrochar play an important role in the adsorption process. These functional groups can interact with nutrients (e.g., NH_4^+ and PO_4^{3-}) in wastewater through mechanisms such as electrostatic adsorption, complexation reaction, or chemical bonding, thus realizing the adsorption and immobilization of nutrients. The type and number of functional groups are affected by preparation conditions, raw material composition, and modification methods. For example, the number of surface functional groups generated can be adjusted by controlling the carbonization temperature. Nepu et al. (2019) found that acidic oxygenated functional groups decreased significantly with increasing pyrolysis temperature, while specific types of functional groups can be introduced by pretreating the feedstock or adding modifiers (e.g., metal salts) to enhance their selective adsorption capacity for specific nutrients. Some experiments have grafted xanthate groups (thiol-containing groups) onto sludge-derived aqueous carbon, providing more heavy metal adsorption sites, while acid leaching experiments have shown that there is a lower risk of secondary leaching of heavy metals, making the modified aqueous carbon suitable for *in situ* immobilization and stabilization of heavy metals (Lin et al., 2024).

11.2.2.3 Chemical Composition and Degree of Carbonization

The chemical composition and degree of carbonization of hydrochar affect its stability and adsorption properties. Hydrochar prepared at low temperatures usually retains more organic matter and has higher adsorption activity, but it has relatively poor structural stability and is prone to degradation during use. Zhang et al. (2022a) found that the H/C and O/C atomic ratios of the hydrochar monotonically decreased with increasing hydrothermal temperatures. X-ray photoelectron spectroscopy (XPS) spectroscopy showed that the carbon atoms on the surface of the hydrochar appeared in the form of C=O bonds, which suggests that high-temperature carbonization can promote dehydration and demethylation. Based on the Fourier-transform infrared spectroscopy results, it was hypothesized that the elevated hydrothermal temperature led to the breakage of the C-C structure of the main chain, and the aliphatic groups attached to the benzene ring were shed, exposing the aromatic substituent sites (Zhang et al., 2022a). These sites are highly reactive and produce more condensed aromatic rings through condensation reactions with similar substituent sites, which improves the degree of aromatization of the hydrochar and increases its stability, but may lead to the reduction of adsorption active functional groups. Therefore, in practical applications, it is necessary to consider the stability and adsorption properties of hydrochar and select appropriate preparation conditions.

11.2.3 Adsorption and Reaction Mechanisms of Hydrochar in Wastewater Treatment

Nutrient removal and recovery by hydrochar in livestock manure wastewater treatment is mainly achieved by two adsorption mechanisms: physical adsorption and chemical adsorption. Table 11.2 presents several research findings on the adsorption and reaction mechanisms of hydrochar and their optimization.

TABLE 11.2
Adsorption and Reaction Mechanisms of Hydrochar and Their Optimization

Raw Materials	Adsorption and Reaction Mechanisms	Optimization Methods	Main Results	References
Sewage sludge	Fitting of Freundlich isotherms, multimolecular layer adsorption	Hydrothermal pretreatment	The strong adsorption of tetracycline by hydrothermal pretreatment may be due to the high specific surface area and pore volume providing more adsorption sites.	Chen et al. (2021)
Sawdust	The activated hydrochar prepared has a microporous structure and monolayer adsorption	KOH activation	The activated hydrochar has formed a microporous structure, and the nitrogen-containing functional groups on the surface play an important role in CO_2 adsorption.	Lilian et al. (2024)
Swine manure and corn cob biomass	Phosphorus adsorption follows quasi-secondary kinetics with chemisorption dominating	H_2O_2 hydrothermal pretreatment, Ca^{2+} and Al^{3+} modified co-precipitation	Catalytic hydrothermal pretreatment enhances inorganic P release, Ca/Al-modified aqueous carbon exhibits high P adsorption capacity.	He et al. (2022b)
Corn cob, biogas residue	Ammonium nitrogen adsorption as monolayer adsorption, chemisorption dominated	KOH activation	Oxygen-containing functional groups as proton donors increased upon KOH activation, promoting ammonium nitrogen adsorption.	Zhang et al. (2020)
Poplar sawdust	Applicable to the Langmuir model, including physical adsorption, surface complexation, electrostatic interactions, and ion exchange	Aging by HNO_3	Alteration of aqueous charcoal polarity, negative surface charge, and SSA, resulting in improved Cd^{2+} adsorption capacity.	Li et al. (2021a)
Bamboo	Electrostatic attraction, reduction, and complexation	NH_4Cl as a nitrogen source and acid catalyst	Ammonia forms amide and protonated amine groups under acidic conditions, which are involved in electrostatic interactions with Cr(VI).	Li et al. (2021b)
Napier grass (*Pennisetum purpureum* Schum)	Cd adsorption was dominated by chemisorption, monolayer adsorption	Hydrothermal carbonization or pyrolysis at different temperatures	The Cd^{2+} adsorption capacity of coke increases and hydrochar decreases at higher temperatures.	Wang et al. (2022)

(*Continued*)

TABLE 11.2 (*Continued*)
Adsorption and Reaction Mechanisms of Hydrochar and Their Optimization

Raw Materials	Adsorption and Reaction Mechanisms	Optimization Methods	Main Results	References
Pine cone (PC)	The adsorption mechanism of Cd(II) and Pb(II) is related to electrostatic attraction, ion exchange, and complexation reaction	Loaded Fe_3O_4 nanoparticles, grafting of xanthate groups	Fe-O and xanthate groups of modified aqueous carbon play a key role in the adsorption process of Cd(II) and Pb(II).	Zhang et al. (2022b)
Wastewater sludge	More consistent with Langmuir isotherm and pseudo-secondary kinetic	KOH activation, HCl treatment	Dual-activated hydrochar structure has surface heterogeneity with both maximum surface area and adsorption capacity.	Sathe et al. (2024)
Rice husk biomass	Dubinin Radushkevich (DR) modeling verified as physisorption, conforming to Langmuir isotherm, monolayer adsorption	Hydrochloric acid (HCl) co-heat treatment	The mechanism of high adsorption to levofloxacin includes electrostatic attraction, hydrogen bonding, pore filling, n-π stacking, π-π stacking, and hydrophobic interactions.	Luo et al. (2023)

11.2.3.1 Physical Adsorption Mechanism

Physical adsorption mainly depends on the pore structure of hydrochar. The pores of the porous hydrochar, which is generated through the HTC process, can accommodate and capture nutrient molecules such as nitrogen and phosphorus in the wastewater. The advantage of this adsorption mechanism is that it does not rely on chemical reactions, so the adsorption process is rapid and reversible. The HTC process undergoes several steps such as hydrolysis, dehydration, decarboxylation, aromatization, and condensation polymerization, and the average pore size of hydrochar synthesized at low temperatures is larger than that synthesized at high temperatures, and the average pore size of activated hydrocarbons is larger than that of pristine hydrocarbons (Zhang et al., 2020). Chen et al. (2021) found experimentally that biochar prepared by direct high-temperature pyrolysis has a lumpy structure. In contrast, the biochar prepared by HTC pretreatment mainly had a spherical morphology. The accumulation of a large number of spherical particles with smooth surfaces and relatively uniform sizes formed a pore structure, which was conducive to the establishment of high porosity and thus enhanced the adsorption performance. By adjusting the pore size distribution of hydrochar, selective adsorption can be realized for nutrient molecules of different sizes, thus improving the efficiency of wastewater treatment.

11.2.3.2 Chemisorption Mechanism

Chemisorption relies on the interaction of functional groups on the surface of hydrochar with nutrient molecules. For example, carboxyl and hydroxyl groups on the surface of hydrochar can electrostatically interact or complex with cations (e.g., NH_4^+) or anions (e.g., PO_4^{3-}) in the wastewater, thereby immobilizing the nutrients on the surface of the hydrochar. Zhang et al. (2020) activated the hydrochar with KOH, and it was determined by XPS analysis that oxygen-containing functional groups (e.g., phenolic hydroxyl groups and noncarbonyl oxygens) that can act as proton donors became more abundant after the KOH activation became more abundant and these protonated groups could interact with ammonium ions to promote ammonium nitrogen adsorption. In addition, by modifying the hydrochar, the chemical affinity of its surface with specific nutrient molecules can be enhanced to improve the adsorption efficiency. For example, modified hydrochar synthesized by modified co-precipitation of Ca^{2+} and Al^{3+} using hydrothermal pretreatment with H_2O_2 can be complexed with phosphates to form monodentate or bidentate inner-sphere surface complexes (He et al., 2022b). XPS and XANES analyses of the spectra showed that metal-O (Ca-O and Al-O) was the main component, which may be related to the substitution of the protonated hydroxyl group by the phosphoric acid group. It has been shown that the surface charge state of hydrochar changes under suitable pH conditions, affecting the adsorption of specific molecules. For example, Li et al. (2021a) aged hydrochar using different mass fractions of nitric acid, and the potential of the samples decreased with the increase of HNO_3, and thus their negative surface charged sites increased, consistent with the decrease in pH. Due to electrostatic interactions, the modified biochar showed superior adsorption capacity compared to pristine biochar.

11.2.3.3 Optimization of Adsorption Properties

The adsorption performance of hydrochar can be further improved by adjusting its preparation process. Some experiments showed that the adsorption capacity of hydrochar for Cd^{2+} decreased with the increase of production temperature, which may be related to the accumulation of minerals and the degradation of oxygen-containing functional groups (Wang et al.,2022). Therefore, we can improve the adsorption performance by optimizing the temperature and time of HTC. Alternatively, modification techniques (e.g., loading of metal oxides) can be introduced to enhance the adsorption capacity. Zhang et al. (2022b) loaded Fe_3O_4 nanoparticles and grafted sulfur-containing groups on pinecone-derived aqueous carbon and found that the negatively charged Fe-O functional groups formed by deprotonation of hydroxyl groups on the surface could be further bound to Pb^{2+} by XPS spectroscopy and thus enhance the adsorption performance. In addition, chemical treatments (e.g., acid washing or alkali treatment) can be applied to the surface of hydrochar to improve the surface area, pore size, and increase the number of active functional groups to enhance the adsorption effect. Sathe et al. (2024) experimentally found that KOH activation resulted in a 17.5-fold increase in the surface area as compared to unactivated hydrochar while alkali activation was followed by leaching out of certain inorganic compounds using HCl treatment (K, Na, P, S, Ca) and metals (Zn, Fe, Cu, Mn), which led to the formation of a porous structure in the carbonaceous skeleton of the activated hydrochar. Luo et al. (2023) co-treated the hydrochar with HCl, which resulted in the production of more aromatic cyclization products on the hydrochar and a significant increase in the amount of C=O in the hydrochar, which may be one of the reasons for the strong adsorption performance.

The application of hydrochar in nutrient recovery from livestock and poultry manure wastewater has great potential. By reasonably optimizing the preparation process and modification method, the adsorption performance of hydrochar can be significantly improved, making it more efficient and economical in wastewater treatment.

11.3 APPLICATION OF HYDROCHAR IN NUTRIENT RECOVERY FROM LIVESTOCK MANURE WASTEWATER

Hydrochar shows various potential applications in the field of wastewater treatment, especially in the recovery of nutrients such as nitrogen and phosphorus from livestock and poultry manure wastewater. In this section, we discuss in detail the application of hydrochar and its mechanisms in nitrogen nutrient adsorption, phosphorus nutrient recovery, and immobilization, as well as the regenerative use and long-term stability of hydrochar.

11.3.1 Nitrogen Nutrient Adsorption and Transformation

Livestock manure wastewater often contains a large amount of ammonia nitrogen (NH_4^+) and organic nitrogen compounds, and these nitrogen substances, if discharged untreated, will not only cause serious pollution of the environment but also lead to eutrophication of the water body and the formation of harmful algal blooms. However, these nitrogen nutrients also have the potential for resource utilization. Hydrochar provides important support for efficient adsorption and immobilization of nitrogen due to its porous structure and active surface functional groups. Table 11.3 presents the nitrogen adsorption and conversion of hydrochar from livestock manure wastewater for nutrient recycling.

TABLE 11.3
Nitrogen Adsorption and Conversion of Hydrochar in Livestock Manure Wastewater for Nutrient Recycling

Raw Materials	Preparation Parameters	Periodicity	Research Progress	References
Cattle manure	170°C, 200°C, and 230°C, 0.1, 0.3, and 0.5 M citric acid and HCl	5, 30, 120 min	Acid addition had little effect on TKN solubilization.	Qaramaleki et al. (2020)
Raw compost leachate, corn cobs	230°C or 260°C, 3 M KOH	0.5 h	The total release of ammonium nitrogen from five intermittent extractions at pH = 8.0 was up to 12.2% of the maximum adsorption capacity (140.3 mg/g).	Zhang et al. (2019)

(Continued)

TABLE 11.3 (*Continued*)
Nitrogen Adsorption and Conversion of Hydrochar in Livestock Manure Wastewater for Nutrient Recycling

Raw Materials	Preparation Parameters	Periodicity	Research Progress	References
Corn stover	220°C, different residence times (1, 2, 4, 12, and 24 h)	4 h	Nitrogen recovery efficiency increased with increasing reaction strength.	Zhang et al. (2020)
Corn stover, biogas	160°C, 190°C, 220°C, 250°C, 280°C	180 min	The adsorption efficiency of TN and NH_4^+ was positively correlated with temperature in the range of 190°C–250°C.	Jiang et al. (2022)
Sewage sludge (SS)	180°C–320°C	30–90 min	Increasing the temperature or prolonging the carbonization time can effectively improve the porosity of aqueous charcoal, but too high a temperature will destroy the microporous structure.	Chen et al. (2023)
Rice straw (RS)	180°C, $MgCl_2$, $AlCl_3$, and ammonium persulfate	12 h	HTC-coupled pyrolysis biochar has a large number of hollow and highly porous structures, which facilitates the adsorption of pollutants.	Tan et al. (2024)
Swine manure (SM) and corn cobs (CC)	180°C–260°C	60–240 min	Co-hydrochars showed the best fertility at 240°C, 120 min, and a mixing ratio of 1:1, with 47.61% nitrogen recovery.	Li et al. (2022)
Sewage sludge (SS) and spent mushroom substrate (SMS)	SS/SMS = 1/0, 3/1°C, 1/1, and 1/3, 220°C	30 min	There was a synergistic effect of SS and SMS in co-HTC, which increased the energy yield, fixed carbon and nutrient recovery of aqueous charcoal.	Shan et al. (2022)
Oilseed rape straw, industrial sewage sludge	160°C–200°C	1–5h	The uptake of NH_4^+ and PO_4^{3-} by co-water carbon was stratified and stable, and the maximum removal efficiency and adsorption capacity of N adsorption were 78.2% and 162.5 mg/g, respectively.	Xu et al. (2024b)
Corn kernel, peanut residue	180°C–260°C	60–240 min	Co-HTC could increase N recovery (54.97%–67.61%) and C/N ratio and showed excellent nitrogen fixation capacity.	Li et al. (2022)

11.3.1.1 Nitrogen Adsorption Mechanism

The adsorption of nitrogen on hydrochar mainly relies on its surface-rich oxygen-containing functional groups (e.g., carboxyl and hydroxyl groups) and its porous structure. In the process of wastewater treatment, these functional groups have electrostatic adsorption and complexation reactions with ammonia and nitrogen ions and immobilize them on the surface and in the pores of hydrochar. At the same time, the porous structure of hydrochar provides space for physical adsorption of nitrogen molecules, which effectively sequesters nitrogen nutrients through capillary adsorption and surface capture. In addition, the larger the specific surface area of hydrochar, the stronger its ability to adsorb nitrogen usually is. Therefore, the preparation of hydrochar with a high specific surface area is essential to enhance the adsorption efficiency of nitrogen nutrients.

11.3.1.2 Influencing Factors of Nitrogen Adsorption Performance

The performance of hydrochar in nitrogen adsorption is affected by several factors, including pH, reaction time, temperature, and the degree of modification of hydrochar.

1. **pH:** pH has a direct effect on the surface charge of hydrochar and the ionic state of ammonia nitrogen. Under neutral or alkaline conditions, the carboxyl and hydroxyl functional groups on the surface of hydrochar are negatively charged, which is conducive to the adsorption of positively charged ammonia nitrogen ions. While under acidic conditions, the charge state of these functional groups changes, which may reduce the adsorption capacity of ammonia nitrogen. However, an experiment was conducted to recover nitrogen from cattle manure by adding different acidic additives and it was found that the recovery of total Kjeldahl nitrogen (TKN) was not affected by pH (Qaramaleki et al., However, an experiment was conducted to recover nitrogen from cattle manure by adding different acidic additives and it was found that the recovery of TKN was not affected by pH (Qaramaleki et al., 2020). This may be due to the fact that the nitrogen present in livestock manure is mainly derived from the breakdown of proteins, and according to Ekpo et al. (2016), further deamination of these amino acids occurs at high temperatures around 500°C, which is well above the temperature range for the preparation of HTC. Therefore, pH has little effect on the nitrogen adsorption performance of hydrochar.
2. **Reaction time and temperature:** Reaction time and temperature are important parameters that determine the efficiency of nitrogen adsorption. Zhang and group members found that the increase in reaction time favorable for the conversion of nitrogen forms from organic nitrogen (e.g., proteins and dissolved organic nitrogen in the feedstocks) to NH_4^+, which realizes the release and recycling of nitrogen from the feedstocks (Zhang et al.,2019). Therefore, the adsorption sites on the surface of the hydrochar can fully contact and react with the nitrogen molecules in the wastewater during a longer reaction time, thus improving the adsorption efficiency. In addition, appropriately increasing the reaction temperature can help increase the

molecular kinetic energy and promote the diffusion of ammonia nitrogen into the pores of the hydrochar, thus enhancing the adsorption effect. Jiang et al. (2022) experimentally found that the recovery of N increased from 46.8% to 140.6% as the temperature increased from 160°C to 250°C and that the adsorption efficiencies of TN and NH_4^+ were positively correlated with the temperature in this temperature range. SEM image analysis of the hydrochar revealed that high temperature promoted carbonization and decomposition of the original fiber structure, and the hydrochar prepared at 250°C had the richest microporous structure, which enhanced the adsorption of nitrogen. However, too high a temperature may lead to desorption phenomena or the generation of other reaction by-products of the hydrochar, while the collapse of the pore wall of the hydrochar destroys the microporous structure when the temperature is too high, leading to a decrease in the adsorption performance (Chen et al., 2023). Therefore, the temperature conditions need to be optimized within a certain range.

3. **Application of modification techniques:** Studies have shown that the adsorption capacity of hydrochar for ammonia nitrogen can be significantly improved by modification techniques (e.g., loading metal oxides or co-HTC). For example, metal modification with iron or zinc can increase the number and type of active sites on the surface of hydrochar and enhance its affinity for ammonia nitrogen. Li et al. (2022) found that co-HTC with the addition of CC effectively reduced the ash content and increased the nitrogen recovery up to 47.61% as compared to SM-based HTC. Shan and group members used sewage sludge (SS) and spent mushroom substrate (SMS) for co-HTC and found that TN and NH_4^+ concentrations decreased as the proportion of SMS added increased, and the AP of the co-HTC could recover 62.03%–64.65% of nitrogen from the feedstock as a fertilizer (Shan et al., 2022). In addition, the modification can modulate the surface charge distribution of hydrochar to maintain high adsorption performance at different pH values.

11.3.1.3 Conversion and Resource Utilization of Nitrogen Nutrients

Nitrogen nutrients adsorbed onto hydrochar can be further converted into organic fertilizer or other valuable compounds through subsequent biodegradation or chemical treatment. It has been found that under specific conditions, ammonia nitrogen adsorbed on hydrochar can interact with carbon sources and microorganisms to be converted into proteins or amino acids, which can be used for soil improvement and plant growth. The co-water charcoal made by Xu et al. (2024a) had a high concentration of alkaline nitrogen in the form of ammonia nitrogen, which when used as a fertilizer resulted in the release of NH_4^+ and H^+ during plant growth, a decrease in the soil pH, a reduction of the possibility of soil hardening and better respiration for plant growth. In addition, by combining hydrochar with other organic fertilizers, the persistence of its nutrient release can be improved, and the resourceful use of wastewater nitrogen nutrients can be realized. For example, hydrochar obtained superior C/N ratios and maturity during the co-heating treatment of CC and peanut residue, which is beneficial for fertilizer application, while the synergistic HTC of the two can further improve the nitrogen recovery rate to achieve resource utilization (Li et al., 2022).

11.3.2 Recovery and Fixation of Phosphorus Nutrients

Phosphorus in livestock manure wastewater mainly exists in the form of phosphate (e.g., PO_4^{3-}), which is one of the important factors leading to eutrophication of water bodies. However, phosphorus is also an essential nutrient in agricultural production, so its recovery and resource utilization are of great significance. Hydrochar has significant advantages in phosphorus recovery, mainly relying on its surface cation exchange sites and porous structure. Table 11.4 presents the performance optimization of hydrochar for phosphorus recovery.

TABLE 11.4
Performance Optimization of Hydrochar for Phosphorus Nutrient Recovery

Adsorption Mode	Targeted Nutrients	Recovery Value	Results	References
Recovery of phosphorus from solid aqueous carbon by HCl leaching and precipitation	Hydroxyapatite	Recovery of phosphorus in the form of crystalline hydroxyapatite was as high as 98%	Hydrochloric acid leaching moves phosphorus from aqueous charcoal to the acidic aqueous phase and extracts almost all of the phosphorus from aqueous charcoal produced at 230°C for 3 h.	Goldfarb et al. (2022)
Binding to metal cations by precipitation or adsorption	Calcium phosphate and magnesium phosphate minerals	Total phosphorus retention of 88% was achieved	$MgFe_2O_4$ modified hydrothermal carbonization achieves excellent phosphorus retention efficiency, but recovery yields less economic benefit than cost.	Sun et al. (2023)
Coordination of metals with high affinity	apatite	Phosphorus recovery close to 100%	CaO-assisted HTC immobilizes phosphorus to calcium phosphate, reducing the bioavailability and mobility of phosphorus in swine manure.	Lang et al. (2018)
Formation of more stable intracellular compounds between phosphate and lanthanum	Phosphate	The maximum adsorption capacity of P reached 72.69 mg/g	Lanthanum-modified aqueous carbon is more resistant to the interference of other anions during the exchange of ligand with phosphate.	He et al. (2022a)
Phosphates form surface complexes through metal-O or metal-OH bond complexation	Phosphate	Maximum adsorption was 81.8 mg/g	Phosphorus adsorption on Ca/Al-modified aqueous carbon follows a quasi-secondary model, with chemisorption predominating. Ca can be used as a reaction site for phosphate precipitation formation.	He et al. (2022b)

(Continued)

TABLE 11.4 (*Continued*)
Performance Optimization of Hydrochar for Phosphorus Nutrient Recovery

Adsorption Mode	Targeted Nutrients	Recovery Value	Results	References
After digestion, total P was extracted by aqua regia	HCl soluble P	Phosphorus recovery was above 80% for all	Phosphorus recovery values were above 80% for all reaction conditions, and the major fraction of inorganic phosphorus in aqueous charcoal was HCl-extractable P.	Khoury et al. (2023)
Al active site binds to phosphate to form aluminum phosphate precipitates	Phosphate	At a temperature of 260°C, the adsorption reached 114.40 mg/g	The adsorption of phosphate by aluminum-modified aqueous carbon mainly depends on the precipitation reaction between Al and phosphate.	Shuang et al. (2023)
PO_4^3 forms new compounds or ionic salts by ion exchange and complexation	Phosphate	The theoretical maximum adsorption capacity is 99.2 mg/g	There is no change in the morphology and surface area of the Co-hydrochar after adsorption of PO_4^{3-} and it can be used as a fertilizer.	Xu et al. (2024b)
Selective recovery of phosphorus from Co-hydrochar formed from concentrate and digestate	P	90% of the P was recovered	The phosphorus content of the aqueous charcoal fraction of the samples increased with increasing HTC temperature and residence time.	Rasoulnia et al. (2024)
Phosphorus is immobilized in aqueous charcoal by precipitation and adsorption with metal ions	P_2O_5		P enrichment in hydrothermal carbon can be achieved by adding metal salts or adjusting pH.	Li et al. (2022)

11.3.2.1 Phosphorus Adsorption Mechanism

The adsorption of phosphorus by hydrochar is mainly realized through the surface cation exchange sites and complexation. During the HTC process, a large number of cation exchange sites are formed on the surface of hydrochar, and these sites can capture phosphate ions in wastewater through electrostatic attraction. As the pH decreases, the hydroxyl and carboxyl groups on the surface of hydrochar are protonated under acidic conditions, and PO_4^{3-} is adsorbed by ligand exchange with the hydroxyl groups (Xu et al., 2024b). Goldfarb et al. (2022) moved phosphorus from aqueous charcoal to the acidic AP by HCl leaching, and the subsequent addition of

NaOH adjusted the pH so that phosphorus was precipitated in the form of crystalline hydroxyapatite. Phosphorus was recovered from the hydrochar and raw fecal samples by filtration. In addition, modification techniques (e.g., metal oxide modification) of hydrochar can enhance its ability to chemisorb phosphorus. For example, iron oxide modification can form highly active iron ion sites on the surface of hydrochar, and these sites can form stable complexes with phosphates, thus improving the adsorption efficiency of phosphorus. Sun et al. (2023) used an in situ synthesis method to modify pig manure-based hydrochar with $MgFe_2O_4$, and the organic phosphorus was degraded during the HTC process, and the multivalent metal cations (Fe^{2+}, Fe^{3+}) and $MgFe_2O_4$ were used in the HTC process. Fe^{3+} and Mg^{2+} significantly enhanced the binding with inorganic phosphorus, trapping more phosphorus into the hydrochar.

11.3.2.2 Adsorption Kinetics and Isothermal Adsorption Modeling

The kinetic performance of hydrochar in the phosphorus adsorption process was affected by several factors, such as the reaction time, the initial phosphorus concentration in the wastewater, and the degree of modification of the hydrochar. The pore structure and surface characteristics of hydrochar under different hydrothermal conditions can explain the changes in adsorption capacity, and high hydrothermal temperatures and long durations lead to structural deactivation, pore clogging, and a decrease in adsorption performance (He et al., 2022a). Experiments showed that the adsorption of phosphorus on hydrochar followed the typical adsorption kinetic law, i.e., the adsorption rate was faster in the initial stage, and then gradually slowed down with time, and finally reached the adsorption equilibrium. Isothermal adsorption models (e.g., Langmuir model and Freundlich model) are commonly used to describe the adsorption behavior of phosphorus on hydrochar and to help predict the adsorption capacity under different conditions. He et al. (2022a) used an isothermal adsorption model for their analysis and found that the Freundlich model is almost the same for the phosphate adsorption process on lanthanum-modified hydrochar and that the constants in the equations $1/n < 1$ indicate that phosphate is easily adsorbed. He et al. (2022b) also used Langmuir and Freundlich models to describe the isothermal data. The highest R^2 value was obtained under the Langmuir model, indicating that the adsorption sites were uniformly distributed on the surface of the Ca/Al-modified hydrochar and that the adsorption of P was approximated as a monolayer adsorption. It was shown that the modified hydrochar was superior to the unmodified hydrochar in terms of phosphorus adsorption amount and adsorption rate, which provided a theoretical basis for the design of an efficient phosphorus recovery system.

11.3.2.3 Optimized Design of Phosphorus Recovery System

For the performance of hydrochar in phosphorus nutrient recovery, researchers have proposed various optimization schemes. For example, Khoury et al. (2023) set different reaction times and temperatures for their experiments and found that the highest P recovery (95%) was observed at the lowest temperature (200°C), while the shortest reaction time (30 min) yielded the highest P recovery from hydrochar and the lowest in the longest reaction time (120 min) group, which suggests that both the temperature and the time during the HTC affect the P recovery from the resulting hydrochar. Therefore, by adjusting the temperature and reaction time of HTC, the pore structure and the density of surface active sites of hydrochar can be altered to enhance the adsorption capacity of phosphorus. In addition, the affinity of hydrochar for phosphate can be improved by loading metal elements such as iron, calcium, and aluminum on the surface of the hydrochar. Shuang et al. (2023) synthesized an aluminum-modified hydrochar for phosphate adsorption using corn stover as a raw material, and the Al element on the hydrochar reacted with phosphate to form aluminum phosphate precipitates attached to the hydrochar, which increased the amount of adsorbed phosphate. co-HTC can eliminate the disadvantages of a single feedstock to achieve high yields and selective recovery of phosphorus from aqueous carbon. Xu et al. (2024a) prepared co-HTC from rapeseed straw and industrial SS, which provided more active adsorption sites for PO_4^{3-} leading to an increase in the removal efficiency, and the adsorbed phosphorus could be released into the soil as a fertilizer afterwards. These optimizations not only improved the phosphorus adsorption capacity of hydrochar but also increased its flexibility in practical applications, allowing it to be adapted to different concentrations and compositions of phosphorus in wastewater.

11.3.3 Recycling of Hydrochar and Its Stability

To enhance the economy and viability of hydrochar, researchers have thoroughly explored the regeneration and utilization of hydrochar. The regeneration technique not only restores the adsorption performance of hydrochar but also produces cost and enhances the feasibility of its use.

11.3.3.1 Regeneration of Hydrochar

Commonly used regeneration includes heat treatment and flushing with chemical solutions. However, regeneration is often challenging for adsorbates with a high affinity for the adsorbent. Because bulky compounds cannot be desorbed from narrow micropores, harsh treatments such as high temperatures and pressures are required to eliminate adsorbates (Dutta et al., 2019). Heat treatment restores the adsorption capacity of hydrochar by heating the adsorbed nutrients such as nitrogen and phosphorus at high temperatures to desorb the adsorbed substances from the pores and surface. Chemical solution rinsing removes adsorbed nutrients by rinsing the surface of the hydrochar with an acidic or alkaline solution. However, the crystal structure of hydrochar may be at high temperatures and acid and alkaline treatments may also negatively affect the stability of hydrochar, leaving behind a large amount of spent solvent. Meanwhile, although these can partially restore the adsorption performance of hydrochar, their adsorption effect may gradually weaken after several regenerations. Therefore, how the structural stability and adsorption activity of hydrochar during regeneration is an important direction of current research.

11.3.3.2 Stabilization of Hydrochar During Regeneration

Hydrochar may cause the collapse of pore structure and surface functional groups due to high-temperature treatment or the corrosive effect of chemical solutions during multiple regeneration (Zhang et al., 2020). To improve the stability of hydrochar, researchers have attempted to modify hydrochar to enhance its resistance to heat and chemical erosion during regeneration. For example, by introducing metal oxides or carbon nanomaterials, a protective layer can be formed on the surface of the hydrochar, with structural loss during regeneration. In addition, regeneration can be improved, such as by using low-temperature heat treatment or mild chemical reagents, to the hydrochar.

11.3.3.3 Evaluation of the Effectiveness of the Application of Recycled Hydrochar

In practical applications, it is crucial to evaluate the adsorption effect and stability of regenerated hydrochar (Xie et al., 2024). Studies have shown that properly regenerated hydrochar still has high efficiency in nitrogen and phosphorus adsorption, and its reuse cycle can be up to several times. However, the fluctuation of the adsorption performance and the stability of the regenerated hydrochar in use still need to be further verified. Therefore, future research should focus on efficient and economical regeneration techniques and evaluate their reliability in use to ensure the economy and availability of hydrochar in wastewater treatment.

Hydrochar has great potential for application in nutrient recovery from livestock and poultry manure wastewater. By optimizing its preparation process and improving its adsorption performance and regeneration technology, hydrochar is expected to become an important material in the field of wastewater treatment and resource utilization. Future research should focus on improving the regeneration capacity and stability of hydrochar, so as to promote its industrialized application.

11.4 EXISTING PROBLEMS AND FUTURE RESEARCH DIRECTIONS

Although hydrochar has shown some potential and advantages in wastewater nutrient recovery, there are still many challenges and limitations in its practical application, which must be effectively addressed in order to promote the widespread application and industrialization of hydrochar technology.

11.4.1 Challenges of Hydrochar in Nutrient Recovery Applications

First, the capacity and efficiency of hydrochar in adsorption of nutrients (e.g., nitrogen and phosphorus) are still limited (Su et al., 2024). Although the modification technology can improve its adsorption performance to a certain extent, there is still much room for improving the nutrient adsorption capacity and selectivity of hydrochar compared with other efficient adsorption materials. In addition, hydrochar may exhibit some lack of selectivity in adsorbing specific nutrients, resulting in its inability to efficiently remove multiple pollutants simultaneously in the same wastewater treatment process. To solve these problems, more targeted modification and optimization of the adsorption mechanism are needed to improve the nutrient recovery efficiency of hydrochar in practical applications.

Second, the relatively high production and application costs of hydrochar are one of the main factors limiting its large-scale promotion (Xu et al., 2024a). The preparation process of hydrochar involves high-temperature and high-pressure conditions as well as specific equipment, with relatively high energy consumption and production costs. Even under laboratory conditions, optimizing the HTC process and modification still requires a large amount of resources, so how to produce the cost, simplify the process, and improve the economic efficiency have become an urgent problem to be solved nowadays. In addition, the chemical reagents and metal salts used in the modification process may also produce costs and pose potential risks, which requires finding a balance in the material preparation process to ensure that hydrochar is balanced between friendliness and economy.

Furthermore, the utilization rate and stability of hydrochar regeneration still need to be further improved (Xie et al., 2024). Although the existing physical and chemical regeneration can restore the adsorption performance of the hydrochar to a certain extent, the pore structure and surface functional groups of the hydrochar may be damaged after repeated regeneration, leading to a decrease in its adsorption efficiency. This problem is especially prominent in practical engineering applications, because the low regeneration efficiency or complex regeneration may lead to high application costs, which affects its economic feasibility. Therefore, more efficient and gentle regeneration is needed in the future.

11.4.2 Future Research Directions

To overcome the above challenges faced by hydrochar in practical applications, future research should focus on the following directions (Liu et al., 2024; Zhang et al., 2022c):

First, novel modification of hydrochar materials is an important way to enhance their adsorption performance and selectivity. By introducing multifunctional modifiers (e.g., nanomaterials, rare earth elements, or composite metal oxides), the selective adsorption capacity of hydrochar for specific nutrients can be enhanced. For example, modified hydrochar loaded with iron or zinc may have higher selectivity in the adsorption of phosphorus and nitrogen, thus improving the efficiency of nutrient recovery. In addition, organic modification strategies can be investigated in the future to enhance the biocompatibility and adsorption capacity of hydrochar by compositing it with natural polymeric materials (e.g., chitosan and lignin). These novel materials not only help improve the adsorption efficiency but also may enable more cost-effective industrial production by optimizing the cost of the production process.

Second, research on the scale-up preparation technology and industrial application strategy of hydrochar is the key to promoting its practical engineering application. In the scale-up preparation process, more efficient and low-cost production processes need to be explored for energy and raw material consumption. For example, the study of low-temperature HTC process or the use of waste biomass and agricultural by-products as raw materials can reduce production cost. Meanwhile, for different types of wastewater, modularized treatment systems can be developed so that hydrochar can be efficiently applied in wastewater treatment projects of different scales. In addition, based on the laboratory research results, the performance optimization strategy of hydrochar under actual engineering conditions can be explored, such as how to control the reaction conditions and the addition of modifiers to adapt to the complexity of the actual wastewater composition.

Finally, to ensure the feasibility of hydrochar in industrial applications, future research should also focus on its regeneration technology and stability. Cost-effective regeneration process, energy consumption and chemical reagent use during the regeneration of hydrochar, and the study of the mechanism of retention of the pore structure and surface functional groups of hydrochar during regeneration are the keys to enhance its regeneration efficiency. In addition, the adsorption performance changes of hydrochar after multiple cycles of regeneration were evaluated by designing experiments to find out the solutions to improve its stability. This will help realize the economic feasibility of hydrochar in large-scale engineering applications.

In summary, although hydrochar has a broad application prospect in the field of wastewater treatment and nutrient recovery, its challenges and limitations still need to be solved through in-depth research and innovative technologies. Through new modified materials, optimization of the scale-up production process, and improvement of the regeneration efficiency, hydrochar is expected to become an important and feasible technological solution in the field of wastewater treatment in the future.

11.5 CONCLUSIONS

This chapter systematically summarizes the potential application of hydrochar in nutrient recovery from livestock manure wastewater and the challenges it faces. The results showed that hydrochar, as a novel material for wastewater treatment, has good nitrogen and phosphorus adsorption capacity, especially when its adsorption efficiency can be significantly enhanced after optimized by adjusting the preparation conditions (e.g., temperature, time, and type of raw materials) as well as the modification techniques (e.g., metal oxide loading). In addition, the porous structure and rich surface functional groups (e.g., carboxyl and hydroxyl groups) of the hydrochar played a key role in the nutrient adsorption process, showing its potential application value in the utilization of wastewater resources.

However, hydrochar still faces a number of problems in practical applications, including its relatively limited adsorption capacity and selectivity, as well as high production and modification costs. Although modification techniques under laboratory conditions can enhance its adsorption performance, the economics of its actual production and application remain key issues that need to be addressed. In addition, the regeneration capacity and stability of hydrochar need to be optimized urgently. Although the current physical and chemical regeneration can restore the adsorption capacity to a certain extent, the adsorption effect of the material gradually decreases after multiple regenerations, and the loss of pore structure and functional groups may also the feasibility of its use.

Future research should focus on modified materials to enhance the selective adsorption capacity of hydrochar while lowering production cost and energy consumption. The adsorption characteristics of hydrochar can be further optimized by introducing nanomaterials, composite metal oxides, or organic polymer modifiers. In addition, research on low-cost and environmentally friendly scale-up preparation processes is one of the keys to promote its engineering application. In addition, the innovation and optimization of regeneration technology will contribute to the adsorption performance of hydrochar after multiple uses and ensure its economic feasibility in large-scale applications.

In summary, while hydrochar has promising applications in wastewater treatment and nutrient recovery, its challenges and limitations require further investigation and technological innovation. With advances in material modification, scale-up production processes, and regeneration efficiency, hydrochar holds potential as a viable technological solution for future wastewater treatment. Future research should focus on enhancing selective adsorption, improving cost-efficiency in production, and developing sustainable regeneration techniques to optimize hydrochar's practical and economic feasibility in large-scale applications.

ACKNOWLEDGMENTS

The research was sustained by a grant from the National Key Research and Development Program of China "Intergovernmental Cooperation in International Science and Technology Innovation" [Grant number 2023YFE0104700] and the National Natural Science Foundation of China [Grant Number 31401944].

REFERENCES

Belete, Y. Z., Mau, V., Spitzer, R. Y., Posmanik, R., Jassby, D., Iddya, A., Kassem, N., Tester, J. W., Gross, A. (2021). Hydrothermal carbonization of anaerobic digestate and manure from a dairy farm on energy recovery and the fate of nutrients. *Bioresource Technology*, 333, 125164.

Bever, C. G., Coronella, C. J. (2024). Carbon sequestration potential of manure-derived hydrochar aided by secondary stabilization. *ACS Sustainable Chemistry and Engineering*, 12(14), 5705–5715.

Chen, W., Zhao, B., Guo, Y., Guo, Y., Zheng, Z., Pak, T., Li, G. (2021). Effect of hydrothermal pretreatment on pyrolyzed sludge biochars for tetracycline adsorption. *Journal of Environmental Chemical Engineering*, 9(6), 106557.

Chen, Y., Tian, L., Liu, T., Liu, Z., Huang, Z., Yang, H., Tian, L., Huang, Q., Li, W., Gao, Y., Zhang, Z. (2023). Speciation and transformation of nitrogen for sewage sludge hydrothermal carbonization-influence of temperature and carbonization time. *Waste Management*, 162, 8–17.

Dutta, T., Kim, T., Vellingiri, K., Tsang, D. C. W., Shon, J. R., Kim, K. H., Kumar, S. (2019). Recycling and regeneration of carbonaceous and porous materials through thermal or solvent treatment. *Chemical Engineering Journal*, 364, 514–529.

Ekpo, U., Ross, A. B., Camargo-Valero, M. A., Fletcher, L. A. (2016). Influence of pH on hydrothermal treatment of swine manure: Impact on extraction of nitrogen and phosphorus in process water. *Bioresource Technology*, 214, 637–644.

Goldfarb, J. L., Hubble, A. H., Ma, Q., Volpe, M., Severini, G., Andreottola, G., Fiori, L. (2022). Valorization of cow manure via hydrothermal carbonization for phosphorus recovery and adsorbents for water treatment. *Journal of Environmental Management*, 308, 114561.

He, L., Chen, Y., Sun, F., Li, Y., Huang, W., Yang, S. (2022a). Controlled release of phosphorus using lanthanum-modified hydrochar synthesized from water treatment sludge: Adsorption behavior and immobilization mechanism. *Journal of Water Process Engineering*, 50, 103319.

He, X., Zhang, T., Niu, Y., Xue, Q., Ali, E. F., Shaheen, S. M., Tsang, D. C. W., Rinklebe, J. (2022b). Impact of catalytic hydrothermal treatment and Ca/Al-modified hydrochar on lability, sorption, and speciation of phosphorus in swine manure: Microscopic and spectroscopic investigations. *Environmental Pollution*, 299, 118877.

Hu Y.C., Su M.R., Wang Y.F., Cui S.H., Meng F.X., Yue W.C., Liu Y.F., Xu C., Yang Z.F. (2020) Food production in China requires intensified measures to be consistent with national and provincial environmental boundaries. *Nature Food*, 1, 572–582.

Inkoua, S., Li, C., Rashid, M., Naeem, M. M., Zhang, S., Gao, W., Gholizadeh, M., Hu, X. (2024). Unveiling drastic influence of cross-interactions in hydrothermal carbonization of spirulina with cellulose, lignin or poplar on nature of hydrochar and activated carbon. *Journal of Environmental Management*, 366, 121713.

Ja Lilian, M., Bissessur, R., Kang, K., He, Q. S., Hu, Y. (2024). Study of KOH-activated hydrochar for CO_2 adsorption. *Journal of Industrial and Engineering Chemistry*, 143, 240–251.

Jiang, H., Deng, F., Luo, Y., Xie, Z., Chen, Y., Zhou, P., Liu, X., Li, D. (2022). Hydrothermal carbonization of corn straw in biogas slurry. *Journal of Cleaner Production*, 353, 131682.

Jin, S.Q., Zhang, B., Wu, B., Han, D. M., Hu, Y., Ren, C. C., Zhang, C. Z., Wei, X., Wu, Y., Mol, A. P. J., Reis, S, Gu, B. J., Chen J. Decoupling livestock and crop production at the household level in China. *Nature Sustainability*, 2021, 4, 48–55.

Karatas, O., Khataee, A., Kalderis, D. (2022). Recent progress on the phytotoxic effects of hydrochars and toxicity reduction approaches. *Chemosphere*, 298, 134357.

Khoury, O., Gaur, R., Zohar, M., Erel, R., Laor, Y., Posmanik, R. (2023). Phosphorus recycling from waste activated sludge using the hydrothermal platform: Recovery, solubility and phytoavailability. *Waste Management*, 169, 23–31.

Kong, Y., Ma, R., Li, G., Wang, G., Liu, Y., Yuan, J. (2022). Impact of biochar, calcium magnesium phosphate fertilizer and spent mushroom substrate on humification and heavy metal passivation during composting. *Science of the Total Environment*, 824, 153755.

Köninger, J., Lugato, E., Panagos, P., Kochupillai, M., Orgiazzi, A., Briones, M. J. I. (2021). Manure management and soil biodiversity: Towards more sustainable food systems in the EU. *Agricultural Systems*, 194, 103251.

Lang, Q., Zhang, B., Liu, Z., Jiao, W., Xia, Y., Chen, Z., Li, D., Ma, J., Gai, C. (2019). Properties of hydrochars derived from swine manure by CaO assisted hydrothermal carbonization. *Journal of Environmental Management*, 233, 440–446.

Li, C. S., Cai, R. R. (2022). Preparation of solid organic fertilizer by co-hydrothermal carbonization of peanut residue and corn cob: A study on nutrient conversion. *Science of the Total Environment*, 838(Part 2), 155867.

Li, C.-S., Cai, R.-R., Hasan, A., Lu, X.-L., Yang, X.-X., Zhang, Y.-G. (2023). Fertility assessment and nutrient conversion of hydrochars derived from co-hydrothermal carbonization between livestock manure and corn cob. *Journal of Environmental Chemical Engineering*, 11(1), 109166.

Li, D., Cui, H., Cheng, Y., Xue, L., Wang, B., He, H., Hua, Y., Chu, Q., Feng, Y., Yang, L. (2021a). Chemical aging of hydrochar improves the Cd^{2+} adsorption capacity from aqueous solution. *Environmental Pollution*, 287, 117562.

Li, S. Y., Teng, H. J., Guo, J. Z., Wang, Y. X., Li, B. (2021b). Enhanced removal of Cr(VI) by nitrogen-doped hydrochar prepared from bamboo and ammonium chloride. *Bioresource Technology*, 342, 126028.

Lin, C., Dong, B., Xu, Z. (2024). Competitive adsorption of heavy metals onto xanthate-modified sludge hydrochar and its solidification as secondary minerals. *Chemosphere*, 356, 141878.

Liu, Q., Zhang, G., Kong, G., Liu, M., Cao, T., Guo, Z., Zhang, X., Han, L. (2023). Valorizing manure waste into green coal-like hydrochar: Parameters study, physicochemical characteristics, combustion behaviors and kinetics. *Renewable Energy*, 216, 119103.

Liu, G. Q., Xu, Q., Abou-Elwafa, S. F., Ali Alshehri, M., Zhang, T. (2024). Hydrothermal carbonization technology for wastewater treatment under the "Dual Carbon" goals: Current status, trends, and challenges. *Water*, 16, 1749.

Luo, Y., Lan, Y., Liang, S., Yu, S., Xue, M., Yin, Z., Shen, F. F., Li, X., Hong, Z., Yan, M., Xie, C., Gao, B. (2024). Rice husk hydrochar prepared by hydrochloric acid assisted hydrothermal carbonization for levofloxacin removal in bioretention columns. *Bioresource Technology*, 393, 130105.

Olszewski, M. P., Nicolae, S. A., Arauzo, P. J., Titirici, M.-M., Kruse, A. (2020). Wet and dry? Influence of hydrothermal carbonization on the pyrolysis of spent grains. *Journal of Cleaner Production*, 260, 121101.

Oumabady, S., Bhardwaj, S. K., Ramasamy, S. P., Dandare, S. U., Sakrabani, R., Doherty, R., Nanukuttan, S., Kumaresan, D. (2024). Thermochemical characterisation of hydrochar from agricultural waste and its efficiency as a supplement with solid fuel. *Biomass and Bioenergy*, 187, 107299.

Qaramaleki, S. V., Mohedano, A. F., Coronella, C. J. (2023). Phosphorus recovery from aqueous product of hydrothermal carbonization of cow manure. *Waste Management*, 168, 301–310.

Qaramaleki, S. V., Villamil, J. A., Mohedano, A. F., Coronella, C. J. (2020). Factors affecting solubilization of phosphorus and nitrogen through hydrothermal carbonization of animal manure. *ACS Sustainable Chemistry and Engineering*, 8(33), 12462–12470.

Rasoulnia, P., Kokko, M., Kinnunen, V., He, C. (2024). Integrated valorization of digestate and concentrate from a centralized biogas plant via hydrothermal co-carbonization. *Chemical Engineering Journal*, 496, 154061.

Saha, N., Xin, D., Chiu, P. C., Reza, M. T. (2019). Effect of pyrolysis temperature on acidic oxygen-containing functional groups and electron storage capacities of pyrolyzed hydrochars. *ACS Sustainable Chemistry and Engineering*, 7(9), 8387–8396.

Sathe, S. M., Chakraborty, I., Dubey, B. K., Ghangrekar, M. M. (2024). Wastewater sludge-derived hydrochar: Effect of operating conditions, activation, and potential use as adsorbent. *Process Safety and Environmental Protection*, 184, 1400–1410.

Shan, G., Li, W., Bao, S., Hu, X., Liu, J., Zhu, L., Tan, W. (2023). Energy and nutrient recovery by spent mushroom substrate-assisted hydrothermal carbonization of sewage sludge. *Waste Management*, 155, 192–198.

Shehata, N., Egirani, D., Olabi, A. G., Inayat, A., Abdelkareem, M. A., Chae, K.-J., Sayed, E. T. (2023). Membrane-based water and wastewater treatment technologies: Issues, current trends, challenges, and role in achieving sustainable development goals, and circular economy. *Chemosphere*, 320, 137993.

Shuang, E., Wang, Q., Sheng, K., Zhang, X. (2024). Synthesis of Al-modified hydrochar from corn stover for efficient phosphate removal. *Journal of Environmental Chemical Engineering*, 12(1), 111725.

Song, C., Shan, S., Yang, C., Zhang, C., Zhou, X., Ma, Q., Yrjälä, K., Zheng, H., Cao, Y. (2020). The comparison of dissolved organic matter in hydrochars and biochars from pig manure. *Science of the Total Environment*, 720, 137423.

Spielmeyer, A., Höper, H., Hamscher, G. (2017). Long-term monitoring of sulfonamide leaching from manure amended soil into groundwater. *Chemosphere*, 177, 232–238.

Springmann, M., Clark, M., Croz, D. M. D., Wiebe, K., Bodirsky, B. L., Lassaletta, L., de Vries, W., Vermeulen, S. J., Herrero, M., Carlson, K. M., Jonell, M., Troell, M., de Clerck, F., Gordon, L. J., Zurayk, R., Scarborough, P., Rayner, M., Loken, B., Fanzo, J., Godfray, H. C. J., Tilman, D., Rockstrom, J., Willett, W. (2018). Options for keeping the food system within environmental limits. *Nature*, 562, 519–525.

Su, X. H., Zhang, T., Zhao, J. Y., Mukherjee, S., Alotaibi, N. M., Abou-Elwafa, S. F., Tran, H. T., Bolan, N. S. (2024). Phosphorus fraction in hydrochar from co-hydrothermal carbonization of swine manure and rice straw: An optimization analysis based on response surface methodology. *Water*, 16, 2208.

Sun, Z., Li, J., Wang, X., Xia, S., Zhao, J. (2024). Enhanced heavy metal stabilization and phosphorus retention during the hydrothermal carbonization of swine manure by in-situ formation of $MgFe_2O_4$. *Waste Management*, 174, 96–105.

Tan, M. J., Li, H., Huang, Z. L., Li, C. Z., Li, X. J., Wu, Z. J., Zhang, X., Qin, X. L., Huang, J., Yang, H. P. (2024). Metal chlorides and ammonium persulfate hydrothermal carbonization for enhanced pyrolysis behavior and biochar properties. *Journal of Analytical and Applied Pyrolysis*, 179, 106469.

Wang, J., Wang, Y., Wang, J., Du, G., Khan, K. Y., Song, Y., Cui, X., Cheng, Z., Yan, B., Chen, G. (2022). Comparison of cadmium adsorption by hydrochar and pyrochar derived from Napier grass. *Chemosphere*, 308, 136389.

Wang, X., Zhang, D., Wang, F., Zheng, X., Yang, X., Zeng, J., Yi, W. (2024). Effects of biogas slurry on hydrothermal carbonization of digestate: Synergistic valorization of hydrochars and aqueous phase. *Journal of Environmental Management*, 362, 121317.

Xie, S., He, X., Alshehri, M. A., Abou-Elwafa, S. F., Zhang, T. (2024). Elevated effect of hydrothermal treatment on phosphorus transition between solid-liquid phase in swine manure. *Results in Engineering*, 24, 102887.

Xiong, W., Luo, Y., Shangguan, W., Deng, Y., Li, R., Song, D., Zhang, M., Li, Z., Xiao, R. (2024). Co-hydrothermal carbonization of lignocellulosic biomass and swine manure: Optimal parameters for enhanced nutrient reclamation, carbon sequestration, and heavy metals passivation. *Waste Management*, 190, 174–185.

Xu, C. L., Li, A., Xu, X., Yang, Z., Wang, G., Cheng, Z., Zhang, S., Li, T., Pu, Y., Lv, G., Cai, J., Zhou, W., Pu, Z., Qi, X. (2024b). Co-hydrothermal carbonization of rape straw and industrial sewage sludge – Applications of hydrochar for N and P adsorptions and for agricultural/horticultural purposes. *Journal of Cleaner Production*, 451, 142154.

Xu, Q., Zhang, T., Niu, Y., Mukherjee, S., Abou-Elwafa, S. F., Nguyen, N. S. H., Al Aboud, N. M., Wang, Y., Pu, M., Zhang, Y., Tran, H. T., Almazroui, M., Hooda, P. S., Bolan, N. S., Rinklebe, J., Shaheen, S. M. (2024a). A comprehensive review on agricultural waste utilization through sustainable conversion techniques, with a focus on the additives effect on the fate of phosphorus and toxic elements during composting process. *Science of the Total Environment*, 942, 173567.

Yao, F., Ye, G., Peng, W., Zhao, G., Wang, X., Wang, Y., Zhu, W., Jiao, Y., Huang, H., Ye, D. (2023). Preparation of activated biochar with adjustable pore structure by hydrothermal carbonization for efficient adsorption of VOCs and its practical application prospects. *Journal of Environmental Chemical Engineering*, 11(2), 109611.

Zhang, L., Wang, Q., Xu, F., Wang, Z., Zhang, G. (2022a). Insights into the evolution of chemical structures in hydrochars from hydrothermal carbonization of PVC. *Journal of the Energy Institute*, 105, 323–333.

Zhang, T., Wu, X., Shaheen, S. M., Zhao, Q., Liu, X., Rinklebe, J., Ren, H. (2020). Ammonium nitrogen recovery from digestate by hydrothermal pretreatment followed by activated hydrochar sorption. *Chemical Engineering Journal*, 379, 122254.

Zhang, T., Wu, X. S., Fan, X., Tsang, D. C. W., Li, G. X., Shen, Y. J. (2019). Corn waste valorization to generate activated hydrochar to recover ammonium nitrogen from compost leachate by hydrothermal assisted pretreatment. *Journal of Environmental Management*, 236, 108–117.

Zhang, X., Zhang, L., Li, A. (2017). Hydrothermal co-carbonization of sewage sludge and pinewood sawdust for nutrient-rich hydrochar production: Synergistic effects and products characterization. *Journal of Environmental Management*, 201, 52–62.

Zhang, Y., Qu, J., Yuan, Y., Song, H., Liu, Y., Wang, S., Tao, Y., Zhao, Y., Li, Z. (2022b). Simultaneous scavenging of Cd(II) and Pb(II) from water by sulfide-modified magnetic pinecone-derived hydrochar. *Journal of Cleaner Production*, 341, 130758.

Zhang, Y. Y., Zhang, T. (2022c). Biowaste valorization to produce advance carbon materials-hydrochar for potential application of Cr (VI) and Cd (II) adsorption in wastewater: A review. *Water*, 14, 3675.

Zhou, C., Zhou, K., Li, H., Xu, X., Liu, B., Li, H., Zeng, Z., Ma, W., Li, L. (2021). Pressure swing adsorption properties of activated carbon for methanol, acetone and toluene. *Chemical Engineering Journal*, 413, 127384.

12 Computational Environmental Chemistry as a Strategy for Remediation

Murielly Fernanda Ribeiro Bihain,
Anna Karla dos Santos Pereira,
and Douglas Henrique Pereira

12.1 INTRODUCTION

Environmental pollution has become increasingly prominent due to industrialization and global development. The continuous contamination of air, water, and soil by chemical substances has heightened the need for research focused on mitigating these problems and developing techniques to remediate the generated waste (Sandoval-Pauker et al. 2024; Wang, Zhang, and Zhu 2024).

Environmental research generally involves the adsorption of different pollutants (Malloum et al. 2023), advanced oxidative processes for wastewater treatment (Dhamorikar et al. 2024), and reaction mechanisms between atmospheric radicals and gaseous pollutants (Warneck 2003), among others (Wang, Zhang, and Zhu 2024; Li and Lopez 2024; Xue et al. 2024). However, with the development of computers with high processing and performance, computational methods were introduced in the remediation process so that it was possible to analyze in a molecular way how processes occur, thus reducing costs and experimentation time (Sandoval-Pauker et al. 2024; Xue et al. 2024; Li, Borthwick, and Liu 2023).

In this context, computational chemistry can be applied to systems across different phases, utilizing theoretical calculations to study molecular properties, interaction mechanisms, and the system's behavior of interest (Li, Borthwick, and Liu, 2023; Morgon and Coutinho 2007). Understanding the molecular-level chemical interactions between pollutants and their surrounding environment makes computational chemistry valuable in environmental remediation research. This chapter introduces various computational chemistry techniques and their applications in addressing environmental pollution.

DOI: 10.1201/9781003536277-12

12.2 COMPUTATIONAL CHEMISTRY

Computational chemistry is a subarea of chemistry that uses computational simulations to elucidate systems involving chemical problems (Hajji et al. 2021). Rooted in quantum mechanics and/or Newtonian equations, this field has gained increasing prominence in the scientific literature (Sandoval-Pauker et al. 2024; Mammino 2023). Figure 12.1a illustrates the number of articles related to computational chemistry from 2014 to 2024, using data exported from the ScienceDirect database. Figure 12.1b provides an overview of the subsections within this field ("ScienceDirect" 2024).

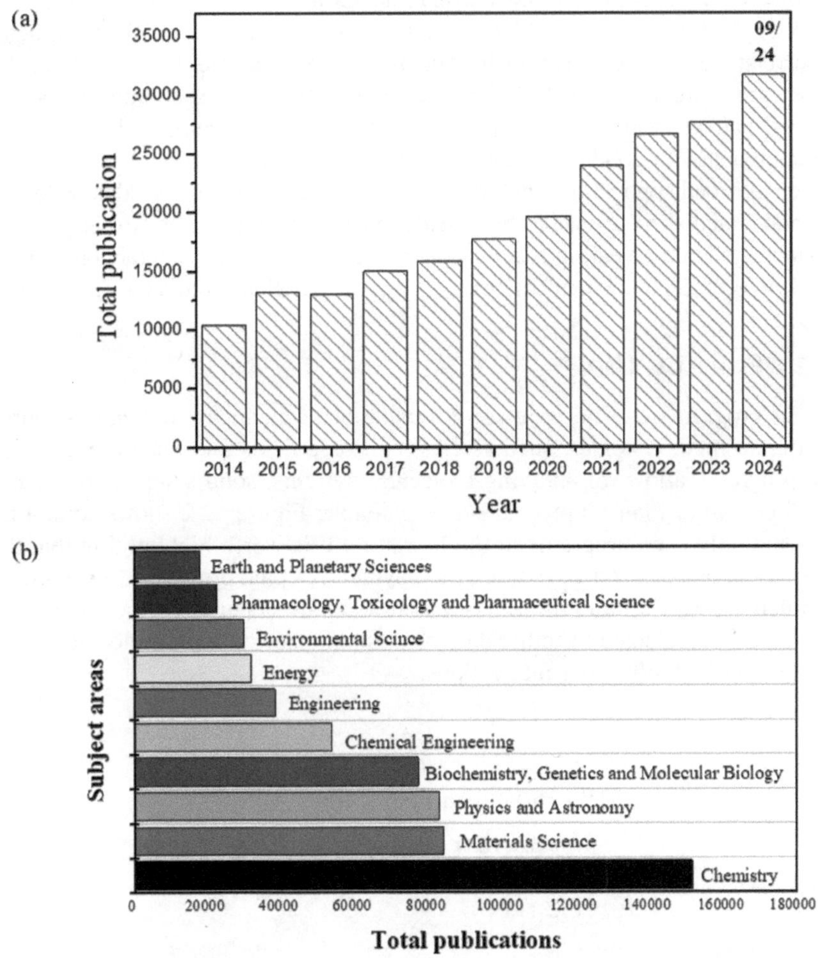

FIGURE 12.1 Total publications with the theme computational chemistry and main subject areas.

As shown in Figure 12.1a, it is possible to observe a growing number of publications over the last few years, highlighting the importance of this line. Another point that deserves to be highlighted is that among the subject areas, environmental chemistry is one of the most important subjects, and, although not highlighted in the subject areas, computational chemistry is found within other subject areas, such as chemistry, materials science, chemical engineering, energy, and engineering.

Computational chemistry includes molecular dynamics methods, semiempirical methods, *ab initio* methods, and density functional theory (DFT) (Sandoval-Pauker et al. 2024). The Hartree–Fock method is the simplest among the *ab initio* methods; however, it ignores electron correlation, unlike DFT, which is based on the Hohenberg–Kohn theorem, which relates electronic energy to electron density and correlation and exchange in its equations. Post-Hartree–Fock methods have more accurate precision for different systems but have higher computational costs (Sandoval-Pauker et al. 2024). Thus, the DFT has become popular because it is a faster approach with good accuracy (Morgon and Coutinho 2007).

Given its broad applicability, computational chemistry has proven valuable for environmental remediation. It enables the elucidation of structural and electronic properties of various systems, revealing reaction mechanisms, absorptive interactions, and the photocatalytic and photochemical processes between pollutants and the environment.

12.2.1 Programs Used

Different programs can be used for computational chemical simulations employing different methodologies, such as classical force fields and quantum chemistry. These programs allow for studying molecular systems, solid states, and dynamics simulations. More than 50 programs are available; Figure 12.2 shows some of the most commonly used programs in the literature. It is worth highlighting that these programs can be open source (not require payment) or paid programs. Each program has numerous features, and the choice of the software depends on the nature of the research being conducted, whether it involves solid-state systems, molecular dynamics, or the study of molecular interactions.

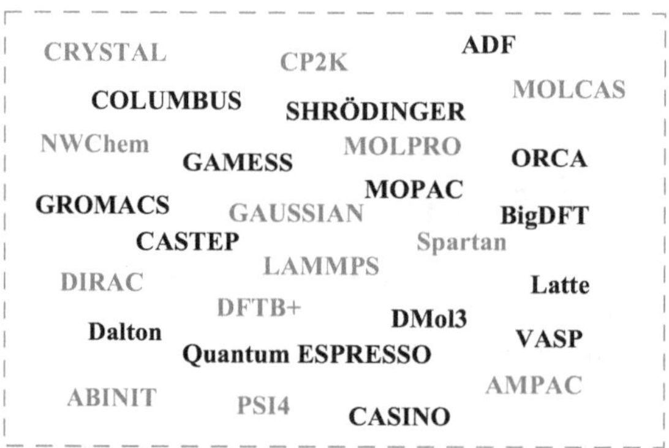

FIGURE 12.2 Schematic representation of some of the programs used in computational chemistry.

12.2.2 Calculated Properties

This work presents the structural properties that can be calculated and the main electronic structure techniques used in environmental chemistry for molecular systems. To represent the simulation data, theoretical calculations were performed at the ωB97XD/6-31+G(d,p) level, with the Gaussian 09 program (Frisch et al., 2009). The programs GaussView (Dennington et al. 2009), AIMALL package (Todd and Keith 2017), Multiwfn (Lu and Chen 2012), VMD (Humphrey, Dalke, and Schulten 1996), and gnuplot were used to generate the structures, topological properties, and noncovalent interaction (NCI) index.

12.2.2.1 Structural and Vibrational Properties

Atomistic model simulations must be as close as possible to an actual situation and consider all chemical details of the molecules. Thus, the structural properties must be well described by any methodology employed in the computational simulations (Batsanov 2017; Poppe et al. 2016).

Once computer simulations correctly describe their structural properties, they can be analyzed. Among the structural properties, it is essential to highlight that the simulations allow evaluation of the length of chemical bonds, the angles of the molecule, dihedrals, and, for solid systems, it is also possible to determine the lattice parameters and equilibrium volume. In conjunction with structural properties, vibrational properties can be generated for chemical systems, providing information about energy minima, infrared (IR) spectra, and data regarding chemical bonds (Batsanov 2017; Poppe et al. 2016).

With information on structural properties, significant gains can be obtained in environmental studies, as it is possible to verify the most stable conformation of the molecule, whether there is the formation of intra-/intermolecular bonds and their respective bond lengths, determine the best angle for interaction as well as the bond length of the interaction, and observe bond breaks in advanced oxidative processes. For solid-state simulations, it is possible to verify which atomic compositions and geometries face different crystalline planes, which alter the photocatalytic properties (Batsanov 2017; Poppe et al. 2016).

Figure 12.3 shows the structure of the emerging contaminant, paracetamol, with its respective bond lengths and the theoretically generated IR spectrum of the molecule.

As observed in Figure 12.3, structural properties can provide a lot of information helping to understand experimental phenomena. It is worth noting that other analyses such as UV-vis and nuclear magnetic resonance (NMR) can also be determined to assist the experimental part.

12.2.3 Electronic Properties

12.2.3.1 Frontier Molecular Orbitals

Understanding the electronic properties of molecules allows us to determine their numerous properties. Frontier molecular orbitals (FMOs) are among the most widely used concepts. Fukui et al. studied the reactivity of aromatic compounds, analyzed

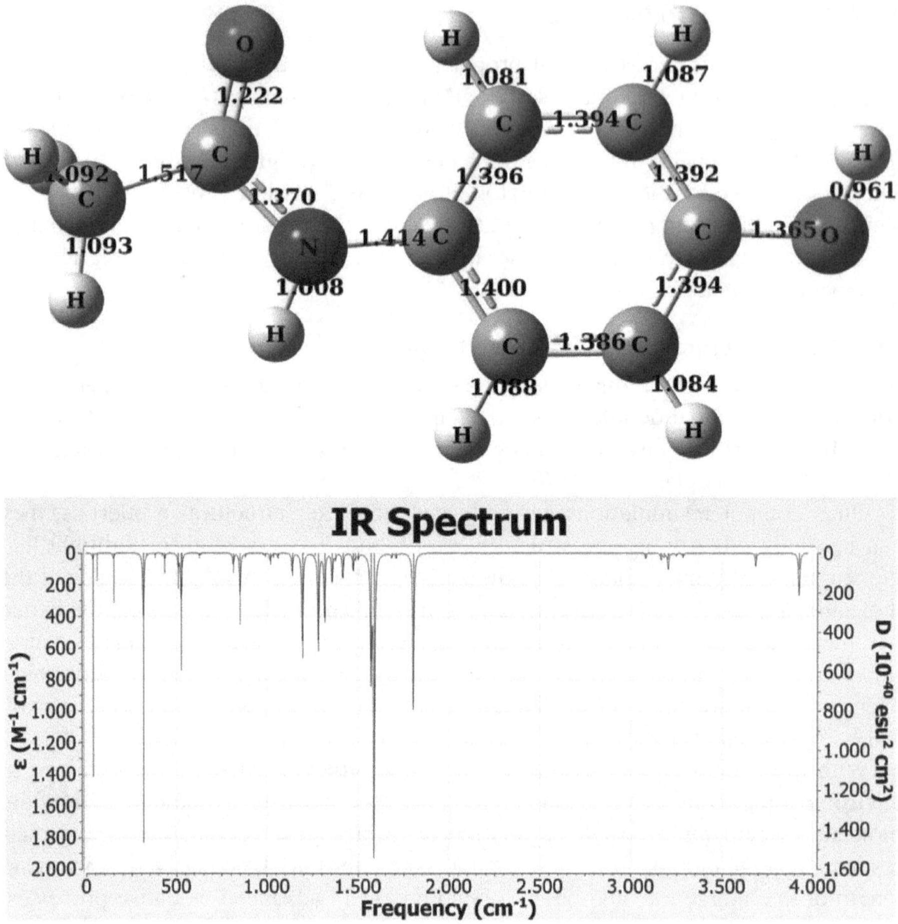

FIGURE 12.3 Chemical structure of paracetamol with bond lengths in Å and IR spectrum. Data obtained at the ωB97XD/6–31+G(d,p) level of theory.

the electron density, and realized that the electron density of the highest occupied molecular orbital (HOMO) and the lowest unoccupied molecular orbital (LUMO) explained the reactivity of the compounds (Pereira et al. 2016). After this interpretation of the frontier orbitals, the HOMO–LUMO concept gained prominence in numerous applications (Pereira et al. 2016).

Several theories have emerged over the years improving the use of FMOs, with an emphasis on the Localized Reactive Orbital, Reactive Hybrid Orbital, Triad, and Frontier Effective for Reaction Molecular Orbital concepts. Although several theories have improved the application of FMOs, HOMO–LUMO orbitals can provide numerous insights into chemical compounds, showing where the electronic density of the orbitals is located, the energy of each orbital, and the gap energy, and thus guiding where reactions will occur (Pereira et al. 2016).

12.2.3.2 Global Reactivity Descriptors

HOMO and LUMO define the chemical stability of a compound. Using the energy levels of HOMO and LUMO, it is possible to determine the molecule's ability to donate (Equation 12.1) or receive an electron (Equation 12.2), respectively. The energy gap (E_{gap}) values in Equation 12.3 allow us to predict the molecule's reactivity. The HOMO–LUMO orbitals and E_{gap} explain how one molecule can interact/react with another (Hamid et al. 2023; Bhatia 2024).

$$IP = -E_{HOMO} \tag{12.1}$$

$$EA = -E_{LUMO} \tag{12.2}$$

$$E_{gap} = E_{LUMO} - E_{HOMO} \tag{12.3}$$

Using DFT, the values of the HOMO–LUMO can be used to determine the reactivity indices of the compounds. Some global indices such as chemical hardness (η), chemical softness (S), chemical potential (μ), electronegativity (χ), and electrophilicity (ω) can be determined using Equations 12.4–12.7.

$$\eta = \frac{(E_{HOMO} - E_{LUMO})}{2} \tag{12.4}$$

$$\sigma = \frac{1}{2\eta} \tag{12.5}$$

$$\mu = \frac{(E_{HOMO} + E_{LUMO})}{2} = -\chi \tag{12.6}$$

$$\omega = \frac{\mu^2}{2\eta} \tag{12.7}$$

η is a reactivity descriptor that quantifies a system's resistance to electron transfer. When values are positive, they indicate a more stable and less reactive system. In contrast, a negative hardness value indicates that the system is more reactive and less stable. For chemical softness, the higher the value of σ, the more reactive the molecule is, and the reverse is observed for low values of σ (Hamid et al. 2023; Bhatia 2024).

Electrophilicity evaluates the tendency of a molecular system to accept electrons. The ω index can be classified as weak ($\omega < 0.8$ eV), moderate (ω between 0.9 and 1.4 eV), and strong ($\omega > 1.5$ eV) (Hamid et al. 2023; Bhatia 2024). Electronegativity (χ) is a measure of the ability of a molecule to attract electrons to itself. As a result, compounds are energetically favored for nucleophilic attack. The chemical potential (μ) is the inverse of electronegativity (Equation 12.6) and measures the tendency of the electron to escape from the electron cloud (Hamid et al. 2023; Bhatia 2024).

12.2.3.3 Fukui Indices

Local descriptors can be used to study the chemical reactivity of compounds and their selectivity at a given site, directing applications such as environmental descriptors. Among these, Fukui indices (Parr and Yang 1984; Morell, Grand, and Toro-Labbé 2005) allow the prediction of where an electrophilic (f^-), nucleophilic (f^+), or radical attack (f^0) will occur. The Fukui indices can be described by Equations 12.8–12.11.

$$\text{Nucleophilic attack: } f_k^+ = q_{N+1}^k - q_N^k \tag{12.8}$$

$$\text{Electrophilic attack: } f_k^- = q_N^k - q_{N-1}^k \tag{12.9}$$

$$\text{Radicular attack: } f_k^0 = (f_k^+ + f_k^-)/2 \tag{12.10}$$

$$\text{Dual descriptor: } f_k^{(2)} = f_k^+ - f_k^- \tag{12.11}$$

where q^k denotes the atomic charge. q_{N+1}^k and q_{N-1}^k are the charges of the molecules in their anionic and cationic forms, respectively. The value of $f_k^{(2)}$ allows a better distinction between nucleophilic and electrophilic attacks in which (1) if $f_k^{(2)} < 0$, the atom is favored by the electrophilic attack, and (2) if $f_k^{(2)} > 0$, the site (atom) is favored by the nucleophilic attack.

It is important to note that higher f values indicate more reactive sites for attack, while negative values suggest that the location is unlikely to undergo an attack. In the environmental context, Fukui function simulations offer valuable insights. When applied to advanced oxidative processes or natural degradation assessments, they can predict where radical attacks are likely to occur on organic contaminants, thereby aiding in the elucidation of reaction mechanisms (Du et al. 2024).

12.2.3.4 Quantum Theory of Atoms in Molecules

Richard Bader and collaborators developed the Quantum Theory of Atoms in Molecules (QTAIM), which allows studying different chemical systems (molecule, crystal, etc.). QTAIM partitions the chemical system into atomic fragments, constructing the topology based on its electron density ($\rho(r)$) (Bader 1994; Bader and Essén 1984; Grabowski 2013).

Through QTAIM, critical points (CPs) where the electron density gradient paths converge can be identified, leading to the classification of nuclear, bond (BCP), ring, and cage critical points. These CPs and the associated parameters derived from QTAIM provide valuable insights for theoretical and experimental chemistry studies (Bader 1994; Bader and Essén 1984; Grabowski 2013). Table 12.1 shows some descriptors generated by QTAIM that can assist in various environmental studies (Bader 1994; Bader and Essén 1984; Grabowski 2013).

12.2.3.5 Noncovalent Interaction (NCI) Index

Studies on chemical systems involving weak interactions between atoms or molecules are essential, as these interactions play a crucial role in determining the properties

TABLE 12.1
Some Descriptors Generated by QTAIM with Their Applicability

Property	Chemical Concept and Description
Electron density ($\rho(r)$)	Bond order/strength - Allows the analysis of the strength of a chemical bond and the bond order
Laplacian of the electron density ($\nabla^2\rho$ (r))	Allows the classification of bonds/interactions in which: i. $\nabla^2\rho$ (r) < 0 and high values of $\rho(r)$ indicate that covalent bond ($\rho(r)$) is > 0.1 a.u. and $\nabla^2\rho$ (r) < 0 ii. $\nabla^2\rho(r) > 0$ and low values of $\rho(r)$ indicate that ionic, hydrogen-bonding, or van der Waals interactions ($\rho(r)$) are < 0.1 and $\nabla^2\rho(r) > 0$
Ellipticity (ε)	π-Character Values of $\varepsilon = 0$ indicate a cylindrical shape of the bond ($\varepsilon \neq 0$ relate to the π character).
Kinetic energy density (G(r))	Helpful in identifying the type of interaction
Potential energy density V(r)	Helpful in identifying the type of interaction It also allows the quantification of the energy of a hydrogen bond: $E_{HB} = 1/2\ V(r)_{BCP}$
Total electron energy density H(r), where H(r) = V(r) + G(r))	Allows the identification of the type of interaction when correlated with $\nabla^2\rho(r)$. i. $\nabla^2\rho < 0$ and H(r)<0 covalent interaction; ii. $\nabla^2\rho < 0$ and H(r)> 0 partially covalent interactions; iii. $\nabla^2\rho > 0$ and H(r)> 0 electrostatic interactions.
Carga atômica q(Ω)	Charges of atoms

and functions of many compounds, including their chemical and biochemical behaviors. Various types of weak interactions are known, such as hydrogen bonds (HB), van der Waals forces, π-π stacking, σ-hole interactions, and metal–ligand bonding (Hajji et al. 2021). Computational chemistry provides a powerful framework for rationalizing experimental results and investigating these interactions. One of the most commonly used tools for analyzing weak interactions is the NCI index, which helps visualize and quantify these subtle forces (Johnson et al. 2010).

NCI analysis is a tool that allows us to expand the results obtained by QTAIM further. It makes it possible to understand the NCI that occurs in molecules, which is determined by electron density and does not involve the sharing of electrons (Hajji et al. 2021; Johnson et al. 2010).

The reduced density gradient (RDG) isosurface function was used to calculate the NCI index using Equation 12.12.

$$\mathrm{RDG} = \frac{1}{2\left(3\pi^2\right)^{1/3}} \frac{\left|\nabla\rho(r)\right|}{\rho(r)^{4/3}} \tag{12.12}$$

where ρ(r) is the electron density and ∇ρ(r) is its first derivative. The NCI is based on the RDG method, which represents the homogeneous deviation of an electron distribution within the structure; in other words, peaks are identified and classified based on the colored RDG isosurface. Thus, it is possible to evaluate and characterize the nature of the interactions as strong attraction or hydrogen bonding, strong repulsion (steric effect), or weak by defining the van der Waals forces (Hajji et al. 2021; Johnson et al. 2010; Jumabaev et al. 2023).

To analyze the nature of the interaction, two parameters are considered: the electronic density of the peaks (λ2), which quantifies the energy associated with the circulation of electrons that are in interaction regions, taking into account both bound and unbound regions, and the electron density (ρ(r)), since these parameters categorize the strength and bond type of the interaction (Hajji et al. 2021; Johnson et al. 2010; Jumabaev et al. 2023).

Values close to zero of λ2 and ρ(r) ($\lambda 2 \approx 0$ and $\rho \approx 0$) show van der Waals interactions, whereas strong repulsive interactions are seen when λ2 have positive values and high densities ($\lambda 2 > 0$ and $\rho(r) > 0$) and hydrogen interactions are characterized by negative values of λ2 and high values of ρ(r) ($\lambda 2 < 0$ and $\rho > 0$) (Hajji et al. 2021; Johnson et al. 2010; Jumabaev et al. 2023).

Furthermore, the visualization of NCI isosurfaces displays a color-coding scheme in which blue represents attractive interactions representing hydrogen bonds, red represents repulsive interactions displaying steric effects, and green represents intermediate interactions involving van der Waals interactions; the darker the color, the stronger the interaction (Hajji et al. 2021; Johnson et al. 2010; Jumabaev et al. 2023).

Figure 12.4 shows the FMO, QTAIM, and NCI analyses for paracetamol. As shown in Figure 12.3, it is possible to observe the orbital densities, through QTAIM and NCI analyses to verify the critical binding positions and the types of NCIs, respectively.

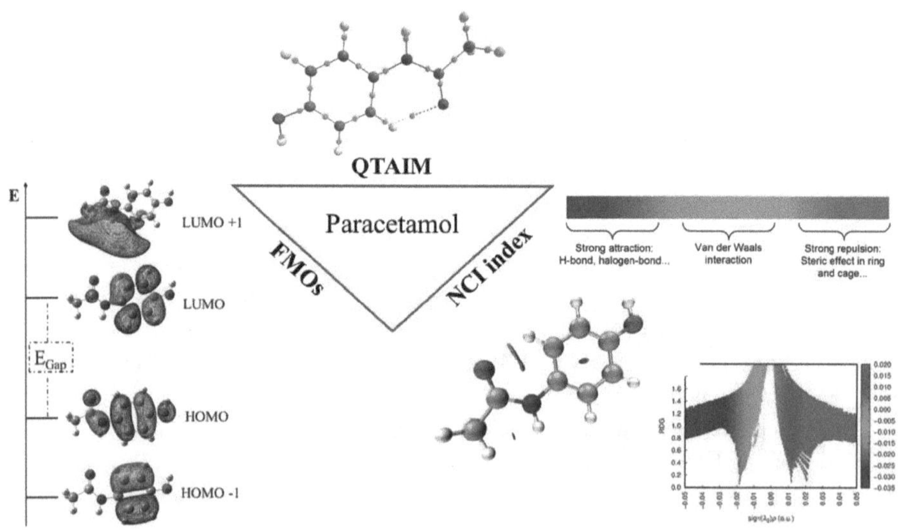

FIGURE 12.4 Frontier molecular orbitals, molecular graph generated by QTAIM, and NCI index result for the paracetamol molecule.

12.2.3.6 Natural Bond Orbital

Analysis of natural bond orbitals is a widely used tool with great applicability in environmental computational chemistry. This theory uses the concept of natural orbitals. It analyzes a many-electron molecular wave function in terms of localized electron-pair bonding units (Glendening, Landis, and Weinhold 2012; Glendening, Hiatt, and Weinhold 2024). Among the analyses that the NBO allows to be carried out, some deserve to be highlighted due to their applicability, which can be highlighted (Glendening, Landis, and Weinhold 2012; Glendening, Hiatt, and Weinhold 2024):

i. The stabilization energy, $E^{(2)}$, of the systems, is shown in Equation 12.13.

$$E_{ij}^{(2)} = q_i \frac{\left\langle i \middle| \hat{F} \middle| i \right\rangle^2}{\varepsilon_j - \varepsilon_i} \tag{12.13}$$

Where ε_j and ε_j are the energies of i and j orbitals and $\left\langle i \middle| \hat{F} \middle| i \right\rangle$ are the elements of the Fock matrix. The results make it possible to evaluate the interaction pathway between an occupied natural orbital i and an unoccupied orbital j. With the values of $E_{ij}^{(2)}$, you can obtain information about which orbitals are interacting, thus providing information about the following parameters, including

ii. stability and reactivity of the molecule;
iii. composition of the orbitals of a chemical bond and its hybridization;
iv. resonance structures;
v. atomic charges;
vi. molecular dipole moment;
vii. bond order;
viii. natural energy decomposition analysis; and
ix. NMR chemical shielding analysis, among other analyses.

12.3 APPLICATION OF COMPUTATIONAL CHEMISTRY IN ENVIRONMENTAL REMEDIATION SYSTEMS

Computational chemistry leverages various properties to elucidate the behavior of atoms and molecules, making it highly applicable to the analysis of pollutants in the environment. These properties are instrumental in studies of adsorption, advanced oxidation processes (AOPs) used in wastewater treatment, and assessments of atmospheric oxidative capacity. Computational chemistry helps to better understand and address environmental pollution challenges through these applications.

12.3.1 Adsorption

Among several techniques for wastewater remediation, adsorption stands out for its effectiveness in removing contaminants; it is versatile and uses few reagents and materials in small quantities. Its efficiency is determined by the porosity of the adsorbent, functional groups present, conditions of the medium in which the interaction occurs, and surface area (Malloum et al. 2023; Satyam and Patra 2024).

Various materials are employed as adsorbents due to their ability to interact with different adsorbates. In wastewater treatment, pollutants such as heavy metals, pesticides, organic contaminants like dyes, and pharmaceuticals are key research targets. These pollutants pose significant risks to human health and ecosystems, making their removal from the environment a critical focus of scientific investigation. Effective adsorbents can mitigate these risks by efficiently capturing and removing these harmful substances from wastewater (Malloum et al. 2023; Satyam and Patra 2024).

Computational chemistry, through molecular simulations, has gained prominence in water treatment research for its ability to predict adsorption energy and characterize the interactions between adsorbate and adsorbent under environmental conditions (Malloum et al. 2023; Cruz et al. 2022; Zhao et al. 2024). It can offer a more ecological alternative by reducing the number of candidate materials to be tested experimentally and avoiding the excessive use of reagents and materials (Satyam and Patra 2024).

Table 12.2 lists some studies that have used computational chemistry to simulate the adsorption process with different adsorbents.

As shown in Table 12.2, different computational analyses were conducted for the adsorption processes. The choice of articles focused on the type of adsorbate and the main current contamination issues, considering wastewater and the atmosphere. These three studies represent only a tiny fraction of the extensive literature that applies computational chemistry to adsorptive processes for wastewater and atmospheric remediation.

Cruz et al. (2022) studied the adsorption of thiamethoxam, a pesticide with carcinogenic potential that is highly soluble in water and has low sorption in soil. The availability of thiamethoxam in the environment as a contaminant is favorable.

TABLE 12.2
Some Adsorption Studies Using Computational Chemistry

Adsorbent	Adsorbate	Theory Level	Computational Analysis	Reference
Graphitic carbon nitride (g-C_3N_4)	Thiamethoxam	ωB97XD/ 6-31+G(d,p)	• FMO; • Global reactivity descriptors • Properties thermodynamics; • QTAIM; • Structural properties.	Cruz et al. (2022)
Polyamide, polyethylene, polypropylene	Cephradine	M06-2X/ 6-311G**	• Adsorption energy; • Structural properties; • QTAIM; • NCI.	G. Zhao et al. (2024)
Cobalt clusters ($Co_{n\,(n=2-7)}$)	Sulfur oxide (SO), nitrogen (NO), and carbon monoxide (CO)	B3LYP/ def2-TZVPPD	• Vibration and bond length; • Adsorption energy ; • Global reactivity descriptors; • Charge transfer.	Shabeeb e Maity (2024)

In this study, the adsorbent, graphitic carbon nitride (g-C_3N_4), formed complexes with the pesticide through hydrogen bonds with energy values that varied between −0.98 and −6.58 kJ/mol, as shown by the results obtained through the QTAIM analysis. Analysis of the thermodynamic properties indicated that the interaction of the complex was spontaneous and exothermic, demonstrating that g-C_3N_4 is a promising adsorbent for removing thiamethoxam (Cruz et al. 2022).

Zhao et al. (2024) employed DFT and molecular dynamics simulations to investigate the adsorption of cefradine, an antibiotic, on various microplastic surfaces—two emerging pollutants that, when combined, may amplify their impact on living organisms. Their DFT-D3 and molecular dynamics calculations revealed that the adsorption of cefradine onto microplastics occurred in two stages: surface adsorption and internal diffusion. The interaction energy was more pronounced with polyamide due to electrostatic and van der Waals effects, resulting in a higher adsorption capacity. QTAIM and NCI analyses further confirmed that the adsorption of cefradine on polyamide was primarily driven by hydrogen bonding and van der Waals forces (G. Zhao et al. 2024). The results demonstrate the importance of theoretical studies as the interaction between two contaminants can further aggravate the harmful effects of pollutants.

Shabeeb and Maity (2024) studied the adsorption of atmospheric pollutants such as SO, NO, and CO, which act as greenhouse gases. The adsorbent used was a small agglomerate of cobalt, Co_{2-7}, as determined by DFT calculations. Through theoretical calculations and computational analyses, the authors evaluated the molecular frontier orbitals and relative energies of optimized structures to assess their stability. They analyzed adsorption energy, reactivity indices, charge transfer via NBO, and bond lengths, revealing that the adsorption of greenhouse gases was favorable. Monocoordinated complexes of CO and NO were more stable, whereas SO formed bidentate complexes. Charge transfer within the clusters was significant for gas activation mechanisms. Additionally, bond length analysis indicated a weakening of NO and CO gas bonds as cluster size increased. The study demonstrated that cobalt clusters effectively adsorb greenhouse gases, acting as reducing agents for these pollutants (Shabeeb and Maity 2024).

Based on the results obtained and the different adsorbates and adsorbents used, it is possible to demonstrate the applicability and importance of computational chemistry in the determination and analysis of these interactions in the environment, showing that for those that are theoretical, it is possible to occur experimentally, and for those that are experimental, complementing the study.

12.3.2 Atmospheric Oxidative Capacity

The atmospheric oxidation capacity refers to the driving force of the atmosphere in carrying out air purification processes through oxidation reactions initiated by atmospheric radicals (Wei et al. 2024), these radicals being (1) the hydroxyl radical (HO), the main daytime atmospheric oxidant due to its high oxidizing power; (2) peroxygen radicals ($HO_2^{\bullet}$ and $RO_2^{\bullet}$) (Warneck 2003), (3) nitrogen trioxide ($NO_3^{\bullet}$) (Yan et al. 2021), the leading nocturnal oxidizing agent, and chlorine radicals (Faxon and Allen 2013). The troposphere, the layer closest to the Earth's crust, is an oxidative environment because of the need for self-purification processes to maintain breathable air quality.

In the literature, it is possible to find simulations of the reaction mechanisms of the oxidation of various gaseous pollutants with these primary radicals. Computational chemistry, through reaction kinetics and the calculation of reaction energies, can assist in identifying and elucidating the reaction mechanisms of the degradation of gaseous pollutants.

Table 12.3 lists some studies that used computational chemistry to simulate atmospheric oxidation reactions.

Several computational analyses were conducted (Table 12.3). In a theoretical–experimental study by Bihain et al. (2023), the authors showed the existence of the carbonate anion radical ($CO_3^{-\bullet}$) in the atmosphere through calculations of the thermodynamic properties (Gibbs energy) of reactions with the primary gaseous pollutants. Consequently, the possibility of the $CO_3^{-\bullet}$ radical reacting with compounds containing sulfur in their structures was identified. Through FMO, they also showed that radicals can react easily in the atmosphere. However, the spin density results showed that $CO_3^{-\bullet}$ is less reactive than HO• (Bihain et al. 2023), confirming what is known experimentally.

Zhao et al. (2024) theoretically studied the degradation of two volatile organic compounds, pentamethylbenzene (5-PeMB) and hexamethylbenzene (6-HeMB), which are pollutants found in petroleum and combustion that can produce photochemical pollution. This study analyzed the kinetics and ecotoxicity of the degradation of

TABLE 12.3
Some Atmospheric Oxidation Reaction Studies That Use Computational Chemistry

Radical	Atmospheric Pollutant	Theory Level	Computational Analysis	Reference
HO	Pentamethylbenzene (5-PeMB) and Hexamethylbenzene (6-HeMB)	M06-2X/6-311++G (3df, 2dp)	• Thermodynamic properties; • Fukui index; Chemical kinetics.	H. Zhao et al. (2024)
Hydroxymethane sulfonic acid	Volatile organic compounds	Methods: B3LYP-D3, M06-2X, PW91PW91, and ωB97X-D Basis set: aug-cc-pVTZ for C, H, and O; aug-cc-pV(T+d)Z to S	• Interaction energy and frequencies; • Geometric parameters; • MEP; • QTAIM; • NCI; • Energy decomposition.	Chen et al. (2024)
$CO_3^{-\bullet}$	Main polluting gases	UMP2/6-311++G (2df,p)	• Thermodynamic properties; • FMO; • Spin density.	Bihain et al. (2023)

these pollutants, showing that reactions of addition and abstraction of hydrogen in this system occur easily in the atmosphere due to low energy barriers. The simulated degradation constants at 298 K were $2.69 \times 10^{-10} cm^3$/molecule s for 5-HePM and $1.28 \times 10^{-10} cm^3$/molecule s for 6-HePM (H. Zhao et al. 2024).

Chen et al. (2024) demonstrated interactions between hydroxymethanesulfonic acid (HMSA) and small organic molecules typical in the atmosphere. Compounds such as formic acid, methanol, dimethyl ether, methyl formate, and acetone are typical precursors of aerosol nucleation. In contrast, sulfur compounds contribute significantly to atmospheric nucleation processes. Therefore, using DFT calculations, the authors investigated HSMA as a precursor for the nucleation of organic compounds. The work showed that these can form a stable ring with approximately six to eight members, which are provided through strong hydrogen bonds, higher negative binding energy, lower enthalpy of formation, and even higher Gibbs energy as HMSA–HMSA and HMSA with methyl formate (Chen et al. 2024).

Through the results obtained, with several demonstrated analyses, it is possible to demonstrate the applicability and importance of computational chemistry in determining and analyzing these interactions in the atmosphere and to elucidate better the oxidative reactions within the atmosphere's self-purification process.

12.3.3 Advanced Oxidation Process

AOPs are considered one of the most effective treatments for wastewater and occur through the radical HO or another radical, which acts as a strong oxidant (Ma et al. 2021; Dhamorikar et al. 2024). AOPs are increasingly popular as they have high oxidation efficiency. A standard AOP is carried out in two basic steps: (1) the formation of strong oxidants and (2) the oxidation of contaminants present in water (Dhamorikar et al. 2024).

Different AOPs vary according to the mechanism by which radicals are introduced into the medium, including photocatalytic oxidation, Fenton reactions, ozonization, electrochemical oxidation, sonochemical oxidation, and AOP based on sulfate radicals (Liang et al. 2024; Dhamorikar et al. 2024). Each of them has its specifications; however, the primary mechanism involves the generation of reactive oxygen species involved in the oxidation of organic pollutants in wastewater. Because they deal with highly reactive free radicals, investigating the reaction mechanisms becomes challenging; therefore, computational chemistry is a method for elucidating the mechanisms of AOPs through theoretical calculations.

Computational chemistry, through DFT calculations, provides functions such as the use of molecular descriptors that predict the active sites of pollutant molecules, such as electrostatic potential, Fukui index, spin density, FMOs, and Gibbs energy for transition state energy barriers to elucidate the reaction mechanism and investigate oxidant activation processes. Table 12.4 summarizes some studies that exemplify the use of computational chemistry in AOPs.

Zhang et al. (2024) investigated the sulfate radical-based advanced oxidation for degradation of carbofuran, a highly toxic pesticide, using DFT calculations. The results indicated sulfate and hydroxyl radicals could undergo addition and abstraction reactions with carbofuran. Thermodynamic and kinetic calculations further revealed

TABLE 12.4
Some AOP Studies Using Computational Chemistry

AOPs	Theory Level	Computational Analysis	Reference	Theory Level
Persulfate based	Carbofuran	M06-2X/6-31+G(d,p) and M06-2X/6-311++G(3df,3pd)	• Fukui index; • Distribution of electrostatic potential; • Gibbs free energy barrier; • Reactive degradation mechanism; • Ecotoxicity.	Zhang et al. (2024)
UV/ Chlorine	Carbamazepine	M062X/6-311++G (d, p) and B3LYP-D3/6-31G (d, p)	• Fukui index; • Molecular frontier orbital; • Reactive degradation mechanism.	Wu et al. (2024)
UV/Fe^{3+} and persulfate	PFOA and PFOS	B3LYP/Lanl2dz for Fe and 6-31G (d,p); M06L/SDD for Fe and 6-311+G (d,p) of remaining atoms	• Reactive degradation mechanism; • Bond lengths.	Lai et al. (2024)

that the radicals were added to the benzene ring; however, the abstraction occurred on the furan ring. The study demonstrated a significant reduction in toxicity, with the sulfate radical exhibiting a more significant initial effect than the hydroxyl radical. The authors concluded that this approach offers a promising method for the degradation of carbofuran (Zhang et al. 2024).

Wu et al. (2024) conducted a study on the degradation of carbamazepine (CBZ), a medicine used for epilepsy and neuropathic pain that has low removal in wastewater treatment, into a micropollutant. Using DFT calculations and experiments, the authors showed that the combined UV/Cl oxidation process significantly degraded the pollutant, with the reactive species HO, $Cl^{\bullet}$, and $ClO^{\bullet}$ being mainly responsible for degradation. Theoretical calculations using the Fukui index showed that the predominant mechanism for hydroxyl and chlorine radicals is through CBZ attack by hydrogen atom transfer, despite also having a radical addition mechanism, which is the $ClO^{\bullet}$ radical that dominates this mechanism in the UV/chlorine degradation process (Wu et al. 2024).

Lai et al. (2024) conducted the first DFT study on the degradation pathways of perfluorooctanoic acid (PFOA) and perfluorooctane sulfonate (PFOS), two persistent organic pollutants that pose significant toxicity to animals, bioaccumulate, and harm human health. The study explored various degradation pathways initiated by sulfate and hydroxyl radicals. Sulfate radicals promote degradation through a single-electron transfer mechanism, whereas hydroxyl radicals act as highly reactive oxidizing agents. Thermodynamic results indicated that persulfate was more favorable for this mechanism, although both degradation reactions were thermodynamically and kinetically favorable. The study also found that while pH can influence the availability of radicals, a low pH is not essential for the persulfate system to function effectively (Lai et al. 2024).

The advanced oxidative processes are diverse. Research using computational calculations has gained prominence in these studies. The articles are a fraction of the literature concerning theoretical studies applied to AOPs using the DFT analysis.

12.4 CONCLUSION

Computational chemistry is a subfield that provides insights into diverse chemical systems across various phases. Its applications in environmental studies have become essential, as it contributes results that may not be fully addressed through experimental methods, ultimately reducing both experimental time and costs, mainly when applied to green chemistry.

Computational chemistry analyzes various properties of systems, including structural, vibrational, and electronic characteristics. These analyses encompass the Fukui index, global reactivity descriptors, FMOs, and thermodynamic and kinetic calculations.

In the domain of environmental remediation, computational chemistry plays an increasingly prominent role, being applied to advanced oxidative processes, atmospheric oxidative processes, and adsorption studies. It helps explain pollutant interactions with the environment and inform the design of remediation systems aimed at environmental protection.

Finally, it concludes with the importance of simulations for environmental chemistry and the countless contributions the area can bring today.

ACKNOWLEDGMENTS

The authors acknowledge financial support from FAPESP -Fundação de Amparo à Pesquisa (Grant 2023/17727-2) and CAPES (Coordination of Improvement of Higher Education Personnel, Brazil, Funding Code 001 CAPES). Anna K. dos S. Pereira acknowledges the PDPG-POSDOC (Programa de Desenvolvimento da Pós-Graduação (PDPG) Pós-Doutorado Estratégico - Postgraduate Development Program (PGDP) Strategic Postdoctoral) No. 88887.798251/2022-00.

REFERENCES

Bader, R. F. W., e H. Essén. 1984. "The characterization of atomic interactions". *The Journal of Chemical Physics* 80 (5): 1943–60. doi:10.1063/1.446956.

Bader, Richard F. W. 1994. *Atoms in Molecules: A Quantum Theory*. Clarendon Press.

Batsanov, A. S. 2017. "X-Ray Diffraction, Small Molecule Applications☆". *Em Encyclopedia of Spectroscopy and* Spectrometry (Third Edition), organizado por John C. Lindon, George E. Tranter, e David W. Koppenaal, 656–66. Oxford: Academic Press. https://api.pageplace.de/preview/DT0400.9780128032251_A28527974/preview-9780128032251_A28527974.pdf.

Bhatia, Manjeet. 2024. "An overview of conceptual-DFT based insights into global chemical reactivity of volatile sulfur compounds (VSCs)". *Computational Toxicology* 29 (março): 100295. doi:10.1016/j.comtox.2023.100295.

Bihain, Murielly Fernanda Ribeiro, Laura Beatriz Camargo Oh, Keila Cardoso Teixeira, Grasiele Soares Cavallini, e Douglas Henrique Pereira. 2023. "Theoretical and experimental evidences for the formation of carbonate anion radical ($CO_3^{-\bullet}$) in the atmosphere". *Atmospheric Pollution Research* 14 (6): 101783. doi:10.1016/j.apr.2023.101783.

Chen, Dong-Ping, Wen Ma, Chun-Hong Yang, Ming Li, Zhao-Zhen Zhou, Yang Zhang, e Zheng-Jun Quan. 2024. "Interaction between hydroxymethanesulfonic acid and several organic compounds and its atmospheric significance". *Journal of Molecular Graphics and Modelling* 130 (julho): 108782. doi:10.1016/j.jmgm.2024.108782.

Cruz, Állefe Barbosa, Raimundo Francisco de Carvalho, Thiago Soares Silva, Renato de Almeida Sarmento, Grasiele Soares Cavallini, e Douglas Henrique Pereira. 2022. "Adsorptive capacity of a g-C3N4 matrix for thiamethoxam removal: A DFT study". *Computational and Theoretical Chemistry* 1215 (setembro): 113816. doi:10.1016/j.comptc.2022.113816.

Dennington, R. D., Keith, T. A., & Millam, J. M. (2009). *Gauss View*. Semichem Inc.

Dhamorikar, Rushikesh S., Vikesh G. Lade, Pratik V. Kewalramani, e Ankush B. Bindwal. 2024. "Review on integrated advanced oxidation processes for water and wastewater treatment". *Journal of Industrial and Engineering Chemistry* 138 (outubro): 104–22. doi:10.1016/j.jiec.2024.04.037.

Du, Hao, Rubin Huo, Aoxiang Liu, Yuxin Hui, Bo Shen, Ningyi Li, Yun Ji, et al. 2024. "Resorcinol-formaldehyde resins/MIL-88A(Fe) S-scheme heterojunctions drive photo-self-Fenton system for enhanced tetracycline degradation and bacterial inactivation: Mechanism insight and DFT calculation". *Applied Catalysis B: Environment and Energy* 357 (novembro): 124283. doi:10.1016/j.apcatb.2024.124283.

Faxon, C. B., e D. T. Allen. 2013. "Chlorine chemistry in urban atmospheres: A review". *Environmental Chemistry* 10 (3): 221. doi:10.1071/EN13026.

Frisch, M. J., Trucks, G. W., Schlegel, H. B., Scuseria, G. E., Robb, M. A.,Cheeseman, J. R., Scalmani, G., Barone, V., Mennucci, B., Petersson, G. A., Nakatsuji, H., Caricato, M., Li, X., Hratchian, H. P., Izmaylov, A. F.,Bloino, J., Zheng, G., Sonnenberg, J. L., Hada, M., … Fox, D. J. (2009).Gaussian 09 Revision D.1., Gaussian Inc.

Glendening, Eric D., Devin M. Hiatt, e Frank Weinhold. 2024. "Natural Bond Orbital Analysis of Chemical Structure, Spectroscopy, and Reactivity: How It Works". *Em Comprehensive Computational Chemistry* (First Edition), organizado por Manuel Yáñez e Russell J. Boyd, 406–21. Oxford: Elsevier. https://www.sciencedirect.com/referencework/9780128232569/comprehensive-computational-chemistry.

Glendening, Eric D., Clark R. Landis, e Frank Weinhold. 2012. "Natural bond orbital methods". *WIREs Computational Molecular Science* 2 (1): 1–42. doi:10.1002/wcms.51.

Grabowski, Sławomir J. 2013. "Non-covalent interactions – QTAIM and NBO analysis". *Journal of Molecular Modeling* 19 (11): 4713–21. doi:10.1007/s00894-012-1463-7.

Hajji, Melek, Nadeem Abad, Mohamed A. Habib, Salima Moftah H. Elmgirhi, e Taha Guerfel. 2021. "Computational chemistry methods for modelling non-covalent interactions and chemical reactivity— An overview". *Journal of the Indian Chemical Society* 98 (11): 100208. doi:10.1016/j.jics.2021.100208.

Hamid, Aabid, Nidhi Deswal, Sourav Pal, e Ram Kinkar Roy. 2023. "Chapter 7- Components of density functional reactivity theory-based stabilization energy: descriptors for thermodynamic and kinetic reactivity". *Em Chemical Reactivity*, organizado por Savaş Kaya, László von Szentpály, Goncagül Serdaroğlu, e Lei Guo, 181–226. Elsevier. https://www.sciencedirect.com/science/article/abs/pii/B9780323902595000135.

Humphrey W, Dalke A, Schulten K. 1996. "VMD: Visual molecular dynamics". *Journal of Molecular Graphics* 14 (1): 33–38.

Janeoo, Shashi, Reenu, Amandeep Saroa, Rakesh Kumar, e Harminder Kaur. 2022. "Computational investigation of bioactive 2,3-diaryl quinolines using DFT method: FT-IR, NMR spectra, NBO, NLO, HOMO-LUMO transitions, and quantum-chemical properties". *Journal of Molecular Structure* 1253 (abril): 132285. doi:10.1016/j.molstruc.2021.132285.

Johnson, Erin R., Shahar Keinan, Paula Mori-Sánchez, Julia Contreras-García, Aron J. Cohen, e Weitao Yang. 2010. "Revealing noncovalent interactions". *Journal of the American Chemical Society* 132 (18): 6498–506. doi:10.1021/ja100936w.

Jumabaev, Abduvakhid, Utkirjon Holikulov, Hakim Hushvaktov, Noureddine Issaoui, e Ahmad Absanov. 2023. "Intermolecular interactions in ethanol solution of OABA: Raman, FTIR, DFT, M062X, MEP, NBO, FMO, AIM, NCI, RDG analysis". *Journal of Molecular Liquids* 377 (maio): 121552. doi:10.1016/j.molliq.2023.121552.

Kurth, S., M. A. L. Marques, e E. K. U. Gross. 2005. "Density-Functional Theory". *Encyclopedia of Condensed Matter Physics*, janeiro: 395–402. https://www.sciencedirect.com/referencework/9780123694010/encyclopedia-of-condensed-matter-physics.

Lai, Yilei, Ying Wang, Shuyu Zhang, e Abing Duan. 2024. "Kinetics and mechanism analysis of advanced oxidation degradation of PFOA/PFOS by UV/Fe3+ and persulfate: A DFT study". *Chemosphere* 357 (junho): 141951. doi:10.1016/j.chemosphere.2024.141951.

Li, Fan, Alistair G. L. Borthwick, e Wen Liu. 2023. "Environmental theoretical calculation for non-periodic systems". *Trends in Chemistry* 5 (6): 410–14. doi:10.1016/j.trechm.2023.03.009.

Li, Jingbai, e Steven A. Lopez. 2024. "Computational Chemistry for Photochemical Reactions". *Em Comprehensive Computational Chemistry*, 658–98. Elsevier. doi:10.1016/b978-0-12-821978-2.00012-x.

Lu, T., & Chen, F. (2012). "Multiwfn: A multifunctional wavefunction analyzer". *Journal of Computational Chemistry* 33 (5), 580–92. doi:10.1002/jcc.22885.

Liang, Jialiang, Peng Zhen, Pengfei Gan, Yunyi Li, Meiping Tong, e Wen Liu. 2024. "DFT calculation of nonperiodic small molecular systems to predict the reaction mechanism of advanced oxidation processes: Challenges and perspectives". *ACS ES&T Engineering* 4 (1): 4–18. doi:10.1021/acsestengg.3c00204.

Ma, Dengsheng, Huan Yi, Cui Lai, Xigui Liu, Xiuqin Huo, Ziwen An, Ling Li, et al. 2021. "Critical review of advanced oxidation processes in organic wastewater treatment". *Chemosphere* 275 (julho): 130104. doi:10.1016/j.chemosphere.2021.130104.

Malloum, Alhadji, Kayode A. Adegoke, Joshua O. Ighalo, Jeanet Conradie, Chinemerem R. Ohoro, James F. Amaku, Kabir O. Oyedotun, et al. 2023. "Computational methods for adsorption study in wastewater treatment". *Journal of Molecular Liquids* 390 (novembro): 123008. doi:10.1016/j.molliq.2023.123008.

Mammino, Liliana. 2023. "Green chemistry and computational chemistry: A wealth of promising synergies". *Sustainable Chemistry and Pharmacy* 34 (setembro): 101151. doi:10.1016/j.scp.2023.101151.

Morell, Christophe, André Grand, e Alejandro Toro-Labbé. 2005. "New dual descriptor for chemical reactivity". *The Journal of Physical Chemistry A* 109 (1): 205–12. doi:10.1021/jp046577a.

Morgon, Nelson H., e Kaline Rabelo Coutinho. 2007. *Métodos de química teórica e modelagem molecular*. Editora Livraria da Física.

Parr, Robert G., e Weitao Yang. 1984. "Density functional approach to the frontier-electron theory of chemical reactivity". *Journal of the American Chemical Society* 106 (14): 4049–50. doi:10.1021/ja00326a036.

Pereira, Douglas Henrique, Felipe A. La Porta, Regis Tadeu Santiago, Danielle R. Garcia, e Teodorico C. Ramalho. 2016. "Novas Perspectivas sobre o Papel dos Orbitais Moleculares de Fronteira no Estudo Da Reatividade Química: Uma Revisão". *Revista Virtual de Química* 8 (2): 425–53.

Poppe, László, József Nagy, Gábor Hornyánszky, e Zoltán Boros. 2016. "Stereochemistry and stereoselective synthesis: An introduction". *Stereochemistry and Stereoselective Synthesis: An Introduction*, setembro. Wiley-VCH Verlag GmbH & Co. KGaA. doi:10.1002/9783527341177.

Sandoval-Pauker, Christian, Sheng Yin, Alexandria Castillo, Neidy Ocuane, Diego Puerto-Diaz, e Dino Villagrán. 2024. "Computational chemistry as applied in environmental research: opportunities and challenges". *ACS ES&T Engineering* 4 (1): 66–95. doi:10.1021/acsestengg.3c00227.

ScienceDirect, 2024. accessed on September 28. https://www.sciencedirect.com/search?qs=computational%20chemistry&date=2014-2024.

Satyam, Satyam, e Sanjukta Patra. 2024. "Innovations and challenges in adsorption-based wastewater remediation: A comprehensive review". *Heliyon* 10 (9): e29573. doi:10.1016/j.heliyon.2024.e29573.

Shabeeb, Muhammed, e Surajit Maity. 2024. "Computational investigation on adsorption and activation of atmospheric pollutants CO, NO and SO on small cobalt clusters". *Chemical Physics* 582 (junho): 112291. doi:10.1016/j.chemphys.2024.112291.

Todd A, Keith T, 2017. AIMAll, Gristmill Software, Overland ParkKS, USA.

Wang, Lan, Wei Zhang, e Zhiqiang Zhu. 2024. "6: Applications in environmental remediation". *Em Full-Spectrum Responsive Photocatalytic Materials*, organizado por Chuanyi Wang, Yanyan Duan, Lan Wang, e Qiuhui Zhu, 157–82. Woodhead Publishing Series in Electronic and Optical Materials. Woodhead Publishing. doi:10.1016/B978-0-443-13631-3.00003-3.

Warneck, Peter. 2003. "Tropospheric Chemistry". *Em Encyclopedia of Physical Science and Technology* (Third Edition), organizado por Robert A. Meyers, 153–74. New York: Academic Press. https://www.sciencedirect.com/referencework/9780122274107/encyclopedia-of-physical-science-and-technology.

Wei, Liyan, Run Liu, Chungui Liao, Shanshan Ouyang, Yanxing Wu, Bin Jiang, Duohong Chen, Tao Zhang, Yan Guo, e Shaw Chen Liu. 2024. "An observation-based analysis of atmospheric oxidation capacity in Guangdong, China". *Atmospheric Environment* 318 (fevereiro): 120260. doi:10.1016/j.atmosenv.2023.120260.

Wu, Duanyang, Shengnan Zhang, Jing Xu, e Wei Cheng. 2024. "Reaction mechanisms of main free radicals (HO•, Cl• and ClO•) with carbamazepine in the UV/Chlorine process". *Journal of Photochemistry and Photobiology A: Chemistry* 456 (novembro): 115869. doi:10.1016/j.jphotochem.2024.115869.

Xue, Qiao, Zhiyue Jiao, Wenxiao Pan, Xian Liu, Jianjie Fu, e Aiqian Zhang. 2024. "Multiscale computational simulation of pollutant behavior at water interfaces". *Water Research* 250 (fevereiro): 121043. doi:10.1016/j.watres.2023.121043.

Yan, Yuhao, Shanshan Wang, Jian Zhu, Yanlin Guo, Guiqian Tang, Baoxian Liu, Xinxin An, Yuesi Wang, e Bin Zhou. 2021. "Vertically increased NO_3 radical in the nocturnal boundary layer". *Science of the Total Environment* 763 (abril). Elsevier: 142969. doi:10.1016/J.SCITOTENV.2020.142969.

Zhang, Chenxi, Youxin Xu, Bingbing Chu, e Xiaomin Sun. 2024. "Mechanism and toxicity assessment of carbofuran degradation by persulfate-based advanced oxidation process". *RSC Advances* 14 (42): 30582–89. doi:10.1039/D4RA05365F.

Zhao, Gaolu, Wanting Li, Chuanhao Xu, Qingsong Qin, Wenjie Fan, Xuehua Li, e Dan Zhao. 2024. "Adsorption mechanism of cefradine on three microplastics: A combined molecular dynamics simulation and density functional theory calculation study". *Science of the Total Environment* 951 (novembro): 175690. doi:10.1016/j.scitotenv.2024.175690.

Zhao, Hui, Shuangjun Wang, Lu Chenggang, Yizhen Tang, e Yaru Pan. 2024. "Theoretical study on degradation of polymethyl substituted benzenes by OH radicals in the atmosphere". *Chemosphere* 352 (março): 141243. doi:10.1016/j.chemosphere.2024.141243.

13 Iron Oxide-Based Technologies for Remediation of Arsenic/Chromium-Contaminated Soils

Shengsen Wang and Muhammad Mahroz Hussain

13.1 OVERVIEW

One of the most challenging contaminants to manage in the soil environment is persistent oxyanion pollutants (POCs), which are harmful to human health and ecosystems, owing to higher stability, availability, and persistence in different environmental scenarios. POCs include arsenic (As), chromium (Cr), selenium (Se), and other metal(liods) also known as class I carcinogens by WHO (WHO, 2011). Their capacity to stay soluble in water and their tendency for minimal degradation give them tenacity, raising the possibility to contaminate groundwater ultimately food chain (Wang et al., 2024b). Arsenic and Cr have very complex biogeochemical cycling under alternate wetting and drying, depending on the soil pH, soil organic matter/carbon (SOM/SOC), microbial action, and redox potential (Hussain et al., 2022). Therefore, addressing POC pollution has emerged as a crucial area of interest to promote environmental restoration and ecologically sound farming.

Moreover, iron oxide (FeOx) gained significant interest among POCs contamination regulation owing to its unique chemical characteristics, cost-effectiveness, and natural occurrence (Singh et al., 2022). Also, Fe-Ox has been known for its large surface area, reactive functional groups, and significant adsorption capacity, including hematite, goethite, and ferrihydrite, which can interact with various oxygenated anionic pollutants (Bashir et al., 2022). Their ability to transform more mobile and poisonous POC forms like arsenite (As^{3+}) or chromate (Cr^{6+}) into mobile, less toxic compounds like arsenate (As^{5+}) or chromite (Cr^{3+}) making them potentially active governing its transformation processes (Rizwan et al., 2019; Brar et al., 2023). Iron oxides are adaptable materials for the restoration of contaminated soils because of their redox-mediated changes as well as their fast adsorption mechanism with oxygenated anions like As and Cr.

By stabilizing other pollutants and enhancing nutrient retention, FeOx not only effectively combats POC also known to enhance soil physicochemical and biochemical

DOI: 10.1201/9781003536277-13

health by providing a strong affinity to bind SOC in the form of mineral-associated organic carbon (Zhao et al., 2024). For instance, FeOx has been thoroughly investigated and shown to have the ability to remediate various pollutants in industrial and agricultural soils by immobilizing Cr and As in soil systems simultaneously (Wang et al., 2024d). Recently, Zhao et al. (2024) found that goethite and birnessite remediate phosphate and cadmium simultaneously by forming P-bridged ternary complexes (≡Fe–P–Cd) and (≡Mn–P–Cd), respectively. Further, the efficacy of FeOx can be increased by the functionalization approach, which can enhance reactivity, selectivity, and stability in challenging environmental circumstances; functionalization approaches such as doping with other metal ions (Amoh et al., 2024), organic coatings (Lee et al., 2024), and surface modifications of polymers (Namasivayam, 2024) have been developed. For instance, Ahmad et al. (2024) demonstrated that mixing biochar (BC) and FeOx produces hybrid materials with improved catalytic and adsorption capabilities. Modifications allow FeOx to have higher remediation ability in soil environments that are diverse, where its performance is influenced by interactions with minerals, organic matter, and microbial communities (Amoh et al., 2024)

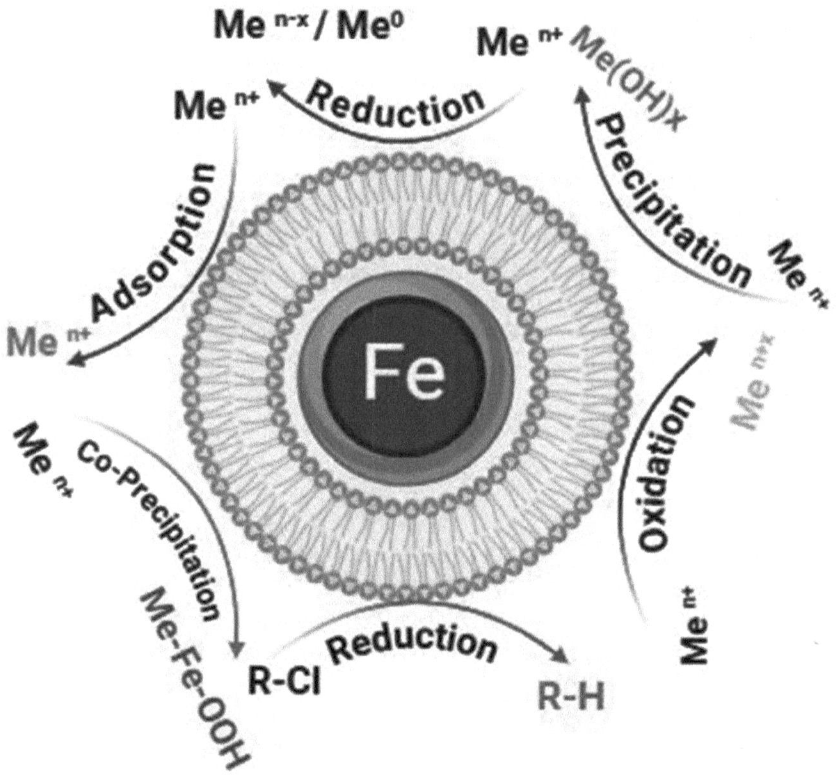

FIGURE 13.1 Iron oxide particles for the remediation of persistent oxyanion pollutants (POCs): Mechanism and pathways owing to their crystallinity and structural stability in the environment under both oxidized and reduced conditions.

by acting as POCs absorber and barrier, which supports this process. Furthermore, iron-based compounds, such as zero-valent iron (ZVI), have demonstrated potential in the reduction of complex organic molecules into less hazardous intermediates or final products via reductive degradation pathways to convert refractory pollutants (Masood ul Hasan et al., 2024).

Iron oxides' use in sustainable remediation techniques is further increased by their interaction with other soil additions like compost and BC; for instance, these can enhance the structural stability and adsorption potential of FeOx (Ahmad et al., 2024). These combinations create a comprehensive strategy for sustainable soil management by enhancing carbon sequestration, lowering greenhouse gas emissions, and improving soil health.

This chapter emphasizes Fe-Ox's vital role in reducing POCs pollution as well as its many uses in resolving agricultural and environmental issues. By emphasizing their functionalization with iron oxides, their interactions with other soil improvers, and their role in regulating pollutant transport, it offers a thorough overview of their potential. To accomplish sustainable soil remediation and guarantee food security in the face of mounting environmental pressures, future research and innovation in this field will be essential.

13.2 FUNCTIONALIZATION AND ENGINEERING METHODS FOR IRON OXIDES/ZERO-VALENT IRON

Given their adsorption and catalytic qualities, iron oxides—such as goethite, magnetite, and hematite—are frequently utilized in soil remediation. However, under some environmental circumstances, aggregation, surface passivation, or low reactivity may restrict their inherent efficiency. To get over these restrictions and increase their efficiency in capturing and immobilizing POCs, functionalization and engineering techniques are being used (Murtaza et al., 2024).

13.2.1 Functionalization of Iron Oxides with Organic Materials

One cutting-edge method to increase FeOx effectiveness in environmental remediation is to functionalize it with organic materials, which can enhance its surface reactivity, stability, and adsorption capability (Chaudhari et al., 2024). Iron oxides can operate in various intricate environmental settings owing to their functionalization and modification. Organic material types utilized for functionalization: (1) humic and fulvic acids are naturally occurring organic compounds that are produced when plant and microbiological matter breaks down. Additionally, these substances increase FeOx dispersion stability in water and soil systems (Altynbay et al., 2024), (2) organic polymers: to functionalize surfaces, synthetic or natural polymers like polyethylene glycol, polydopamine, or alginate are frequently utilized. The polymer improves the FeOx interaction with impurities by stabilizing it, making it more hydrophilic, and preventing it from aggregating. Through improved adsorption and catalytic degradation, iron oxide functionalized with polymers can efficiently reduce organic pollutants including pesticides and polycyclic aromatic hydrocarbons (PAHs) (de Farias et al., 2024), (3) biochar and organic compounds derived from compost: biochar, with high labile C source which is synthesized from pyrolyzing biomass. It produces composites with higher adsorption potential and redox-assisted reactions combined with FeOx because of organic acids and phenolic compounds produced by compost-derived materials. These composites work very well to improve soil health and immobilize POCs (Ahmad et al., 2024), and (4) surfactants and organic ligands: FeOx surface charge and hydrophobicity can be changed by surfactants (like cetyltrimethylammonium bromide) and ligands (like citrate and ethylenediaminetetraacetic acid) (Yodsanit et al., 2021); improving the soil buffering capacity by increasing iron oxide FeOx affinity for hydrophobic organic pollutants.

By modifying FeOx surface chemistry, it can interact with contaminants in a more efficient way, functionalizing FeOx with organic molecules via the addition of functional groups including hydroxyl, carboxyl, and amine groups to improve pollutant adsorption and stability (Aftabtalab et al., 2022). This surface alteration enhances particle dispersion and generates new reactive sites, both of which are essential for immobilizing As and Cr. For example, FeOx surfaces and organic molecules form covalent or hydrogen bonds, which stabilize the materials and improve their selectivity against particular oxyanions pollutants like As^{5+} (Li et al., 2024). Furthermore, by improving hydrophobic and electrostatic interactions, the changed surface increases the FeOx ability to absorb organic contaminants from soil systems, including hydrocarbons and persistent pesticides (Mostafa et al., 2024).

The capacity of FeOx that is functionalized using organic materials that can stimulate redox transformations with both organic and inorganic pollutants can be more effective in remediation process. The FeOx core and the organic enrichment might also enhance the electron transport capabilities by changing POCs redox state, e.g., As^{3+} and Cr^{6+} into less mobile and available species (As^{5+} or Cr^{3+}) due to its redox-assisted mechanism (Kumarathilaka et al., 2018). Furthermore, the Fenton-type oxidation reaction can generate hydroxyl radicals, which can assist the oxidation mechanism to convert the complex bound pollutant structure (Kumar and Chopra, 2015). Owing to the FeOx advanced sorption and oxidation mechanism, these functionalized materials can remediate the As- and Cr-contaminated soils. Singaraj et al. (2019) reported that humic acid-functionalized iron oxide can remediate higher Cr concentration (24.13 mg/g) from water compared to its unfunctionalized FeOx.

Moreover, organic functionalized-FeOx can also enhance soil health and its physicochemical sustainability by forming stable organic molecules, higher molecular weight C compounds, particularly the enrichment with surface functional groups. As reported by Cai et al. (2023), the addition of typha biochar enriched with magnetic $FeCl_3$ can remediate As^{5+} and Cr^{3+} owing to its structural stability and surface functional group that ultimately enhances its remediation potential by keeping the natural biogeochemical cycle undisturbed. Moreover, the decomposition of organic materials leads to more C sources for the microbiota that can aid in the bioremediation of the POCs like As and Cr via methylation and thiolation processes (Hossain et al., 2020; Breuninger et al., 2024). However, more detailed work is still required to develop better insight into the mechanistic approach of these organic functionalized FeOx toward sustainable application and remediation of contaminated soil.

13.2.2 Functionalization of Iron Oxides with Inorganic Materials (Metals and Nonmetals)

Iron oxide functionalization with metals and nonmetals is another emerging aspect to remediate the soil contaminated with POCs particularly As and Cr; however, not much attention has been given to this due to variable limitations. Different processes like doping, coating, enrichment, and hybridization are mostly employed to enhance the reactivity, sorption capabilities, and catalytic effectiveness in terms of electron donor and acceptors and stability, ultimately enhancing the soil remediation potential (Mostafa et al., 2024; Murtaza et al., 2024).

13.2.2.1 Functionalization with Metals

Among the above-mentioned functionalization methods, metal doping and hybridization are the most promising ones as these can alter the FeOx electrical, magnetic, and catalytic abilities and also can regulate the mobility and availability of transition metals and metalloids like As and Cr from soil. The major metals that have been used for the FeOx doping and hybridization are manganese (Mn), cobalt (Co), nickel (Ni), copper (Cu), or zinc (Zn). These metals can change the redox potential of the materials; thus, their addition can increase the FeOx potential to remediate As and Cr (Ersan et al., 2023). Recently, He et al. (2024) reported that Mn-doped FeOx can remediate 100% of Ce(VI) from Cr-contaminated soil via enhanced sorption and increased stable complexes via inner layer complexes. Another possible mechanism is by enhancing the electrostatic interaction and electron transfer, which is confirmed by the zeta potential and electrochemical analytical approach in this study. More of the functionalization studies with other POCs have been given in Table 13.1.

Iron oxide has also been functionalized using noble metals like gold (Au), silver (Ag), lanthanum (La), and platinum (Pt) other than the above-mentioned metals. Noble metals facilitate the Fenton-type reaction and advanced oxidation process (AOP) and serve as electron mediators, e.g., lanthanum (La)-doped magnetic biochar and FeOx complexes reduce As mobility by 39%–95% via different mechanisms chemical modifications, enhanced retention, soil fraction partitioning, and stable complexes within the soil aggregates (Bagherifam et al., 2022). Furthermore, the formation of bimetallic complexes comprises mixing two metal complexes together to form a strong reactive substance that can sequester a higher amount of POCs, but the application of these bimetallic complexes in soil still remains the least explored. These combinations can increase the FeOx metals' relative bonding strength, diatomic surface dimensions, surface energy, and electron transfer that will enhance the electric and magnetic effect based on the two mixed metals such as Fe-Cu, Fe-Zn, or Fe-Mn, which is another viable strategy (Allam et al., 2021). For instance, the bimetallic-based compounds such as hydroxyethyl cellulose–graphene oxide (HEC–GO) and HEC–GO/Fe–Zn mole ratio (2:1) remediate doxycycline from contaminated water, i.e., 97% and 95%, respectively (SefidSiahbandi et al., 2023); however, bimetallic composites application in soil is not elucidated.

Combining the FeOx with hydroxides and oxides of other metals can also increase its efficacy in terms of soil remediation and stabilization, e.g., TiO_2–FeOx and

TABLE 13.1
Functionalize Biochar and Its Potential to Remediate Different Persistent Oxyanion Pollutants (POCs) from Soil under Varying Experimental Conditions

Functionalize FeOx	POCs	Remediation Potential	Experimental Type	Reference
Rice Hull-biochar/FeOx	Lead(II) and Arsenate	90.21 mg/g; 90.21 mg/g	Lab-scale batch experiment	Kim et al. (2024)
Biochar-supported zero-valent iron	Total Arsenic	64.2%	Lab incubation experiment	Wang et al. (2024c)
Biochar-supported sulfidized nZVI	Cadmium	32.04%	Lab incubation experiment	Xu et al. (2024a)
Typha biochar with phosphoric acid modification and ferric chloride impregnation	Hexavalent chromium	55.5 mg/g	Controlled condition greenhouse experiment	Cai et al. (2024)
Fe-modified biochar (N–Fe and S–Fe)	Cadmium	79%	Soil incubation experiment	Si et al. (2024)
Rice biochar (RBC) and doped with iron oxide (IO) nanoparticles	Arsenic	60.5%	Pot culture with recommended fertilizer application	Majumdar et al. (2023)
BC combined with ferrihydrite (FH) or goethite (GT)	Cadmium	<80.5% for both doped iron oxide composites	Pot culture experiment	Liu et al. (2023b)
Fe-oxide loaded biochars (FBCs) and humic acid (HA)	Lead and Cadmium	96.8–151 mg/g; 29.1–82.3 mg/g	Incubation experiment (lab scale)	Wang et al. (2023)
Fabrication of amino- or hydroxyl-functionalized Fe/Ni-biochar	Lead and Cadmium	242.15/229.52 mg/g	Soil incubation experiment	Wang et al. (2024a)

SiO_2–FeOx composites. For instance, Yin et al. (2024a) reported that the addition of TiO_2/α-FeOOH enhances the degradation of PAHs in soil by 3.0 and 4.8 times compared to original materials via electron capacity making more sorption holes by broadening the spectrum range. Similarly, it also enhances the Z-type of electron shift and the contributing factors include the $\bullet O^{2-}$, H^+, and $^1O_2$ known to be for TiO_2/α-FeOOH photocatalysis. However, limited studies have been conducted on the immobilization of As and Cr under varying environmental conditions.

Functionalization with Nonmetals: Not only the metal functionalization, FeOx can also be enriched, doped, and functionalized with the nonmetals like fluorine (F), phosphorus (P), sulfur (S), and nitrogen (N), which can increase its efficiency in terms of remediation and regenerative sustainable approaches. The nonmetal functionalization can improve its surface chemistry, adding more active function group sites that can increase the FeOx reactivity and sorption ability in both aquatic and terrestrial environments (Hu et al., 2019). Additionally, phosphate enrichment and silicon dioxide (SiO_2) have also been used based on the characteristic removal of the POCs from soil. The mechanism of lowering the bioavailability, particularly by the application of phosphate, is also known to be the elemental isomorph for As^{5+} in soil under a reducing environment (Hussain et al., 2021). Furthermore, silica-FeOx is known to be hydrophilic in the mode of action ultimately improving the ability to remediate As and Cr under varying environmental conditions (Garg et al., 2023). Also, FeOx that are enriched with carbon-based compounds, i.e., GO and carbon nanotubes, can enhance the conductivity capabilities, sorption, and catalytic qualities by altering the redox and pH of the soil as described by Farooq and Jalees (2020). As Su et al. (2017) reported that many enrichment materials have been used for the remediation of As and Cr; however, GO either oxidized or reduced via redox reaction of graphite can enhance the As immobilization under aquatic condition owing to its two-dimensional crystallinity and higher surface area. In a recent study, biobased graphene and magnetite composites were synthesized for the removal of As, which showed the As removal of 93.5% from the soil due to surface functional groups and supportive elemental composition (Akhtar et al., 2024).

13.3 IRON OXIDE-BASED REMEDIATION FOR POC-CONTAMINATED SOILS

Iron oxide has been known for its large surface size, stability, and conversion of metastable phases into more stable phases under different soil conditions because of microbial activity, soil pH, SOC, and other geoenvironmental factors, particularly temperature and moisture content. Multiple mechanisms are associated with the POCs immobilization in soil which regulates the FeOx stability and efficiency along with POCs remediation.

13.3.1 Iron Oxides for As and Cr Remediation in Soil

Iron oxide is commonly observed as a promising approach to remediate As and Cr from contained soils due to its high reactivity, higher surface area, and capacity to engage in redox and sorption processes (Akhtar et al., 2024) via surface complexation, precipitation, and redox reactions. Iron oxide minerals (functionalized and nonfunctionalized) might interact with As^{3+} and Cr^{6+} via surface complexation and surface sorption (Chemi- or physical sorption) based on the species and soil redox conditions. Arsenate may bind with the ferrihydrite (Fang et al., 2024) and hematite (Cui et al., 2020) because of their hydroxylated surfaces through inner layer complexes. Similarly, Cr mobility might be rendered via the FeOx-mediated redox process and then by adsorption or precipitation of $Cr(OH)_3$.

To increase iron oxide's durability and efficacy in variable environments, sophisticated methods have been devised, such as functionalizing it through organic and inorganic changes. The efficiency of FeOx can be enhanced if used in combination with biochar-iron oxide composites, which have been reported to decrease As and Cr bioavailability in plants and increase its sorption potential by managing soil biogeochemical cycles (Ahmad et al., 2024). FeOx-BC not only immobilizes As and Cr but also enhances the soil health by stimulating the SOC dynamics in soil as mentioned by many researchers in Cr immobilization and enhancing the SOC of soil by improving the soil health (physical, biological, and biochemical) (Cai et al., 2023). These are the two most common mechanisms in Cr immunization as FeOx can form stable complexes and, on the other hand, the organic matter can act as an electron donor or acceptor to mediate the Cr species interconversion.

Despite the effectiveness of FeOx-Cr and As complexes, still research is warranted for their enduring stability in shifting soil conditions. The purpose of immobilized As and Cr can be impacted by changes in soil parameters like pH, redox potential, and microbial activity, which may result in their remobilization that should be investigated in detail. To maximize its use and guarantee the long-term restoration of soils contaminated with As and Cr, more field studies and investigations into the incorporation of FeOx into sustainable soil management techniques are essential.

13.3.2 Iron Oxide-Biochar Composites for As and Cr Remediation in Soil

Iron oxide coupled with BC has been reported to remediate As and Cr from varying soil condition via different mechanisms. These composites have a high large surface area, porosity, and functional groups of biochar with the adsorption and redox characteristics of FeOx that simultaneously can remediate the soil contaminated with As and Cr (Song et al., 2021) through surface complexation, precipitation, and formation of stable covalent bond/hydrogen bonding owing to porosity of the structural combination. In the geochemical phase of soil, the main As chelators are Fe-oxides, and specifically As–Fe-containing minerals such as pharmacosiderite, arseniosiderite, and scorodite. Thus, this serves as a controller for applying Fe-modification to boost biochar's ability to bind As (Fendorf et al., 2010).

Iron oxide–biochar composites for Cr have the dual advantage of immobilizing the reduced species and converting Cr^{6+} to the less hazardous Cr^{3+}. While the FeOx component offers a site for the adsorption and precipitation of immobilized Cr^{3+}, the biochar component serves as an electron donor and encourages redox processes that transform Cr^{6+} into Cr^{3+} (Xu et al., 2023). Here, Xu et al. (2023) reported that high electron donating efficiency was found with Fe-Mn BC enrichment that can reduce up to 78% of the Cr^{6+} from the solution phase; the proposed result demonstrated that oxygen might be removed during the process of pyrolysis that can restrict the formation of stable Fe-Cr complexes which later on can be taken via Fe-Mn BC enrichment to enhance its efficiency. A recent study by Lyu et al. (2024) explored the potential of magnetic FeOx and BC composites that reduce the As availability by 85% under reduced conditions and it was 70% under oxic settings and also the results were supported via advanced-integrated approaches like X-ray photoelectron analysis reassured the As^{3+} and As^{5+} re-oxidation from ferrite complexes. FeOx-BC composites have higher remediation potential; however, they are still dependent on attributes such as soil pH, redox conditions, and contaminant composition. The efficiency of the redox transformation might be reduced due to interference from high quantities of organic matter or competing ions in the soil.

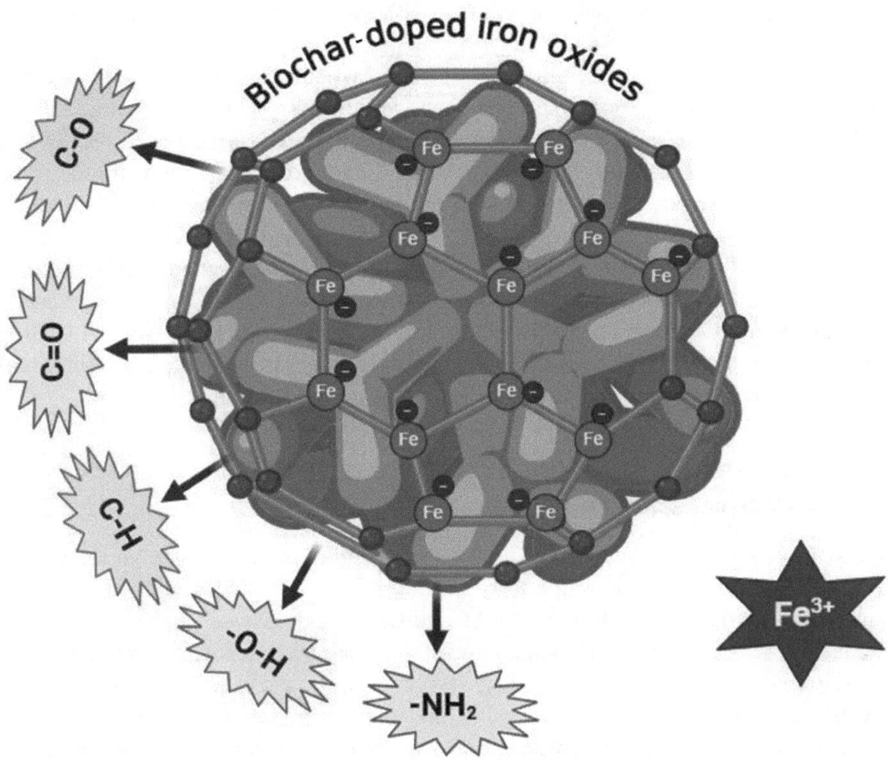

FIGURE 13.2 Biochar-doped iron oxides in environmental application for the remediation of persistent oxyanion pollutants (POCs): Associated functional groups and role of Fe in the remediation under changing redox conditions.

13.3.3 Iron Oxide-Organic Fertilizer Compounds for As and Cr Remediation in Soil

The past few decades have focused on the use of organic fertilizer (OF) to enhance soil health and crop productivity by the addition of OF with FeOx possibly be a feasible alternative approach to remediate As and Cr from contaminated soils (Garg et al., 2023; He et al., 2024). OFs have a higher amount of C and N in them, particularly in a higher C/N ratio that can supplement As and Cr remediation as mentioned above in the N-doped POCs removal. Higher N content can aid in the As remediation as mentioned by Hussain et al. (2021) with more sustainable rice production by altering the redox transformation of the As under paddy conditions. However, the Cr removal by the OF has not been yet well explored. The possible mechanism can be the formation of stable methylated and thiolated Cr and As complexes that have a more sustainable ratio in the environment; Breuninger et al. (2024) explained the formation of the thiolated As complexes in the different water and soil regime settings. The composites application of OF and FeOx also remained partially explored, however, as electron donors, OF aids in the transformation of Cr^{6+} to Cr^{3+}, which is less mobile and less poisonous (Cai et al., 2023). Moreover, the OF can also increase the soil microbial population that plays a crucial role in the reduction of the As and Cr both under reduced and oxidized conditions. Particularly, the role of algae in the degradation and stabilization of As has been well understood (Hussain et al., 2019), while the role of the bacterial action remained partially explored based on the cell structure and role of vacuoles. The FeOx and OF composites can also stimulate the cation exchange capacity and soil pH by altering the formation of organic acid under reduced conditions that can bind As and Cr in more stable complexes. The stability of the immobilized POCs and the persistence of SOC regulate FeOx and OF application and environmental implications. Arsenic and Cr release can be governed by breakdown of organic materials, competing ion interactions, and shifts in redox conditions. Furthermore, overuse of OFs releases nutrients that encourage microbial activity, which might alter contaminant stability and redox balance.

13.3.4 Other Iron Oxide-Bearing Composites for As and Cr Remediation in Soil

In the past few decades, the FeOx and industrial and municipal waste together have gained a lot of attention that reflects exercise nutrients and more binding sites for the sustainable removal of the POCs like As and Cr from the wastewater and contaminated soils. The application of such composites gained more researcher's prospective due to low cost, environmental stability, and green approach to reuse the waste materials like coal fly ash, red mud, steel slag, and agriculture straw waste (Żhang et al., 2024). FeOx-containing composites, like iron oxide-clay and polymer-iron oxide blends, are also known to be more effective in removing As and Cr from the contaminated soil and water owing to their large surface area, cation exchange capacity, and higher sorption capabilities (Su et al., 2017). Clay minerals and FeOx composites comprising bentonite, kaolinite, and montmorillonite can remediate <90% of As and Cr from water as reported by many researchers due to internal spherical complexation and electrostatic interactions (Ugwu and Igbokwe, 2019). Arsenic can be desorbed from FeOx under more favorable conditions, and these clay minerals' composites enhance the sorption capacity of FeOx. As explained by Abbaslou et al. (2023), the addition of sepiolite enhances the As immobilization by improving the soil physicochemical attributes along with the minimum desorption from the soil. Another approach includes the polymer enrichment with FeOx that improves the reusability, stability, environmental implications, and adaptability, and this includes the functionalization with chitosan, polyaniline, or polyacrylamide (Xu et al., 2024b). These polymers add extrafunctional groups like amine and hydroxyls along with some C=C that improve the sorption potential of the As and Cr as reported by Xu et al. (2024b) that the addition of chitosan polymers in a combination with FeOx can remediate 78% of As with Langmuir as the best-fit model for the As-contaminated soil. Other possible mechanism is that these polymers can enhance the FeOx aggregation and reduce its leaching; along with providing the coordination, chelation, and electrostatic interaction to remediate the contaminated soil (Kloster et al., 2020). Furthermore, even in alkaline soils, the polymer covering helps buffer the pH of the soil, assuring ideal conditions for chromium reduction. Kloster et al. (2020) reported that the polyaniline–FeOx composites have been effectively employed in lab experiments to adsorb chromium and arsenic from mining-affected soils at the same time. Future studies should focus on improving these composites' yields, assessing their long-term stability, and optimizing them for economical field deployment in order to sustainably address a range of soil pollution issues.

13.4 ROLE OF IRON OXIDES ON TRANSPORT OF POCs IN SOILS

Iron oxide plays an important role in regulating the transport of POCs in soil due to its abundance in the soil and also its mode of action under different soil environmental conditions. The oxyanion pollutants including arsenic (As), chromium (Cr), selenium (Se), and phosphate in soils have complex biogeochemical behavior, which is governed by the SOC, microbes, Fe/Mn oxides in soil and, most importantly, alkalinity and acidity of the soil (Hossain et al., 2020). While FeOx owing to its strong reactivity, surface area, and adsorption capacity can manage the transport of Cr and As via different processes like adsorption, surface complexation, and redox reactions, these minerals serve as organic barriers that immobilize pollutants, lowering their mobility and potential environmental hazards (Suda and Makino, 2016). As Cr^{6+} can be reduced to Cr^{3+} via microbial action or the other geoenvironmental factors in the presence of FeOx however, As can be firmly bound to FeOx that can restrict its availability even at a wide range of pH (von der Heyden and Roychoudhury, 2015).

13.4.1 Cotransport of Colloidal Fe Oxide-POCs

Colloidal FeOx and POCs cotransport mainly because of their vigorous sorption and complexation mechanisms where they can act as a carrier for POCs and form complexes that impact their transport in soil systems because of their vast surface area and reactivity (Hassellov and von der Kammer, 2008). Depending on the contaminants' characteristics and the surrounding environment, these interactions can either increase or decrease POC mobility, which is further governed by the electrostatic interactions including charged functional groups (Dam et al., 2019). Negatively charged surfaces are known to be attracted highly to the positively charged FeOx surface in acidic soil where the function of FeOx as POC transporters and immobilizing agents is contingent on soil pH and ionic strength. A competitive adsorption of dissolved organic matter might lower the affinity between POC and iron oxide colloids in alkaline or highly organic soils, changing their transit kinetics (Dam et al., 2019).

13.4.2 Influential Environmental Factors

Transformation and stability of FeOx-assisted POCs in soils are regulated by microbial activity, organic matter, pH, and ionic strength and ability to adsorb POC, which are influenced by its surface charge, which is largely determined by pH (Aftabtalab et al., 2022). FeOx has a positive surface charge under acidic environments, which can either improve the adsorption of cationic pollutants or repel negatively charged contaminants (Parkinson, 2016). Further, FeOx surface becomes more negatively or neutrally charged in neutral to alkaline conditions, which improves the adsorption of negatively charged oxygen anions like phosphate and As^{5+}. Another important element influencing the FeOx bilayer's capacity to adsorb POC is ionic strength in which compression of bilayer in high ionic strength soils reduces electrostatic interactions, which in turn lessens the adsorption of organic and heavy metal pollutants (Cornell and Schwertmann, 2003). Thus, the stability of iron oxide-POC complexes and their mobility in the soil system are determined by pH and ionic strength, which in turn affects the movement of pollutants in the environment.

The mobility of POC linked to FeOx is tangentially impacted by SOC and its associated acids like fulvic and humic acids, which can compete with impurities for adsorption sites on the oxide (Parkinson, 2016). However, SOC might enhance contaminant retention by changing the surface characteristics of FeOx, which increases its capacity to sorb POCs, or by generating organometallic complexes (Ashraf et al., 2016). Another important factor in changing how FeOx–POC interactions behave is microbial activity, e.g., anaerobic microbial reduction of FeOx can increase the mobility of adsorbed pollutants like As by immobilizing via precipitation or surface complexation on FeOx surfaces (Bashir et al., 2022). This is further complicated by the temperature and moisture as the FeOx kinetics can be influenced by these two factors under varying conditions of the soil (Yin et al., 2024b).

13.4.3 Fe Oxide-Based Permeable Reactive Barriers for Groundwater Remediation

A resourceful and operative groundwater remediation technique, FeOx-based permeable reactive barrier (PRB) primarily uses adsorption, precipitation, and redox reaction processes to immobilize or detoxify POCs like As and Cr (Thiruvenkatachari et al., 2008). Iron oxide adsorbs and traps various contaminants, including heavy metals and persistent organic pollutants, when contaminated groundwater flows past these barriers mitigating the migration of pollutants to groundwater. This process of adsorption aids in binding Cr and As to the FeOx surface and is regulated by electrostatic interactions, surface complexation, and hydrogen bonding (Budania and Dangayach, 2023). Iron oxide might also have a coprecipitation process in which As and Cr precipitate out of the solution and are immobilized by FeOx particles (Sakr et al., 2023). Besides sorption and co-precipitation methods, FeOx can also act as a catalytic function in stimulating redox processes, which can control the redox transformation of As and Cr making them less available and mobile by enhanced binding on the FeOx surface (Sakr et al., 2023). For example, magnetite and hematite can catalyze the reduction of As, Cr, and also other contaminants like insecticides and chlorinated solvents, into less hazardous by-products. Also, ZVI and magnetite PRBs have been used to remediate Cr- and As-contaminated groundwater with high removal efficiency of up to 98% (Masood ul Hasan et al., 2024). These barriers offer an environmentally benign auxiliary for conservative practices like pumping and treatment systems, and they have shown themselves to be a sustainable and economical solution, especially in regions with significant levels of groundwater contamination. PRBs can be used in groundwater remediation to address a greater variety of contaminants and offer long-term solutions to the problems associated with groundwater contamination as research into their materials and design advances.

13.5 SOIL MANAGEMENT PRACTICES FOR POCs REMEDIATION AND FOOD SAFETY

In addition to using iron oxides to reduce POC contamination, sustainable soil management techniques also enhance soil health and food safety. The main tactics for managing organic matter, water, and nutrients are highlighted in this section.

13.5.1 Water Management

Particularly in flooded or waterlogged areas like rice fields, water management techniques—in particular, the control of redox conditions—have a significant impact on how As and Cr behave in soil systems (Hussain et al., 2021). Under the flooded condition, the oxygen diffusion can be restricted and soil thus has alternate redox conditions that can regulate the transformation of As can Cr speciation (Hussain et al., 2021). The reduced environment can impact the dissolution and efficiency of the FeOx. As reported by Liu et al. (2023a), under alternate wetting and drying conditions, goethite and ferrihydrite converted to a more labile fraction losing its structural stability and crystallinity. Further, FeOx has an important role in the remediation of As and Cr as it can serve as electron donors that oxidize Fe^{2+} to Fe^{3+} in response to changes in the redox potential, which impact the movement of As under such conditions. The conversion of FeOx can also result in the release of the bind As onto the FeOx minerals structure as reported by Fendorf et al., (2010). To overcome this, controlled irrigation practice is always recommended even in the case of rice crops that can maintain the stability and redox condition in soil, which ultimately can regulate the mobility of As and Cr (Hussain et al., 2019).

13.5.2 Nutrient Management

Iron oxides and POCs in soil interaction are significantly impacted by the addition of nutrients and their management in soil via enhanced fertilization and other management approaches like tillage operations. The mobility and availability of the FeOx and POCs can be regulated by the nitrogen and phosphorus addition under different soil environments that can stimulate the soil's chemical and biological attributes. For instance, the addition of excessive phosphorus can increase the formation of stable Fe-P complexes (Xiao et al., 2023), which are known to be more stable and can also serve as a P source upon degradation of these complexes. These complexes can also act as strong remediation potential for As and Cr as phosphate can act as an elemental isomorph for As^{5+} (Hussain et al., 2022). Moreover, by encasing POC in crystalline structures, phosphorus-induced FeOx precipitation can stabilize it while decreasing its bioavailability. Phosphate fertilizers and FeOx (such as goethite or hematite) might reduce As mobility in contaminated soils, by making stable Fe–PO_4–As complexes and supplying vital nutrients for plant growth (Mokadem et al., 2023)

Iron oxide has been used for regulating N in the environment in both ways to remediate it and to increase its efficiency by minimizing its losses. Ammonia-based fertilizers have the potential to improve the solubility of the FeOx by increasing the chances to release POCs in the soil solution causing acidification of soil that favors more mobility and availability of As and Cr. However, no comprehensive work has been done on such a mechanism and remained partially elucidated. These mechanisms also impact the soil microbiota and its associated pathways to remediate As and Cr from the soil. However, nitrate-based fertilizer has contrasting behavior that can increase the stability of the FeOx in soil, thus making stable complexes with As and Cr making them less mobile and bioavailable. This phenomenon can also be altered with the soil extracellular enzymes like leucine aminopeptidase by breaking the protein in the soil and β-glucosidase that can help in breaking the glucose into simpler C compounds, which can impact the soil physicochemical and microbial attributes (Cheng et al., 2023). Moreover, more research is still warranted to explore the impact of N/P fertilizers in combination with FeOx to remediate As and Cr from contaminated soil and its impact on its respective speciation.

13.5.3 Soil Organic Matter (C vs Fe vs As and Cr)

As a mediator of intricate environmental processes, SOM is crucial in determining how FeOx and POCs interact in soil as SOM provides many surface functional groups like hydroxyl, carboxyl, and phenolic groups that can regulate the POCs binding and surface kinetics (Chaudhari et al., 2024). Owing to its specific character, FeOx can form stable organic complexes that can reduce the As and Cr mobility and availability under different kinds of soil types and conditions. Furthermore, SOM changes the structure of the soil, promotes aggregation, and produces a microenvironment that increases FeOx capacity to immobilize contaminants (de Farias et al., 2024).

Moreover, SOM boosts carbon sequestration and indirectly aids in FeOx–POC complex stabilization. The persistence of carbon in the form of humic compounds and organic molecules adsorbed on FeOx is encouraged by soils that are high in organic matter, which ultimately improve the sorption of As and Cr to interact with FeOx forming organic–mineral interactions (Cheng et al., 2023). Also, SOM reassures the microbial activity to mediate Fe cycling and redox changes as its associated reduction of Fe^{3+} in FeOx under anoxic conditions can stimulate POCs regulation, but the coexistence of SOM provides alternate electron acceptors or donors, thereby attenuating the release and enhancing the long-term stability of the POCs-Fe complexes (Zhang et al., 2024).

Biochar can increase the microbial activity, thus stimulating the interaction of SOM and FeOx due to its labile soil carbon fraction creating more reaction surfaces. Biochar, which is high in stable organic matter, has been demonstrated to improve FeOx capacity to absorb POCs (Cheng et al., 2023). Hussain et al. (2022) reported that biochar has higher soil labile carbon, which helps in the As sequestration in the soil organic bound soil fraction, thus reducing its mobility and transformation.

13.6 FUTURE PROSPECTS AND RESEARCH GAPS

This chapter concluded that multipurpose FeOx composites which can mitigate POCs functionalized with charcoal, polymers, and other components can be an effective approach to remediate the soil without disturbing its natural biogeochemical cycle. The use of iron oxide-based amendments controls soil contamination by increasing plant productivity, nutrient cycling, and soil microbial health. Iron oxide coatings that are biodegradable and environmentally benign are made to reduce environmental hazards when applied. To improve the effectiveness of POCs removal, combining FeOx with technologies like electroremediation, bioremediation, or phytoremediation should be used in conjunction with AOPs to remediate POCs, which remained still partially elucidated. To dynamically optimize the use of FeOX in soil sensor-based technologies should be used to pathway performance of remediation. Iron oxides can contribute significantly to the development of efficient and ecologically friendly soil remediation systems by addressing gaps and utilizing developments in materials science and sustainable practices.

REFERENCES

Abbaslou, H., Poorangha, P., Bakhtiari, S., Abolpour, B., Shahrashoub, M., (2023). Remediation of artificially Arsenic Contaminated clay Soil (Sepiolite) by extraction and its effect on soil properties. *Anthropogenic Pollution* 7, 8–16.

Aftabtalab, A., Rinklebe, J., Shaheen, S.M., Niazi, N.K., Moreno-Jiménez, E., Schaller, J., Knorr, K.-H., (2022). Review on the interactions of arsenic, iron (oxy)(hydr)oxides, and dissolved organic matter in soils, sediments, and groundwater in a ternary system. *Chemosphere* 286, 131790. https://doi.org/10.1016/j.chemosphere.2021.131790

Ahmad, A., Khan, Z.U.H., Sabahat, S., Sun, J., Shah, N.S., Khan, Z.U., Muhammad, N., Mir, S., Rahim, A., Nadeem, M., (2024). Innovations in metal oxides-biochar nanoparticles for dye removal. *Nano-Structures & Nano-Objects* 39, 101269.

Akhtar, M.S., Jutt, D.S.R., Aslam, S., Nawaz, R., Irshad, M.A., Khan, M., Khairy, M., Irfan, A., Al-Hussain, S.A., Zaki, M.E.A., (2024). Green synthesis of graphene oxide and magnetite nanoparticles and their arsenic removal efficiency from arsenic contaminated soil. *Scientific Reports* 14, 23094. https://doi.org/10.1038/s41598-024-73734-9

Allam, B.K., Musa, N., Debnath, A., Usman, U.L., Banerjee, S., (2021). Recent developments and application of bimetallic based materials in water purification. *Environmental Challenges* 5, 100405. https://doi.org/10.1016/j.envc.2021.100405

Altynbay, N., Tastambek, K., Akimbekov, N., Digel, I., Tagayev, K., Kamenov, B., Liu, X., (2024). A comprehensive review on enhancing saline soil reclamation efficiency through bacterial solubilization of low-rank coal. *Engineered Science* 30, 1147.

Amoh, P.O., Samy, M., Elkady, M., Shokry, H., Mensah, K., (2024). Surface modification of toner-based recyclable iron oxide self-doped graphite nanocomposite to enhance methylene blue and tetracycline adsorption. *Journal of Environmental Management* 357, 120786.

Ashraf, A., Bibi, I., Niazi, N., Ok, Y.S., Murtaza, G., Shahid, M., Kunhikrishnan, A., Li, D., Mehmood, T., (2016). Chromium(VI) sorption efficiency of acid-activated banana peel over organo-montmorillonite in aqueous solutions. *International Journal of Phytoremediation* 19, 605–613. https://doi.org/10.1080/15226514.2016.1256372

Bagherifam, S., Brown, T.C., Fellows, C.M., Naidu, R., Komarneni, S., (2022). In situ stabilization of arsenic in soil with organoclay, organozeolite, birnessite, goethite and lanthanum-doped magnetic biochar. *Pedosphere* 32, 764–776. https://doi.org/10.1016/j.pedsph.2022.06.008

Bashir, H., Bibi, I., Hussain, M.M., Niazi, N.K., Iqbal, J., (2022). Nanoparticulate Iron Oxide Minerals for Arsenic Removal from Contaminated Water, in: Niazi, N. K., Bibi, I., Aftab, T. (Eds.), *Global Arsenic Hazard: Ecotoxicology and Remediation*. Springer, pp. 459–479.

Brar, B., Marwaha, S., Poonia, A.K., Koul, B., Kajla, S., Rajput, V.D., (2023). Nanotechnology: a contemporary therapeutic approach in combating infections from multidrug-resistant bacteria. *Archives of Microbiology* 205, 62.

Breuninger, E.S., Tolu, J., Aemisegger, F., Thurnherr, I., Bouchet, S., Mestrot, A., Ossola, R., McNeill, K., Tukhmetova, D., Vogl, J., (2024). Marine and terrestrial contributions to atmospheric deposition fluxes of methylated arsenic species. *Nature Communications* 15, 9623. https://doi.org/10.1038/s41467-024-53974-z

Budania, R., Dangayach, S., (2023). A comprehensive review on permeable reactive barrier for the remediation of groundwater contamination. *Journal of Environmental Management* 332, 117343.

Cai, Y., Ran, Z., Cang, Y., Chen, X., Shaaban, M., Peng, Q.-A., (2023). Efficient removal of Cr(VI) and As(V) from an aquatic system using iron oxide supported typha biochar. *Environmental Research* 225, 115588. https://doi.org/10.1016/j.envres.2023.115588

Cai, Y., Yang, J., Ran, Z., Bu, F., Chen, X., Shaaban, M., Peng, Q.-a., (2024). Optimizing Typha biochar with phosphoric acid modification and ferric chloride impregnation for hexavalent chromium remediation in water and soil. *Chemosphere* 354, 141739. https://doi.org/10.1016/j.chemosphere.2024.141739

Chaudhari, D.S., Upadhyay, R.P., Shinde, G.Y., Gawande, M.B., Filip, J., Varma, R.S., Zboril, R., (2024). A review on sustainable iron oxide nanoparticles: synthesis and application in organic catalysis and environmental remediation. *Green Chemistry* 26, 7579–7655.

Cheng, L., Liang, H., Yang, W., Yang, T., Chen, T., Gao, D., (2023). The biochar/Fe-modified biocarrier driven simultaneous NDFO and Feammox to remove nitrogen from eutrophic water. *Water Research* 243, 120280. https://doi.org/10.1016/j.watres.2023.120280

Cornell, R.M., Schwertmann, U., (2003). *The Iron Oxides: structure, Properties, Reactions, Occurrences, and Uses*. Wiley-vch Weinheim.

Cui, H., Zhang, X., Wu, Q., Zhang, S., Xu, L., Zhou, J., Zheng, X., Zhou, J., (2020). Hematite enhances the immobilization of copper, cadmium and phosphorus in soil amended with hydroxyapatite under flooded conditions. *Science of the Total Environment* 708, 134590.

Dam, T.T., Hoang, N.T., Nguyen, A.T., Nguyen, H.X., Nguyen, P.M., Nguyen, K.M., Nguyen, M.N., (2019). Colloidal dynamics of freshly formed iron oxides under the influence of silicic acid. *Journal of Environmental Quality* 48, 670–676.

de Farias, M.B., Elgueta, N.M., Camparotto, N.G., Prediger, P., Vieira, M.G.A., (2024). Comparative efficiency of conventional and green functionalized carbon-based materials for the uptake of priority produced water contaminants. *Journal of Cleaner Production* 447, 141458.

Ersan, G., Brienza, M., Mulchandani, A., Apul, O.G., Garcia-Segura, S., (2023). Trends on arsenic species removal by metal-based nanoadsorbents. *Current Opinion in Environmental Science & Health* 34, 100478.

Fang, M., Sun, Y., Zhu, Y., Chen, Q., Chen, Q., Liu, Y., Zhang, B., Chen, T., Jin, J., Yang, T., (2024). The potential of ferrihydrite-synthetic humic-like acid composite as a soil amendment for metal-contaminated agricultural soil: Immobilization mechanisms by combining abiotic and biotic perspectives. *Environmental Research* 250, 118470.

Farooq, M.U., Jalees, M.I., (2020). Application of magnetic graphene oxide for water purification: heavy metals removal and disinfection. *Journal of Water Process Engineering* 33, 101044.

Fendorf, S., Nico, P.S., Kocar, B.D., Masue, Y., Tufano, K.J., (2010). Arsenic chemistry in soils and sediments, in: Singh, B., Gräfe, M. (Eds.), *Developments in Soil Science*. Elsevier, pp. 357–378.

Garg, R., Garg, R., Khan, M.A., Bansal, M., Garg, V.K., (2023). Utilization of biosynthesized silica-supported iron oxide nanocomposites for the adsorptive removal of heavy metal ions from aqueous solutions. *Environmental Science and Pollution Research* 30, 81319–81332.

Hassellov, M., von der Kammer, F., (2008). Iron oxides as geochemical nanovectors for metal transport in soil-river systems. *Elements* 4, 401–406.

He, K., Lai, Y., Wang, S., Gong, L., He, F., (2024). Mechanochemical synthesis of manganese-modified microscale zerovalent iron for efficient Cr(VI) removal: performance and mechanism. *ACS ES&T Engineering* 4, 2553–2562. https://doi.org/10.1021/acsestengg.4c00316

Hossain, M., Mestrot, A., NORTON, G., Deacon, C., ISLAM, M., MEHARG, A., (2020). Arsenic dynamics in paddy soil under traditional manuring practices in Bangladesh. *Environmental Pollution* 268, 115821. https://doi.org/10.1016/j.envpol.2020.115821

Hu, Y., Peng, X., Ai, Z., Jia, F., Zhang, L., (2019). Liquid nitrogen activation of zero-valent iron and its enhanced Cr(VI) removal performance. *Environmental Science & Technology* 53, 8333–8341. https://doi.org/10.1021/acs.est.9b01999

Hussain, M.M., Bibi, I., Ali, F., Saqib, Z.A., Shahid, M., Niazi, N.K., Hussain, K., Shaheen, S.M., Wang, H., Shakil, Q., Rinklebe, J., (2022). The role of various ameliorants on geochemical arsenic distribution and CO2-carbon efflux under paddy soil conditions. *Environmental Geochemistry and Health*. https://doi.org/10.1007/s10653-021-01196-3

Hussain, M.M., Bibi, I., Niazi, N.K., Shahid, M., Iqbal, J., Shakoor, M.B., Ahmad, A., Shah, N.S., Bhattacharya, P., Mao, K., Bundschuh, J., Ok, Y.S., Zhang, H., (2021). Arsenic biogeochemical cycling in paddy soil-rice system: Interaction with various factors, amendments and mineral nutrients. *Science of the Total Environment* 773, 145040. https://doi.org/10.1016/j.scitotenv.2021.145040

Hussain, M.M., Bibi, I., Shahid, M., Shaheen, S.M., Shakoor, M.B., Bashir, S., Younas, F., Rinklebe, J., Niazi, N.K., (2019). Chapter Two – Biogeochemical cycling, speciation and transformation pathways of arsenic in aquatic environments with the emphasis on algae, in: Duarte, A. C., Reis, V. (Eds.), *Comprehensive Analytical Chemistry*. Elsevier, pp. 15–51. https://doi.org/10.1016/bs.coac.2019.03.007

Kim, S.H., Park, J.-I., Lee, S., An, H.-R., Kim, H., Son, B., Seo, J., Kim, C., Jeong, Y., Choi, K., Jeong, S., Lee, H.U., (2024). Enhancing adsorption efficiency for environmentally-friendly removal of As(V) and Pb(II) using a biochar-iron oxide composite. *Applied Surface Science* 667, 160348. https://doi.org/10.1016/j.apsusc.2024.160348

Kloster, G.A., Valiente, M., Marcovich, N.E., Mosiewicki, M.A., (2020). Adsorption of arsenic onto films based on chitosan and chitosan/nano-iron oxide. *International Journal of Biological Macromolecules* 165, 1286–1295. https://doi.org/10.1016/j.ijbiomac.2020.09.244

Kumar, V., Chopra, A., (2015). Toxicity of chromium in agricultural crops with respect to its chemical speciation: A review. *World Applied Sciences Journal* 33, 944–969.

Kumarathilaka, P., Seneweera, S., Meharg, A., Bundschuh, J., (2018). Arsenic speciation dynamics in paddy rice soil-water environment: sources, physico-chemical, and biological factors – A review. *Water Research* 140, 403–414. https://doi.org/10.1016/j.watres.2018.04.034

Lee, J., Kim, C., Schmucker, D., Lee, S.S.S., Liao, S., Cápiro, N.L., Pennell, K.D., Fortner, J.D., (2024). Delineating the role of surface grafting density of organic coatings on the colloidal stability, transport, and sorbent behavior of engineered nanoparticles. *Environmental Science: Nano* 11, 578–587.

Li, M., Zhang, X., Zhang, Y., Xu, X., Liu, Y., Zhang, Y., He, Z., Wang, J., Liang, Y., (2024). Effect of interaction between dissolved organic matter and iron/manganese (hydrogen) oxides on the degradation of organic pollutants by in-situ advanced oxidation techniques. *Science of the Total Environment* 918, 170351.

Liu, Q., Huang, Y., Zhou, Y., Chen, Z., Luo, J., Yan, X., (2023a). Impacts of wet-dry alternations on cadmium and zinc immobilisation in soil remediated with iron oxides. *Journal of Environmental Management* 326, 116660. https://doi.org/10.1016/j.jenvman.2022.116660

Liu, Y., Wang, L., Liu, C., Ma, J., Ouyang, X., Weng, L., Chen, Y., Li, Y., (2023b). Enhanced cadmium removal by biochar and iron oxides composite: Material interactions and pore structure. *Journal of Environmental Management* 330, 117136. https://doi.org/10.1016/j.jenvman.2022.117136

Lyu, P., Li, L., Huang, J., Ye, J., Zhu, C., (2024). Magnetic biochar-supported layered double hydroxide for simultaneous remediation of As and Cd in soil: Effectiveness, retention durability, and insight into a new immobilization mechanism. *Journal of Cleaner Production* 434, 140136. https://doi.org/10.1016/j.jclepro.2023.140136

Majumdar, A., Upadhyay, M.K., Giri, B., Karwadiya, J., Bose, S., Jaiswal, M.K., (2023). Iron oxide doped rice biochar reduces soil-plant arsenic stress, improves nutrient values: An amendment towards sustainable development goals. *Chemosphere* 312, 137117. https://doi.org/10.1016/j.chemosphere.2022.137117

Masood ul Hasan, I., Niazi, N.K., Bibi, I., Younas, F., Al-Misned, F., Shakoor, M.B., Ali, F., Ilyas, S., Hussain, M.M., Qiao, J., Lüttge, A., (2024). Enhanced capacity of thiol-functionalized sugarcane bagasse and rice husk biochars for arsenite sorption in aqueous solutions. *Environmental Science and Pollution Research* 31, 52293–52305.10.1007/s11356-024-34661-4

Mokadem, N., Ngie, A., Viljoen, G., (2023). The Role of water quality on the water-agriculture-health nexus: an assessment of surface water in Vaalharts, South Africa. *American Journal of Public Health* 23, 1–19.

Mostafa, S.M., Fathy, M.M., Khalil, W.M., Mohamed, N., (2024). Chitosan functionalized iron oxide nanocomposites as a novel carbonic anhydrase nano-inhibitor. *Inorganic Chemistry Communications* 166, 112662.

Murtaza, G., Ahmed, Z., Valipour, M., Ali, I., Usman, M., Iqbal, R., Zulfiqar, U., Rizwan, M., Mahmood, S., Ullah, A., (2024). Recent trends and economic significance of modified/functionalized biochars for remediation of environmental pollutants. *Scientific Reports* 14, 217.

Namasivayam, S.K.R., (2024). Eco friendly, green route method for the preparation of poly ethylene glycol (PEG) mediated surface modified iron oxide nanoparticles (PEG-IONps) with potential biological activities. *Environmental Quality Management* 33, 967–978.

Parkinson, G.S., (2016). Iron oxide surfaces. *Surface Science Reports* 71, 272–365.

Rizwan, M., Ali, S., Ali, B., Adrees, M., Arshad, M., Hussain, A., ur Rehman, M.Z., Waris, A.A., (2019). Zinc and iron oxide nanoparticles improved the plant growth and reduced the oxidative stress and cadmium concentration in wheat. *Chemosphere* 214, 269–277.

Sakr, M., El Agamawi, H., Klammler, H., Mohamed, M.M., (2023). A review on the use of permeable reactive barriers as an effective technique for groundwater remediation. *Groundwater for Sustainable Development* 21, 100914.

SefidSiahbandi, M., Moradi, O., Akbari–adergani, B., Aberoomand Azar, P., Sabar Tehrani, M., (2023). The effect of Fe–Zn mole ratio (2:1) bimetallic nanoparticles supported by hydroxyethyl cellulose/graphene oxide for high-efficiency removal of doxycycline. *Environmental Research* 218, 114925. https://doi.org/10.1016/j.envres.2022.114925

Si, T., Yuan, R., Qi, Y., Zhang, Y., Wang, Y., Bian, R., Liu, X., Zhang, X., Joseph, S., Li, L., Pan, G., (2024). Enhancing soil redox dynamics: Comparative effects of Fe-modified biochar

(N–Fe and S–Fe) on Fe oxide transformation and Cd immobilization. *Environmental Pollution* 347, 123636. https://doi.org/10.1016/j.envpol.2024.123636

Singaraj, S.G., Mahanty, B., Balachandran, D., Padmaprabha, A., (2019). Adsorption and desorption of chromium with humic acid coated iron oxide nanoparticles. *Environmental Science and Pollution Research* 26, 30044–30054. https://doi.org/10.1007/s11356-019-06164-0

Singh, Y.D., Panda, M.K., Sarma, M.K., Ningthoujam, R., Babu, P.J., Das, M., Patra, J., (2022). Nanotechnology for Sustainable Bioenergy Production, in: Arakha, M., Pradhan, A. K., Jha, S. (Eds.), *Bio-Nano Interface*. Springer, pp. 339–355.

Song, J., Huang, Z., Gamal El-Din, M., (2021). Adsorption of metals in oil sands process water by a biochar/iron oxide composite: Influence of the composite structure and surface functional groups. *Chemical Engineering Journal* 421, 129937. https://doi.org/10.1016/j.cej.2021.129937

Su, H., Ye, Z., Hmidi, N., (2017). High-performance iron oxide–graphene oxide nanocomposite adsorbents for arsenic removal. *Colloids and Surfaces A: Physicochemical and Engineering Aspects* 522, 161–172. https://doi.org/10.1016/j.colsurfa.2017.02.065

Suda, A., Makino, T., (2016). Functional effects of manganese and iron oxides on the dynamics of trace elements in soils with a special focus on arsenic and cadmium: a review. *Geoderma* 270, 68–75. https://doi.org/10.1016/j.geoderma.2015.12.017

Thiruvenkatachari, R., Vigneswaran, S., Naidu, R., (2008). Permeable reactive barrier for groundwater remediation. *Journal of Industrial and Engineering Chemistry* 14, 145–156.

Ugwu, I.M., Igbokwe, O.A., (2019). Sorption of heavy metals on clay minerals and oxides: a review. in Edebali, S. (Eds.), *Advanced Sorption Process Applications* 2019, 1–23. InTech Open.

von der Heyden, B.P., Roychoudhury, A.N., (2015). Application, chemical interaction and fate of iron minerals in polluted sediment and soils. *Current Pollution Reports* 1, 265–279.

Wang, H., Chen, Q., Xia, H., Liu, R., Zhang, Y., (2024a). Enhanced complexation and electrostatic attraction through fabrication of amino- or hydroxyl-functionalized Fe/Ni-biochar composite for the adsorption of Pb(II) and Cd(II). *Separation and Purification Technology* 328, 125074. https://doi.org/10.1016/j.seppur.2023.125074

Wang, J., Shaheen, S.M., Swertz, A.-C., Liu, C., Anderson, C.W., Fendorf, S., Wang, S.-L., Feng, X., Rinklebe, J.r., (2024b). First insight into the mobilization and sequestration of arsenic in a karstic soil during redox changes. *Environmental Science & Technology*. https://doi.org/10.1021/acs.est.4c02989

Wang, S., Li, W., Ding, C., Zhang, J., Zhang, N., Li, Y.C., Gao, B., Wang, B., Wang, X., (2024c). Biochar-supported zero-valent iron enhanced arsenic immobilization in a paddy soil: the role of soil organic matter. *Biochar* 6, 26. https://doi.org/10.1007/s42773-024-00318-1

Wang, X., Jia, P., Hua, Y., Xu, H., Xi, M., Jiang, Z., (2023). Natural organic matter changed the capacity and mechanism of Pb and Cd adsorptions on iron oxide modified biochars. *Separation and Purification Technology* 314, 123625. https://doi.org/10.1016/j.seppur.2023.123625

Wang, Y., Yang, Y., Shi, J., An, W., Lyu, T., Zhang, P., (2024d). Processes and mechanisms in remediation of aqueous chromium contamination by sulfidated nano-scale zerovalent iron (S-nZVI): Experimental and computational investigations. *Journal of Hazardous Materials* 469, 134031.

WHO Guidelines., (2011). Guidelines for Drinking-Water Quality. 4, 315–318.

Xiao, Y., Liu, F., Shi, H., Hou, L., Qin, G., Yuan, C., Lou, X.W., (2023). Construction of Ultrastable Ultrathin Black Phosphorus Nanodisks Hybridized with Fe3O4 Nanoclusters and Iron (V)-Oxo Complex for Efficient Potassium Storage. *Advanced Materials* 23, 2301772.

Xu, Y., Cao, S., Chen, X., Li, J., Liu, H., Gao, Y., Wen, S., Guo, J., Shi, X., Xue, W., (2024a). Enhanced immobilization of cadmium in contaminated paddy soil by biochar-supported sulfidized nanoscale zero-valent iron. *Journal of Soils and Sediments* 24, 259–274. https://doi.org/10.1007/s11368-023-03618-4

Xu, Y., Huang, X., Liu, C., Kong, D., Qian, G., (2024b). Chitosan-modified iron fillings materials for remediation of arsenic-contaminated soil. *Chemical Engineering Journal* 486, 150261.

Xu, Z., Sun, M., Xu, X., Cao, X., Ippolito, J.A., Mohanty, S.K., Ni, B.-J., Xu, S., Tsang, D.C.W., (2023). Electron donation of Fe-Mn biochar for chromium(VI) immobilization: Key roles of embedded zero-valent iron clusters within iron-manganese oxide. *Journal of Hazardous Materials* 456, 131632. https://doi.org/10.1016/j.jhazmat.2023.131632

Yin, B., Li, J., Guo, W., Dong, H., Zhang, G., Xin, Y., Zhang, G., Chen, Q., (2024a). Photocatalytic degradation of fluoranthene in soil suspension by TiO2/α-FeOOH with enhanced charge transfer capacity. *Environmental Science and Pollution Research* 31, 20621–20636. https://doi.org/10.1007/s11356-024-32501-z

Yin, Y., Wang, Y., Ding, C., Zhou, Z., Tang, X., He, L., Li, Z., Zhang, T., Wang, X., (2024b). Impact of iron and sulfur cycling on the bioavailability of cadmium and arsenic in co-contaminated paddy soil. *Journal of Hazardous Materials* 465, 133408.

Yodsanit, N., Wang, B., Zhao, Y., Guo, L.-W., Kent, K.C., Gong, S., (2021). Recent progress on nanoparticles for targeted aneurysm treatment and imaging. *Biomaterials* 265, 120406.

Zhang, Y., Fu, P., Ni, W., Zhang, S., Li, S., Deng, W., Hu, W., Li, J., Pei, F., Du, L., Wang, Y., (2024). A review of solid wastes-based stabilizers for remediating heavy metals co-contaminated soil: Applications and challenges. *Science of the Total Environment* 920, 170667. https://doi.org/10.1016/j.scitotenv.2024.170667

Zhao, W., Xu, Y., Gu, L., Zhu, M., Yang, P., Gu, C., Liu, Z., Feng, X., Tan, W., Huang, Q., (2024). Elucidating phosphate and cadmium cosorption mechanisms on mineral surfaces with direct spectroscopic and modeling evidence. *Environmental Science & Technology* 58, 20211–20223.

Index

Note: **Bold** page numbers refer to tables and *italic* page numbers refer to figures.

C

M

T

U

V

W

X

Z

FSC
www.fsc.org
MIX
Papier | Fördert gute Waldnutzung
FSC® C083411

Kolibri 360 GmbH
Ferdinand-Jühlke-Straße 7
99095 Erfurt, Deutschland
produktsicherheit@kolibri360.de